Universitext

Marcel Berger

Geometry I

Translated from the French by M. Cole and S. Levy

With 426 Figures

Springer-Verlag Berlin Heidelberg New York
London Paris Tokyo

Marcel Berger
Institut des Hautes Etudes Scientifiques
35, route de Chartres
91440 Bures-sur-Yvette
France

Michael Cole
Academic Industrial Epistemology, 17 St. Mary's Mount
Leyburn, North Yorkshire DL8 5JB
England

Silvio Levy
Department of Mathematics, Princeton University
Princeton, NJ 08544, USA

Title of the original French edition: Géométrie (vols. 1–5)
Published by CEDIC and Fernand Nathan, Paris
© CEDIC and Fernand Nathan, Paris 1977

Mathematics Subject Classification (1980): 51-XX, 52-XX, 53-XX

ISBN 3-540-11658-3 Springer-Verlag Berlin Heidelberg New York
ISBN 0-387-11658-3 Springer-Verlag New York Berlin Heidelberg

© Springer-Verlag Berlin Heidelberg 1987
Printed in Germany

Printing and bookbinding: Druckhaus Beltz, Hemsbach/Bergstr.
2141/3140-543210

Table of contents

Volume I

Volume II

Introduction

This book arose essentially from two sources: a course called "Mathématiques Élémentaires Approfondies", taught in 1972–1973 and 1973–1974 at the University of Paris VII; and the author's fifteen years of experience in preparing the geometry part of the orals in the Agrégation de Mathématiques, a competition to select the best high school teachers in France.

The main objectives pursued in this book, more precisely, in the framework of elementary geometry, are the following:

— to emphasize the visual, or "artistic", aspect of geometry, by using figures in abundance;

— to accompany each new notion with as interesting a result as possible, preferably one with a simple statement but a non-obvious proof;

— finally, to show that this simple-looking mathematics does not belong in a museum, that it is an everyday tool in advanced mathematical research, and that occasionally one encounters unsolved problems at even the most elementary level.

Here are some particularities of this book that derive from the general principles above. Main definitions are, whenever possible, followed by non-trivial (and sometimes new) examples. Figures abound: at the beginning, especially, each geometric reasoning is illustrated by a diagram. (Such was the general practice fifty years ago, but pictures have all but disappeared from modern geometry books. One reason seems to be that authors think that readers keep pencil and paper next to them and draw figures as they go along, or else draw mental pictures. But the author's experience from university examinations shows that students aren't likely to draw pictures, either on paper or in their heads. Thus one of the aims of this book is to teach the reader to make systematic use of figures as he reads.)

Notes are also common, referring to both the historical development and the current, often very advanced, applications of the ideas introduced. This is meant to convey to the reader a feeling that the elementary mathematics that he is studying is an integral part of the living, continuing corpus of mathematical knowledge. The notes are backed by an extensive bibliography.

References in brackets are to book and page number, except in the case of [B–G] and Bourbaki, which use a system similar to the one adopted in this book. Other exceptions are always clearly labeled.

There are many internal references as well. The author believes that most readers will not read this book sequentially, from beginning to end. They will pick it up to find such and such a theorem, or a discussion of such and such a subject; or they will be "hooked" by a figure or result while leafing through the book, and then will want to read about the background of that figure or result. Either way their task will be facilitated by the many internal references and by the two indexes at the end, the second of which attempts to include all references to a given entry, not just the primary one.

We will not dwell on the contents for long, but it's worth mentioning a few results included here that are often absent from comparable books, or relegated to exercises: the fundamental theorem of affine geometry, the classification of crystallographic groups, the classification of regular polytopes in arbitrary dimension, Cauchy's theorem on the rigidity of convex polyhedra, the discussion of polygonal billiards, Poncelet's theorem on polygons inscribed in a conic, the Villarceau circles on the torus, Clifford parallelism, the isoperimetric inequality in arbitrary dimension, the theorems of Helly and Krasnosel'skii, the simplicity of the orthogonal group, the theorems of Witt and Cartan–Dieudonné, and complete expositions of spherical, elliptic and hyperbolic geometry.

This book can be used in different ways. Besides consulting it as a source for problems and for its extensive bibliography, teachers can use it as a textbook for several courses: a freshman-level course on Euclidean geometry (based on chapters 8 and 9), a higher-level course on the sphere and hyperbolic geometry (based on chapters 18 and 19), a higher-level course on convexity from the geometric point of view (based on the whole or parts of chapters 11 and 12).

One last important point. Most of the exercises are more difficult than those in comparable books, since the text itself already contains a huge number of ideas for exercises and applications. In particular, the reader often has to find out what to show. This will require from him a certain amount of maturity, a will to take risks and to develop his originality. Many of the
* exercises, marked with a star, are completely solved in the companion volume [B–P–B–S].

This work owes much to the whole of the mathematical community, both students and colleagues. In particular, I've profited a lot from lectures by candidates to the *Agrégation*, which showed me the right and the wrong way to teach things; I've also leant heavily on many colleagues for advice and references. Since I owe so much, I would not dare include a necessarily incomplete list of acknowledgements; I ask you all to accept my warm, global thanks.

I would like to thank Silvio Levy for his rapid and excellent translation.

M. BERGER

It was in 1897, I was then a third-year student at the École Normale, and Joseph Caron, a teacher whose name I bring up with pleasure, had assigned us a very difficult exercise in descriptive geometry. It involved the intersection of two tori arranged in such a way that, to find an arbitrary point of the intersection, one would use sections of the tori by certain spheres bitangent to each of them.

Joseph Caron taught descriptive geometry in Paris schools; he also held the chair of geometric drawing at the Sorbonne and the École Normale. He has written a treatise on descriptive geometry and a manual on topographic projections (a reaction against programs that he condemns); he has published in the *Bulletin de la Société Mathématique de France* an article about the construction of a third-order surface having its twenty-seven lines real (a plaster cast of this surface is part of the higher geometry collection of the University of Paris); above all, he has contributed to instilling the love of geometry into numerous students, at a time when many eminent scholars, endowed with great geometric talent, make a point of never disclosing the simple and direct ideas that guided them, subordinating their elegant results to abstract general theories which often have no application outside the particular case in question. Geometry was becoming a study of algebraic, differential or partial differential equations, thus losing all the charm that comes from its being an art.

Henri Lebesgue
[LB1, 209–210]

Chapter 0

Notation and background

0.1. Set theory

If A and B are subsets of E, we *denote* their *difference set* by

$$A \setminus B = \{ x \in E \mid x \in A \text{ and } x \notin B \}.$$

If $f : E \to F$ is a map and $A \subset E$, we *denote* by $f|_A$ the *restriction* of f to A. The identity map from X into itself will be *denoted* by Id_X, and the *cardinality* of X by $\#X$.

If $\{x_i\}_{i=1,\dots,n}$ is a family, we write $\{x_1, \dots, \hat{x}_i, \dots, x_n\}$ for the same family without the element x_i.

0.2. Algebra

We *denote* by $\mathbf{N}$ the non-negative integers, by $\mathbf{Z}$ the ring of integers, by $\mathbf{R}$ the field of real numbers, by $\mathbf{C}$ the field of complex numbers and by $\mathbf{H}$ the skew field of quaternions (see 8.9). The non-negative (resp. non-positive) real numbers are *denoted* by $\mathbf{R}_+$ (resp. $\mathbf{R}_-$), and the strictly positive (resp. negative) reals by $\mathbf{R}_+^*$ (resp. $\mathbf{R}_-^*$). If K is a field, K^* *denotes* the set $K \setminus 0$.

If E and F are sets endowed with the same kind of algebraic structure, $\mathrm{Hom}(E; F)$ and $\mathrm{Isom}(E; F)$ *stand for* the set of homomorphisms and isomorphisms, respectively, from E into F. However, if E and F are vector spaces, we *write* $L(E; F)$ instead of $\mathrm{Hom}(E; F)$ for the set of linear maps from E into F, and $\mathrm{GL}(E)$ for $\mathrm{Isom}(E; E)$, the so-called *linear group* of E. When two vector spaces are considered, they are vector spaces over the same field, unless we explicitly say otherwise.

If a vector space E is the *direct sum* of two subspaces A and B, we *write* $E = A \oplus B$. The *algebraic dual* of a vector space E is *denoted* by E^*.

We *denote* by I the *identity matrix*, and by ${}^tA$ the *transpose* of the matrix A.

If X is a set we *denote* by S_X the *group of permutations* (that is, bijections) of X under composition of maps: $fg = f \circ g$. When $X = \{1, 2, \ldots, n\}$ we *write* S_n for S_X (the *symmetric group* of order n), and $\mathcal{A}_n$ for the *alternating group* of order n, defined as the subgroup of S_n formed by even permutations. The *Klein group* $\mathcal{V}_4$ is the product $\mathbf{Z}_2 \times \mathbf{Z}_2$ of the group with two elements with itself. The *dihedral group* $\mathcal{D}_{2n}$, where n is an integer, is the extension by $\mathbf{Z}_2$ of the cyclic group $\mathbf{Z}_n$, with relation $ab = ba^{-1}$, where a (resp. b) is a generator of $\mathbf{Z}_n$ (resp. $\mathbf{Z}_2$).

Binomial coefficients are *denoted* by $\binom{n}{p}$ (see 1.5.2).

All fields are assumed commutative, unless we state otherwise.

0.3. Metric spaces

Let X be a metric space with distance d. The *diameter* $\operatorname{diam}(A) \in \mathbf{R}_+ \cup \infty$ of a subset $A \subset X$ is defined as $\operatorname{diam}(A) = \sup \{ d(x, y) \mid x, y \in A \}$. The *distance* $d(A, B)$ between two subsets $A, B \subset X$ is the non-negative number

$$d(A, B) = \inf \{ d(x, y) \mid x \in A, y \in B \}.$$

We also *write* $d(x, A) = d(\{x\}, A)$. The distance $d(A, B)$ should not be confused with $\delta(A, B)$ (introduced in 9.11).

The *open ball* of radius r around a point a is *denoted* by $U(a, r) = \{ x \in X \mid d(a, x) < r \}$, and the corresponding *closed ball* by $B(a, r) = \{ x \in X \mid d(a, x) \leq r \}$. Balls are also defined around sets: we put

$$U(A, r) = \{ x \in X \mid d(x, A) < r \},$$
$$B(A, r) = \{ x \in X \mid d(x, A) \leq r \}.$$

We also *write* $U_X(\cdot, \cdot)$ and $B_X(\cdot, \cdot)$ if necessary.

If X, Y are metric spaces, we *denote* by $\operatorname{Is}(X; Y)$ the set of *isometries* from X into Y, i.e., the set of maps $f : X \to Y$ such that

$$d(f(x), f(y)) = d(x, y) \qquad \text{for } x, y \in X.$$

In particular, we *write* $\operatorname{Is}(X) = \operatorname{Is}(X; X)$.

0.4. General topology

We shall use several times the fact that the intersection of a decreasing family of compact sets is non-empty (see a proof in 11.7.3.2).

0.5. Hyperbolic trigonometry

The *hyperbolic cosine, sine* and *tangent* of a real number t are *defined* by the formulas

$$\cosh t = \frac{e^t + e^{-t}}{2}, \qquad \sinh t = \frac{e^t - e^{-t}}{2}, \qquad \tanh t = \frac{\sinh t}{\cosh t}.$$

The inverse of the restriction of the hyperbolic cosine to the positive real axis is called the *principal hyperbolic arc-cosine*; it is a map Arccosh : $[1, \infty[\rightarrow$ $\mathbf{R}_+$.

0.6. Lebesgue measure; integration theory

In certain parts of the text we shall use results from Lebesgue measure and integration theory, especially the following fundamental theorems: the dominated convergence theorem, Fubini's theorem, and the change of variable theorem for C^1 maps in open sets of $\mathbf{R}^n$. We shall also need the *characteristic function* χ_K of a set K, and the *image* of a measure under a map. A general reference for these topics is [RU].

Chapter 1
Group actions: examples and applications

In this chapter we define group actions, transitivity, stabilizers, homogeneous spaces and faithful actions. These notions are abundantly illustrated with examples from algebra and particularly from geometry. These examples will be encountered again in later chapters.

The last two sections, 1.7 and 1.8, are devoted respectively to crystallographic groups and finite groups of rotations around a fixed point in space. These two topics make full use of the language introduced in the previous sections. They were chosen for their esthetic interest and also because, in spite of the elementary nature of the questions posed, their solution presents a certain degree of difficulty. Section 1.8 is closely connected with the theory of regular polyhedra in three-dimensional Euclidean space (section 12.5).

The idea of considering the geometry of a space as the study of properties invariant under a group of transformations is due to Felix Klein, who formulated it in his famous "Erlangen program" (see, for example, [GG1, p. 367]).

1.1. Group actions

1.1.1. DEFINITION. *Let G be a group and X a set. A G-action on X is a homomorphism $\phi : G \to S_X$. If ϕ is a G-action on X, we say that G acts or operates on X (by ϕ).*

1.1.2. NOTATION. The action ϕ is sometimes written (G, X, ϕ) or even (G, X). If ϕ is fixed, we write $g(x)$ for $\phi(g)(x)$, where $g \in G$ and $x \in X$.

It follows from the definition that the map $x \to g(x)$ is a bijection for every $g \in G$, and that $g(h(x)) = (gh)(x)$ for all $g, h \in G$, $x \in X$. Also $e(x) = x$, where e is the identity of G, and $\phi(g^{-1}) = (\phi(g))^{-1}$.

1.2. Examples

1.2.1. G is a subgroup of S_X; this is the most frequent case. For example, G can be defined as the subgroup of S_X satisfying certain conditions.

1.2.2. Put $A = \{1, \ldots, n\}$, $S_A = S_n$ (the symmetric group). Let $G = S_n$. Then G acts on A; but it also acts in a natural way on $X = P_{n,p} = \{ P \subset A \mid \#P = p \}$, the set of subsets of A with p elements $(0 \leq p \leq n)$.

1.2.3. For a given $g \in S_n$, put $G = \{ g^k \mid k \in \mathbf{Z} \} \subset S_n$. Then G acts on $X = \{1, \ldots, n\}$.

1.2.4. If X is a vector space, its linear group

$$G = \mathrm{GL}(X) = \{ f : X \to X \mid f \text{ is linear and bijective} \}$$

acts on X.

1.2.5. Let E be a Euclidean vector space (chapter 8), and put

$$G = O(E) = \{ f \in \mathrm{GL}(E) \mid f \text{ is an isometry} \}.$$

Then there is a natural $O(E)$-action on $X = E$. But $O(E)$ also acts on each grassmannian (manifold)

$$X = G_{E,p} = \{ V \subset E \mid V \text{ is a } p\text{-dimensional vector subspace of } E \},$$

and again on the set X of orthonormal bases of E.

1.2.6. Let $X = G$ be a group. Then G acts on itself in several important ways:

$$\phi(g)(h) = gh \qquad \text{(left translations)};$$
$$\phi(g)(h) = hg \qquad \text{(right translations)};$$
$$\phi(g)(h) = ghg^{-1} \qquad \text{(inner automorphisms)}.$$

1.2.7. Let X be an affine plane (chapter 2), $\mathrm{GA}(X) = G$ the group of affine bijections of X (called the affine group of X), and $\Gamma(X)$ the set of conics of X (a conic is a subset of X given by a certain type of equation, see 16.7). Then $\mathrm{GA}(X)$ acts on $\Gamma(X)$.

1.2.8. Let X be a metric space, with distance d. The isometry group $G = \mathrm{Is}(X)$, defined in 0.3, acts on X in a natural way.

1.2.9. Put $G = \mathbf{R}$, $X = S^3 = \{\, x \in \mathbf{R}^4 \mid \|x\| = 1 \,\} \subset \mathbf{R}^4$. Identifying $\mathbf{R}^4$ with $\mathbf{C}^2$, we define an operation $(\mathbf{R}, S^3, \phi)$ by

$$\phi(t)(z, z') = (e^{it}z, e^{it}z').$$

This example is of great importance in geometry: see 4.3, 6.2, 18.8.1.

1.2.10. For other examples, see 4.5.9, 8.8, 9.5, 18.10.

1.3. Faithful actions

1.3.1. DEFINITION. *The action (G, X, Φ) is called faithful if Φ is injective* (in other words, if only $e \in G$ maps to the identity Id_X).

This always happens when $G \subset \mathcal{S}_X$ (cf. 1.2.1). If G is not faithful, one can consider the associated faithful action $(G/\operatorname{Ker}\phi, X, \phi)$.

In section 1.2, all actions are faithful, with the following exceptions: in 1.2.9, we have $\operatorname{Ker}\phi = 2\pi\mathbf{Z}$; and in 1.2.6, the action of inner automorphisms is faithful if and only if the center Z_G of G consists only of the identity.

1.4. Transitive actions

1.4.1. DEFINITION. *The action (G, X, ϕ) is called transitive if for every x, y in X there exists $g \in G$ such that $g(x) = y$.*

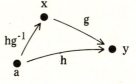

Figure 1.4.1

In practice, it suffices to check that some fixed $a \in X$ can be mapped to any $x \in X$ by some element $g \in G$. For if we have $g(a) = x$ and $h(a) = y$, then $y = (hg^{-1})(x)$.

1.4.2. EXAMPLES. We study the examples in 1.2 from the point of view of transitivity:

— 1.2.2 is transitive;

— 1.2.3 is transitive if and only if the permutation g is cyclic;

— 1.2.4 is not transitive. Is it transitive on $X \setminus 0$?

— 1.2.5 is not transitive on E (why?), but is transitive on any $G_{E,p}$ (why?);

— 1.2.6: translations are transitive, but inner automorphisms are not (every point is fixed if G is abelian!);

— 1.2.7 is not transitive. For example, ellipses cannot be taken into parabolas (see 15.3.3.2);

— 1.2.9 is not transitive—see 18.8.

1.4.3. DEFINITION. *The action (G, X) is called simply transitive if for every $x, y \in X$ there exists a unique $g \in G$ such that $g(x) = y$.*

1.4.4. EXAMPLES.

1.4.4.1. Proposition. *If G is an abelian group, any faithful transitive action is simply transitive.*

Proof. If for some $x \in X$ and $g, h \in G$ we have $g(x) = h(x)$, then $g(y) = h(y)$ for any $y \in G$, since

$$g(y) = g\big(k(x)\big) = k\big(g(x)\big) = k\big(h(x)\big) = h\big(k(x)\big) = h(y). \qquad \square$$

1.4.4.2. Left translations (cf. 1.2.6) are simply transitive.

1.4.4.3. The group $O(E)$ does not act simply transitively on $G_{E,p}$ for any $0 \leq p \leq \dim E$, but does act simply transitively on the set of orthonormal bases.

1.4.5. GENERALIZATION. *We say that (G, X) is p-transitive $(p \in \mathbf{N})$ if G acts transitively on p-tuples of distinct points of X.*

See examples in 2.3.3.5, 4.5.10, 4.6.9, 6.1.1, 9.1.6, 9.1.7, 9.6.2, 9.7.1, 18.5.5, 18.8.4, 18.10.6, 19.4.5.1.

1.5. Stabilizers; homogeneous spaces

1.5.1. DEFINITION. *If (X, G) is a group action and $x \in X$, the stabilizer or isotropy group of x is the subgroup $G_x = \{ g \in G \mid g(x) = x \}$.*

Stabilizers in some sense tell us how far a group is from acting simply transitively: just notice that $g(x) = h(x)$ if and only if $h^{-1}g$ is in the stabilizer of x.

1.5.2. EXAMPLES. In 1.2.2, for $X = A$ and $x = 1$, we have a natural isomorphism $G_1 \cong S_{n-1}$. For $X = S_{n.p}$, we have

$$G_{\{1,\dots,p\}} \cong S_p \times S_{n-p}.$$

These isomorphisms will allow us to count the number of permutations, $\# S_n = n!$, and of combinations, $\# P_{n.p} = \binom{n}{p}$. See 1.5.8.

For inner automorphisms (1.2.6) we have the following equality:

$$G_g = \{ h \in G \mid hg = gh \},$$

the set of elements of G commuting with g.

In 1.2.5, for $X = G_{E,p}$, we have the isomorphism $G_1 \cong O(p) \times O(n - p)$ (where $n = \dim E$ and $O(p) = O(\mathbf{R}^p)$).

1.5.3. One always has

$$G_{g(x)} = gG_x g^{-1}.$$

In other words, $G_{g(x)}$ and G_x are conjugate subgroups of G, and in particular are always isomorphic. Example 1.4.4.1 is an immediate consequence of this remark.

1.5.4. DEFINITION. *The set X is called a homogeneous space under G (for the action ϕ) if (G, X, ϕ) is transitive.*

Let X be homogeneous under (G, X, ϕ), and fix $x \in X$. Define $\theta : G \to X$ by $\theta(g) = g(x)$. Since $g(x) = h(x)$ is equivalent to $h^{-1}g \in G_x$, we can pass to a map $\underline{\theta} : G/G_x \to X$, where G/G_x is the quotient of G by the equivalence relation $g \sim h \iff h^{-1}g \in G_x$:

$$G \xrightarrow{p} G/G_x$$
$$\theta \searrow \quad \swarrow \underline{\theta}$$
$$X$$

(Observe that, in general, G/G_x is not a group.) By construction, $\underline{\theta}$ is bijective. Thus we have, in the set-theoretical sense,

1.5.5 $$X \overset{\text{set}}{\cong} G/G_x.$$

This equation is invaluable, for it reduces the study of the set X to an algebraic problem, namely, the study of the pair (G, G_x). See, for example, [B–H], [WF].

1.5.6. COROLLARY. *If G is finite, so is X. Moreover, for every $x \in X$, we have $\#X = (\#G)/(\#G_x)$.*   □

1.5.7. It is important to notice that if G and X are topological spaces and $\phi : G \times X \to X$ is a continuous map, the topological space X is not, in general, homeomorphic to G/G_x (with the quotient topology). See 1.9.1 for a counterexample.

1.5.8. Using an elementary reasoning (the so-called "shepherd principle": to compute the number of sheep in a flock, count the number of legs and divide by four, cf. [BIO, 179]), one deduces from 1.5.2 and 1.5.6 that $\#A = n = (\#S_n)/(\#S_{n-1})$, and hence, by recurrence, that $\#S_n = n!$. Thus

$$\#S_{n,p} = \frac{\#S_n}{(\#S_p)(\#S_{n-p})} = \frac{n!}{p!\,(n-p)!}.$$

1.5.9. From 1.5.2 and 1.5.5 one deduces the expression for the grassmannian $G_{E,p}$ as a homogeneous space $G_{n,p} = O(n)/(O(p) \times O(n-p))$. See some applications in [HU, chapter 18]; see also 14.3.7.

1.6. Orbits; the class formula

We now study non-transitive actions.

1.6.1. DEFINITION. *Let (G, X) be a homogeneous space. The orbit $O(x)$ of $x \in X$ is defined as*

$$O(x) = \{ g(x) \mid g \in G \}.$$

Thus orbits are just equivalence classes under the equivalence relation

$$x \sim y \Longleftrightarrow \text{there exists } g \in G \text{ such that } g(x) = y.$$

In particular, we can form the space of orbits X/G. From 1.5.5 we deduce that, for all $x \in X$,

1.6.2
$$O(x) \stackrel{\text{set}}{\cong} G/G_x.$$

1.6.3. COROLLARY. *If G is finite, so is $O(x)$ for every x, and*

$$\#O(x) = (\#G)/(\#G_x).$$

Proof. This follows from 1.6.2 and 1.5.6. $\qquad\square$

1.6.4. EXAMPLES. In 1.2.3, the orbits are called *cycles of the permutation g.* Their study leads to a complete classification of conjugate elements in S_n.

In 1.2.5, for $X = E$, the orbits are the spheres $S(0, r)$ centered at the origin (where $r \in \mathbf{R}_+$).

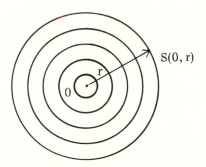

Figure 1.6.4

In 1.2.9 the structure of the set of orbits is fascinating. It is straightforward to see that the orbits are all circles (like $S^1 = \mathbf{R}/2\pi\mathbf{Z}$), but less so to visualize how they all fit together to form the famous Hopf fibration: any two of the circles are linked, and the quotient space is homeomorphic to the sphere S^2 (see 4.3.6). See 18.8 and 18.9 for different approaches to this very important example.

1.6.5. NOTE. Given an action (G, X), the problem of finding the orbits and parametrizing the set X/G can often be described as a *classification problem*. We shall see numerous examples of this: 2.7.5.11 deals with the examples in 1.2.6 and 1.4.2; sections 8.6 and 8.7 deal with angles; 13.1.4 with the classification of quadratic forms; 15.2, with affine quadrics; see also 18.6 and 13.10.

1.6.6. COROLLARY (THE CLASS FORMULA). *Let X and G be finite, and consider a G-action on X. If $A \subset X$ is a subset which intersects each orbit in exactly one point, we have*

$$\#X = \sum_{x \in A} \frac{\#G}{\#G_x}. \qquad \square$$

The set A is called a *section* of the map $p : X \to X/G$ because it is the image $s(X/G)$ of a section $s : X/G \to X$ of p, i.e., a map such that $p \circ s = \mathrm{Id}_{X/G}$. One can say that A *parametrizes* the orbits.

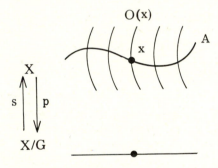

Figure 1.6.6

1.6.7. APPLICATION TO p-GROUPS

1.6.7.1. Definition. *A finite group G is called a p-group if $\#G = p^m$, where p is a prime and $m \in \mathbf{N}^*$.*

1.6.7.2. Proposition. *Every p-group has a non-trivial center.*

Proof. Consider the action of G on itself by inner automorphisms (1.2.6), call Z_G the center of G, and look at the induced G-action on $X = G \backslash Z_G$ (cf. 1.4.2, 1.2.6). If A parametrizes X/G we have, by 1.6.6,

$$\#X = \sum_{x \in A} \frac{\#G}{\#G_x},$$

whence

$$\#G = p^m = \#Z_G + \sum_{x \in A} \frac{\#G}{\#G_x}.$$

But $\#G_x$ is a power of p, since it divides p^m, and $\#G/\#G_x = p^{m(x)}$, where $m(x) \geq 1$. Then p divides $\#G$ and $\#X$, whence also $\#Z_G$. $\qquad \square$

1.6.7.3. Notes. The proof of 1.6.7.2 is one of the keys to showing that every finite skew field is commutative; see, for example, [AN, 37].

A considerable refinement of the same argument also works in studying Sylow subgroups of a finite group of order $p^m q$ (where p is a prime not dividing q); see [SE1, 139].

1.6.8. We shall now conclude this chapter of generalities with two more elaborate studies involving plane and spatial geometry, respectively. Proofs will only be sketched; the interested reader can fill in the details himself or find them in the references given.

1.7. Tilings and crystallographic groups

1.7.1. A tourist visiting the Alhambra in Spain—just as the reader who leafs through the next few pages—is bound to develop an interest in plane figures with regularly repeated motifs covering the whole plane. The motifs admit an infinite number of variations, but there is only a finite number of ways of reproducing them—seventeen (see figures 1.7.4.1 through 1.7.4.5 and 1.7.6.1 through 1.7.6.12). At the Alhambra only thirteen of these patterns are present; two of the remaining ones are found in the town of Toledo and the last two are represented in the native art of various tribes. The fact that animal or human motifs are not depicted in the Alhambra (unlike figures 1.7.4.8 and 1.7.6.13, for example) is solely due to the Islamic ban on such representations. It may in fact be argued that this limitation on the development of pictorial motifs steered the artists' creativity towards new ways of repeating the available ones, accounting for the surprisingly high number (for the time) of patterns which were used.

The aim of this section is to give an axiomatization of these "regularly repeated motifs" and sketch a proof of the existence of only seventeen patterns (but see also 1.7.7.8).

1.7.2. We begin our axiomatization by assuming that the artist fills his surface with copies of one *standard tile*. To simplify matters, we require at first that he use his tiles in any position, but without turning them over—maybe the tiles are decorated on one side only (see figures 1.7.6.1 through 1.7.6.14).

Let E be the Euclidean plane, and $P \subset E$ a subset in the shape of the standard tile. We want to express the fact that P and its copies fill the whole plane without leaving any gaps; thus axiom (CG2) in 1.7.3. This by itself, however, is insufficient to obtain only "regular" tilings such as those at the Alhambra, since there exist "non-regular" tilings, like the one in figure 1.7.2.1.

This example, due to Voderberg ([VG1], [VG2]), is obtained from a nine-sided standard tile and its images under orientation-preserving isometries of the affine plane. What makes it work is the tiles' remarkable property that,

Figure 1.7.2.1 (Source: [VG1])

when two of them are juxtaposed with only two vertices in common, the gap left between them can be filled with one or two other tiles.

For that matter, the reader himself can construct a non-regular tiling by considering the squares formed by joining adjacent points with integer coordinates in $\mathbf{R}^2$, and randomly dividing each square into two vertical or horizontal rectangles (figure 1.7.2.2).

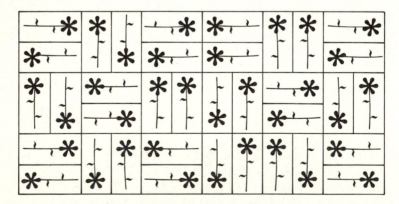

Figure 1.7.2.2

On the fascinating problem of "necessarily non-periodic tilings", the reader may consult [RN1], [RN2] (cf. 19.6.12), [RN3], [GA]. See also 1.9.16 and figure 1.9.16.2.

In order to restrict ourselves to "regular" tilings, we must introduce a group of isometries of E, as in axiom (CG1) below. In figures 1.7.2.1 and 1.7.2.2 no such group can be found.

1.7.3. AXIOMS FOR TILINGS. *A tiling consists of a connected compact subset P of the Euclidean plane E and a subgroup G of the group $\mathrm{Is}^+(E)$ of orientation-preserving isometries of E, such that the interior $\overset{\circ}{P}$ of P is non-empty and the following conditions are satisfied:*

CG1)
$$\bigcup_{g \in G} g(P) = E$$

CG2)
$$g(P) = h(P) \text{ whenever } g(\overset{\circ}{P}) \cap h(\overset{\circ}{P}) \neq \emptyset$$

The group G is called a crystallographic group.

1.7.4. Our aim here is to prove that, up to conjugation in the linear group of E, there exist only five crystallographic groups G, corresponding to figures 1.7.4.1 through 1.7.4.5. (These figures depict *fundamental tilings*, which means that they satisfy the following axiom, stronger than (CG2): $g(\overset{\circ}{P}) \cap h(\overset{\circ}{P}) \neq \emptyset$ implies $g = h$.)

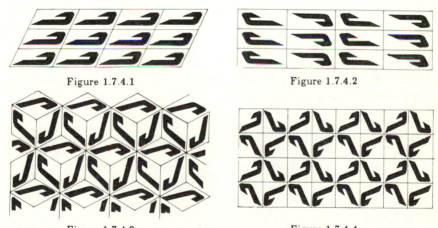

Figure 1.7.4.1

Figure 1.7.4.2

Figure 1.7.4.3

Figure 1.7.4.4

Figure 1.7.4.5
(Source: [BD])

While the number of crystallographic groups is finite, the tiles themselves admit of infinite variations (figures 1.7.4.6, 1.7.4.7 and 1.7.4.8).

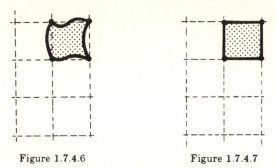

Figure 1.7.4.6   Figure 1.7.4.7

M. C. Escher, *Perpetual Motion*, 1953. Watercolor, 305 × 230 mm.
Escher Foundation–Haags Gemeentemuseum (The Hague)

Figure 1.7.4.8 (See [ER])

1.7.5. PROOF OF 1.7.4.

We shall use freely, sometimes without giving the exact reference, facts from plane affine Euclidean geometry found in chapter 9. Let $\vec{E}$ be the vector space underlying E, and $\text{Is}^+(E) \to \text{GL}(\vec{E})$ the homomorphism whose kernel $T(E)$ is the group of translations of E (see 2.3.3.4 if necessary). Recall (cf. 9.3.4) that a map $f \in \text{Is}^+(E) \setminus T(E)$ is necessarily a *rotation* by some *angle* around its unique fixed point, called its *center*. The first key idea in the proof is the following:

1.7.5.1. *The group G acts discretely on E.* This means that, for any $a \in E$, the orbit $G(a)$ is discrete in E, i.e., it consists solely of isolated points. (We recall that a point x of a metric space X is called *isolated* if there exists $\epsilon > 0$ such that $B(x, \epsilon) \cap X = \{x\}$.) In particular, the intersection of any orbit with a compact set is finite, since a discrete compact set is finite.

To prove the action is discrete, remark first that every compact set of E contains only a finite number of distinct tiles $g(P)$. This comes from (CG2) and the fact that every sequence of points in a compact set has a convergent subsequence. Next, observe that the stabilizer of a tile $P' = g(P)$, defined as $G_{P'} = \{g \in G \mid g(P') = P'\}$, is always finite. In fact, by 9.8.6.1, there exists $a' \in E$ such that

$$G_{P'} \subset G_{a'} = \{g \in G \mid g(a') = a'\}.$$

Choose another tile P'' such that the associated point a'' given by 9.8.6.1 is different from a'. Since only the identity leaves both a'' and a' fixed (see 9.1.6 if necessary), the cardinality of $G_{P'}$ is equal to the number of tiles $g(P'')$, for $g \in G_{P'}$. Since these tiles are distinct and contained in a circle of fixed radius, there are only finitely many of them (figure 1.7.5.1).

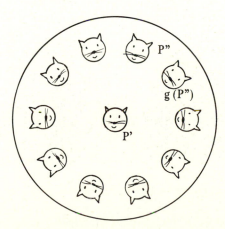

Figure 1.7.5.1

Now for the proof that every orbit $G(a)$ is discrete. Since G is a group of isometries, it is enough to show that a is isolated in $G(a)$. Take an arbitrary $\eta > 0$. The number of points of $G(a) \cap B(a, \eta)$ lying in any given tile is finite, by the paragraph above, and the number of tiles intersecting $B(a, \eta)$ is also finite, since they are all contained in the compact disc $B(a, \eta + \delta)$, where δ is the diameter of a tile. Thus $G(a) \cap B(a, \eta)$ is a finite set, and we can find ϵ such that $G(a) \cap B(a, \epsilon) = \{a\}$.   □

Now consider the subgroup $\Gamma = G \cap T(E)$ of translations of G. The second key point in the proof is the following property:

1.7.5.2. *The group Γ is a lattice, i.e., there exists a basis $\{\vec{u}, \vec{v}\}$ of $\vec{E}$ such that Γ consists exactly of translations by vectors in $\mathbf{Z}\vec{u} + \mathbf{Z}\vec{v}$.*

Proof. We start by proving, by contradiction, that Γ contains at least two linearly independent translations. Assume first that $\Gamma = \mathrm{Id}_E$, that is, G contains only rotations; if two such rotations r, s had different centers, their commutator $rsr^{-1}s^{-1}$ would be a non-trivial translation, by 1.7.5.0. Thus the rotations of G all have the same center ω, and $\bigcup_{g \in \Gamma} g(P)$ is contained in a disc of fixed radius, contradicting (CG1) (figure 1.7.5.2.1).

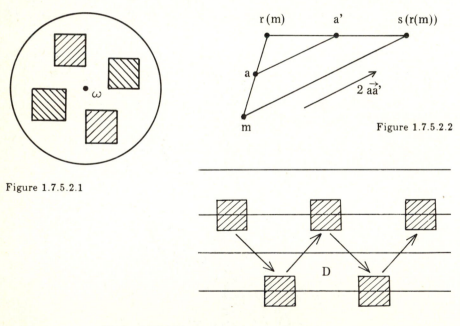

Figure 1.7.5.2.1

r(m)   a'   s(r(m))

a

$2 \overrightarrow{aa'}$

m   Figure 1.7.5.2.2

D

Figure 1.7.5.2.3

Now suppose that the directions of elements of Γ are all parallel. Take $r \in G \backslash \Gamma$ and a translation $t \in \Gamma$ by a vector $\vec{\xi}$. Then rtr^{-1} is a translation by the vector $r(\vec{\xi})$, and, since $r(\vec{\xi})$ must be parallel to $\vec{\xi}$, we conclude that r is a reflection through some point. All elements of $G \backslash \Gamma$ are thus reflections through points; but if two reflections r, s have centers a, a', then sr is a translation by the vector $2\overrightarrow{aa'}$ (figure 1.7.5.2.2). Thus the centers of all elements of $G\backslash\Gamma$ are on a line D parallel to the direction of the translations; this implies that $\bigcup_{g\in G} g(P)$ is contained in a strip centered on D (figure 1.7.5.2.3), again contradicting (CG1).

This shows that Γ contains translations by two linearly independent vectors. There remains to show that there exist $\vec{u}, \vec{v}$ such that

$$\Gamma = \{\text{ translations by } \vec{w} \mid \vec{w} \in \mathbf{Z}\vec{u} + \mathbf{Z}\vec{v}\}.$$

By discreteness (1.7.5.1), we can choose a translation in Γ by a vector $\vec{u}$ with minimal norm $\|\vec{u}\|$, then another translation in Γ by $\vec{v} \notin \mathbf{Z}\vec{u}$ with minimal norm $\|\vec{v}\|$. We will show that $\vec{u}$ and $\vec{v}$ are the desired vectors.

Let Q be the parallelogram

$$Q = \{\, a + t\vec{u} + s\vec{v} \mid t, s \in [0, 1]\,\},$$

for a fixed $a \in E$. Since G is a group, and

$$\Gamma \supset \{\text{ translations by } \vec{w} \mid \vec{w} \in \mathbf{Z}\vec{u} + \mathbf{Z}\vec{v}\},$$

the images $g(Q), g \in \Gamma$, fill E. Thus, if there is a point in the orbit of a under Γ that is not in $a + \mathbf{Z}\vec{u} + \mathbf{Z}\vec{v}$, there is also one, which we call y, inside Q (figure 1.7.5.2.4).

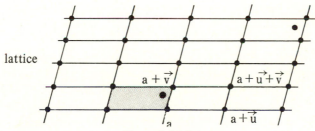

lattice

Figure 1.7.5.2.4

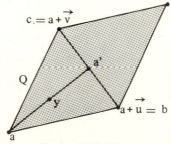

Figure 1.7.5.2.5

We claim that the distance from y to one of the vertices of Q is strictly less than $\|\vec{u}\|$ or $\|\vec{v}\|$; to see this, assume, without loss of generality, that y is in the interior of the triangle $\{a, b = a + \vec{u}, c = a + \vec{v}\}$, and observe that the line $\langle a, y \rangle$ intersects the side $\langle b, c \rangle$ at a', and we have $d(a, y) < d(a, a')$ (figure 1.7.5.2.5). But then $d(a, a') < \big(d(a, b) + d(a, c)\big)/2 = (\|\vec{u}\| + \|\vec{v}\|)/2$, proving our claim, and contradicting the choice of $\vec{u}$ and $\vec{v}$. $\square$

1.7.5.3. Now let $G' = G \backslash \Gamma$. If G' is empty, we're in the situation of figure 1.7.4.1; otherwise, the elements of G' are all true rotations, and they must be of *finite order* by (CG2). Assume first that they all have order two; then we're in the situation of figure 1.7.4.2. Suppose now that $r \in G'$ has order $\alpha \geq 3$, and assume also its angle is $2\pi/\alpha$, which is always possible. Call its center a. Let $b \neq a$ be the center of a rotation in G' such that $d(b, a)$ is minimal, and let s be a rotation of center b, order $\beta \geq 3$ and angle $2\pi/\beta$. Put $t = (rs)^{-1}$, so that $rst = \mathrm{Id}_E$. Now 9.3.6 shows that the center c of t is such that the angles of the triangle formed by a, b, c are half the angles of r, s, t, respectively (figure 1.7.5.3.1).

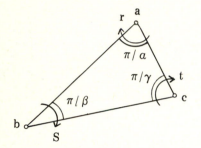
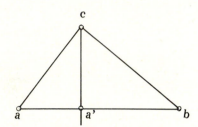

Figure 1.7.5.3.1   Figure 1.7.5.3.2

We now show that the choice of b implies that the angle of t is $2\pi/\gamma$, where γ is the order of t. In fact, if t had angle $2\pi n/\gamma$, $n \geq 2$, there would exist a rotation $t' \in G'$ with center t and angle less than $2\pi n/\gamma$. Then, applying the result from 9.3.6 mentioned above, it would follow that $(rt')^{-1} \in G'$ has center a', a point closer to a than b is, contradicting the choice of b (figure 1.7.5.3.2).

Since the sum of angles of a triangle is π, we obtain the basic condition

$$\frac{1}{\alpha} + \frac{1}{\beta} + \frac{1}{\gamma} = 1.$$

But $\alpha \geq 3$, $\beta \geq 3$, $\gamma \geq 2$ are all integers, so the only possibilities are the following:

1.7.5.4

	α	β	γ
case I	3	3	3
case II	2	4	4
case III	2	3	6

We easily deduce 1.7.4, by showing that cases I, II and III correspond to the groups associated with figures 1.7.4.3, 1.7.4.4 and 1.7.4.5, respectively.

1.7.6. We now lift the restriction that tiles cannot be turned over; in other words, the tiles are decorated on both sides. This corresponds to working in $\mathrm{Is}(E)$ instead of $\mathrm{Is}^+(E)$. The considerations in 1.7.5.1 and 1.7.5.2 still hold, but the possibilities for G increase to seventeen. The additional twelve groups are associated with figures 1.7.6.1 through 1.7.6.12.

Figure 1.7.6.1

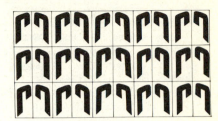

Figure 1.7.6.2

Figure 1.7.6.3

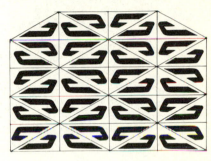

Figure 1.7.6.4

Figure 1.7.6.5

Figure 1.7.6.6

Source: [BD]

1.7.7. NOTES.

1.7.7.1. The first rigorous proof that the number of crystallographic groups is equal to seventeen seems to be due to Fedorov (1891). For more historical or physical information, see [FT2, 38], [CR1, 279], and [H–C, chapter II].

Fedorov's prime interest was in crystallography. The most recent and complete reference on plane and spatial crystallographic groups is [BT]; [CR2] contains some very perceptive and precise historical remarks. See especially [SC1]; also 1.9.14, [SC2] and [B–B–N–W–Z].

Figure 1.7.6.7

Figure 1.7.6.8

Figure 1.7.6.9

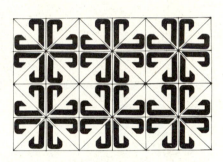

Figure 1.7.6.10

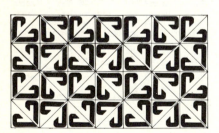

Figure 1.7.6.11

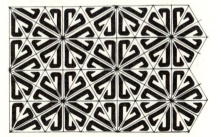

Figure 1.7.6.12

Source: [BD]

1.7.7.2. The problem of tilings can be extended to three-dimensional Euclidean space, and also to other, non-Euclidean, spaces, for example, spheres and hyperbolic spaces. Tilings of the sphere S^2 are treated in the next section (especially 1.8.6), and hyperbolic tilings in 19.6.12.

1.7.7.3. For tilings of three-dimensional Euclidean space, see [BT]. The complexity of the classification of crystallographic groups increases fast with the

M. C. Escher—Sketch for a periodic tiling by horsemen
India ink and watercolor. Escher Foundation–Haags Gemeentemuseum, The Hague.

Figure 1.7.6.13

Figure 1.7.6.14. The tile P.

dimension: there are 230 crystallographic groups in three dimensions, 4783 in four. The fact that their number is finite in any dimension is non-trivial and is known as Bieberbach's theorem; the proof is based on a generalization of 1.7.5.2, but is much harder. See [WF, 100].

1.7.7.4. It is a little easier to classify the crystallographic groups (in arbitrary dimension) that act with no fixed points, i.e., such that the stabilizer of any point is trivial. The interest in such groups lies in that the quotient space E/G is locally homeomorphic to E, and is thus a differentiable manifold. In two dimensions, only two groups satisfy this condition: the group of figure 1.7.4.1, which has as quotient the two-torus $\mathbf{R}^2/\mathbf{Z}^2$, and the group of figure 1.7.6.3, whose quotient is the famous Klein bottle. For a classification in arbitrary dimension, see [WF, chapter 3]; in three dimensions there are six such groups ([WF, 117]).

1.7.7.5. On tilings and marginally related, but equally captivating, subjects, see [FT1] and [FT2].

1.7.7.6. For pictures similar to 1.7.4.8 and 1.7.6.13, see [FT2, plates I, II and III], [ER] and [MG], which is entirely devoted to Escher's use of plane crystallographic groups.

1.7.7.7. See also 1.8.7, 1.9.4, 1.9.9, 1.9.12.

1.7.7.8. There are finer classifications, which lead to more types of tilings. See [B–P–B–S, 7], as well as the two articles [GR–SH1] and [GR–SH2], which discuss in detail two such classifications (one with 81 types, the other with 91).

1.8. Tilings of the two-sphere and regular polyhedra

1.8.1. One way of summarizing 1.7.5.3 is by saying that a rotation of finite order with center a gives rise to a regular tiling of a circle centered at a (figure 1.8.1). Thus, tilings of the plane lead to the study of finite subgroups of the group of rotations around a fixed point, which is isomorphic to $O^+(2)$ (see section 8.3). Analogously, the study of tilings of three-dimensional Euclidean space leads to the study of finite subgroups of $O^+(3)$, or, which is the same, of tilings of the sphere S^2. In this section we will only deal with the latter question; the reader can consult [BT] for the 230 crystallographic groups in three dimensions.

 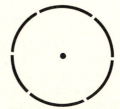

Figure 1.8.1

1.8.2. THEOREM. *Up to conjugation in $O(3)$, finite subgroups of $O^+(3)$ fall into five types: two families, indexed by an integer $n \geq 2$, and three exceptional groups.*

We begin with a description of the mathematical approach to this problem. Then, in 1.8.3.4, we show how regular polyhedra fit in with our result and lead to a visual understanding of the subject. For more on regular polyhedra, see 12.5 and 12.6.

1.8.3. OUTLINE OF PROOF

1.8.3.1. The point of departure is a clever use of the class formula (1.6.6). Another method, using regular polyhedra and Euler's formula, is given in 12.7.4. In section 12.6 we shall classify regular polytopes in arbitrary dimension.

Let $G \subset O^+(3)$ be a finite subgroup, and consider its action restricted to the sphere S^2. We introduce the set

$$\Gamma = \big\{ (g, x) \in (G \setminus e) \times S^2 \mid g(x) = x \big\},$$

which projects onto the set

$$X = \{ x \in S^2 \mid g(x) = x \text{ for some } g \in G \setminus e \}$$

of non-trivial fixed points of G (here e denotes the identity element). The latter can also be seen as the set of rotational axes of $G \setminus e$ (see 8.4.7.1). We will compute $\#\Gamma$ in two ways, first by summing over g, and then over x.

Since each $g \in G \setminus e$ has exactly two distinct fixed points (see 8.4.7.1), we have $\#\Gamma = 2(\#G - 1)$. To calculate $\#X$, we parametrize the orbits of the G-action on $X \subset S^2$ by a section A, as in 1.6.6. For $x \in A$, we have (cf. 1.6.2)

$$\#O(x) = \frac{\#G}{\#G_x};$$

but $\#G_x$ is constant for $y \in O(x)$, by 1.5.3. We call this constant ν_x; it is none other than the order of a rotation generating G_y. Thus $\#(G_y \setminus e) = \nu_x - 1$. Since $G_y \setminus e = \{ g \in G \mid (g, y) \in \Gamma \}$, we conclude that

$$\#\Gamma = \sum_{x \in A} (\nu_x - 1)\#O(X) = \sum_{x \in A} (\nu_x - 1)\frac{\#G}{\#G_x} = \sum_{x \in A} \frac{\#G(\nu_x - 1)}{\nu_x},$$

whence

1.8.3.2
$$2 - \frac{2}{\#G} = \sum_{x \in A} \left(1 - \frac{1}{\nu_x}\right).$$

1.8.3.3. If all the ν_x are large, each quantity $1 - 1/\nu_x$ is close to 1, so the number of summands must be two or less, otherwise the sum would be greater than 2. Sharpening this observation a little bit, it is easy to conclude that the only possible cases are the ones listed in the tables below (where n is an arbitrary integer ≥ 2):

2 orbits	ν_1	ν_2	$\#G$
case I	n	n	n

3 orbits	ν_1	ν_2	ν_3	$\#G$
case II	2	2	n	$2n$
case III	2	3	3	12
case IV	2	3	4	24
case V	2	3	5	60

1.8.3.4. Thus there are only five possible cases for the cardinality of the orbits and the orders of the associated stabilizers. There remains to show that all possible cases occur, and that each case corresponds, up to conjugation in $O(3)$, to a unique group G.

The uniqueness of G is not very difficult but requires some care; see [AS]. For existence, observe that:

— Case I is realized by the cyclic group of order n generated by a rotation of order n in $\mathbf{R}^3$; the two orbits consist of the two points of S^2 lying on the common rotation axis (figure 1.8.3.1).

— Case II corresponds to the subgroup of $O^+(3)$ leaving invariant a regular n-sided polygon drawn on a plane of $\mathbf{R}^3$ and centered at the origin. This group, called the *dihedral group* of order $2n$ (see section 0.2), contains n rotations (by angles $2\pi k/n$) around the axis perpendicular to the plane of the polygon, plus n reflections through the lines joining the center of the polygon to its vertices and to the midpoints of the sides (figure 1.8.3.2—watch out for a small distinction between the cases n even and n odd). The two points of S^2 situated on the axis of the polygon form a single orbit. A second one consists of the vertices of the polygon (if it was drawn inscribed in S^2), and the third, of the radial projections of the midpoints of the sides on S^2.

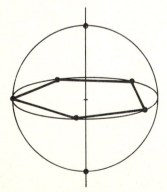

Figure 1.8.3.1

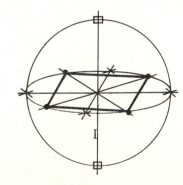

Figure 1.8.3.2

— Cases III, IV, V are realized by the subgroups of $O^+(3)$ leaving invariant a regular tetrahedron, cube and dodecahedron, respectively, all centered at the origin (figures 1.8.4.3, 1.8.4.4 and 1.8.4.5). The existence of the tetrahedron and cube is trivial. The same cannot be said for the existence of the dodecahedron; we refer the reader to 12.5.5.

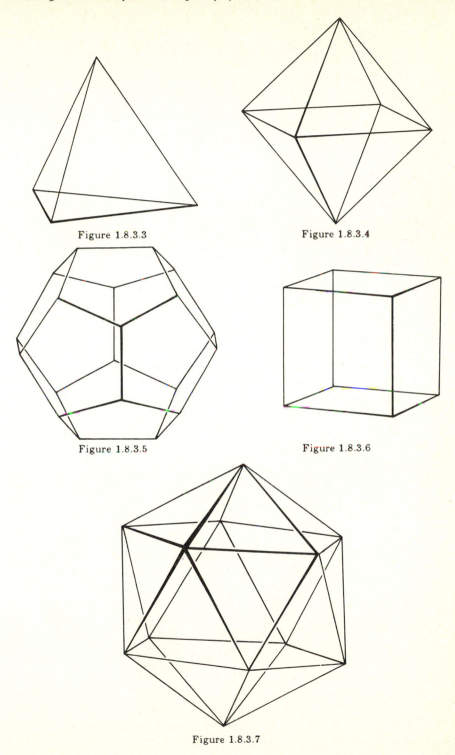

Figure 1.8.3.3

Figure 1.8.3.4

Figure 1.8.3.5

Figure 1.8.3.6

Figure 1.8.3.7

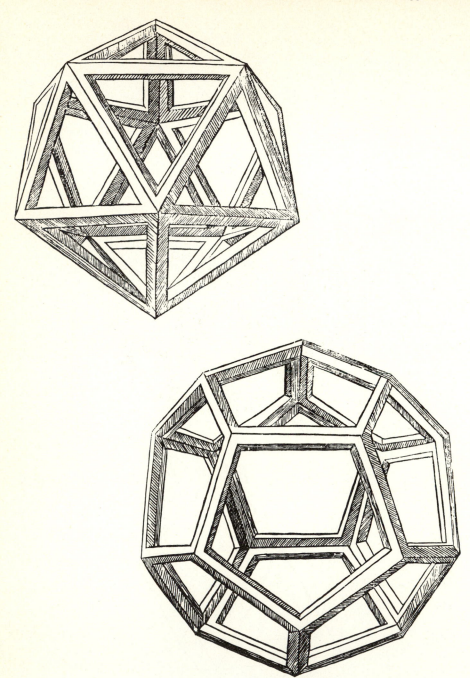

Icosahedron and dodecahedron drawn by Leonardo da Vinci
for Fra Luca Pacioli's *De Divina Proportione*.

Figure 1.8.3.8

1.8.4. We remark (cf. 12.5.4 and 12.5.5) that the groups of the cube and the regular octahedron are the same, as are those of the regular dodecahedron and icosahedron. Also, the tetrahedron group is isomorphic to $\mathcal{A}_4$, the group of the cube to $\mathcal{S}_4$, and the octahedron group to $\mathcal{A}_5$; see 12.5.5.6.

1.8.5. It is a relatively simple task to investigate the finite subgroups of $O(3)$, an extension analogous to that of 1.7.6. We find, associated with case V, a group of order 120, whose corresponding tiling of S^2 is shown in figure 1.8.5.

Figure 1.8.5 (Source: [CR1])

1.8.6. When studying the tilings of S^2 (as well as those of the hyperbolic plane, cf. 19.2), the key point, the analysis in 1.7.5.2, remains valid. Specifically, one considers three "rotations" r, s and t such that $rst = \mathrm{Id}$. In all three geometries, Euclidean, spherical and hyperbolic, the angles of such rotations are of the form $2\pi/\alpha$, $2\pi/\beta$, $2\pi/\gamma$, where α, β, γ are integers; these angles, moreover, are twice those of the triangle formed by the centers of the three rotations. In the Euclidean case we always have $\dfrac{1}{\alpha} + \dfrac{1}{\beta} + \dfrac{1}{\gamma} = 1$. In the spherical case,

1.8.6.1
$$\frac{1}{\alpha} + \frac{1}{\beta} + \frac{1}{\gamma} > 1,$$

by 18.3.8.4, and in the hyperbolic case $\dfrac{1}{\alpha} + \dfrac{1}{\beta} + \dfrac{1}{\gamma} < 1$ by 19.5.4.

Formula 1.8.6.1 can be taken as the starting point for the classification of the types of finite subgroups of $O^+(3)$; the number of choices for α, β, γ is clearly finite. On the other hand, the equation $\dfrac{1}{\alpha} + \dfrac{1}{\beta} + \dfrac{1}{\gamma} < 1$ is satisfied by all but a few triples of integers (α, β, γ), and in 19.6.2 we shall see that all such triples do indeed give rise to a hyperbolic tiling. Thus, in contrast with the Euclidean and spherical cases, the number of hyperbolic tilings is infinite. For references and additional material, see also 19.6.12.

1.8.7. GROUP PRESENTATIONS. The dihedral group $\mathcal{D}_{2n}$ (cf. 0.2 and 1.8.3.4) can be defined by two generators r, s, satisfying the two relations

1.8.7.1 $r^n = s^2 = e, \qquad rs = sr^{-1}.$

The three exceptional finite subgroups of $O^+(3)$, corresponding to regular polyhedra (1.8.3.4), can also be defined by two generators r, s satisfying the relations

$$r^p = s^q = (rs)^2 = e,$$

where $p = 3$, $q = 3$ for the regular tetrahedron, $p = 3$, $q = 4$ for the cube and $p = 3$, $q = 5$ for the dodecahedron. Observe that the existence of a finite group with each of these presentations is not obvious.

Plane crystallographic groups can be presented in a similar way. The one corresponding to figure 1.7.4.1, for example, has two generators r, s and the relation $rs = sr$. We leave the others to the reader.

Another way of describing the dihedral group is by using two generators r, s that satisfy only the relations

1.8.7.2 $r^2 = s^2 = (rs)^n = e.$

The presentation of the symmetry groups of polyhedra in $O(3)$ is similar: three generators r, s, t, satisfying the relations

1.8.7.3 $r^2 = s^2 = t^2 = (rs)^p = (st)^q = (tr)^2 = e,$

where p and q are as above. The order of these groups is 24, 48 and 120, respectively; see 12.5.4.1, 12.5.4.2 and 12.5.5.6.

Observe the close similarity between 1.8.7.2 and 1.8.7.3; in fact, these are all cases of more general groups, defined as discrete subgroups of the group of isometries of an affine Euclidean space, whose generators are reflections through hyperplanes. The interest in such groups derives from the fact that they occur outside the bounds of Euclidean geometry; in particular, they play an essential role in the study of semisimple Lie groups, a field that has recently experienced a spectacular growth. The reader interested in this subject or in the details of the examples above should consult [C–M] (in particular pp. 38–51) and [BI4] (especially the historical note, pp. 234–240).

1.9. Exercises

1.9.1. Study the orbits of the **Z**-action on the circle $S^1 \subset \mathbf{R}^2 \cong \mathbf{C}$ given by $\phi(n)(z) = e^{i\alpha n}z$, according to the nature of the real number α. If α is irrational, can this be used to construct the counterexample mentioned in 1.5.7?

1.9.2. Study the same question for the **R**-action on S^3 given by $\Phi(t)(z, z') = (e^{it}z, e^{i\alpha t}z')$, $\alpha \in \mathbf{R}$, where the notation is the same as in 1.2.9.

1.9.3. Fill in the details of the proof of theorem 1.8.2.

* **1.9.4.** For each of the five crystallographic groups containing only proper motions, then for all seventeen, find the following: the order of the stabilizers; the structure of the group; the different types of orbits; a presentation for the group (cf. 1.8.7).

1.9.5. Same question for the finite subgroups of $O^+(3)$ (resp. $O(3)$) acting on S^2.

1.9.6. Draw fundamental tilings of S^2 under the action of groups of type I, II, III and IV (see 1.8.3.3 and figure 1.8.5).

Figure 1.9.6 (Source: [CR1])

1.9.7. Show that in 1.7.7.4 we do indeed obtain as quotients the torus and the Klein bottle.

1.9.8. Show that there exists an infinite number of distinct tilings of the circle S^1, of the sphere S^3 (see for example 18.8.1), and, in general, of any sphere S^{2n+1}.

1.9.9. Make a comparative critical study of the classification of crystallographic groups in [GR, 72–84], [H–C, 70–81] and [WL, 22–115].

1.9.10. Compute $\#G_{E,p}$ (cf. 1.2.5) when E is a vector space of finite dimension n over a finite field of k elements.

1.9.11. Determine all compact subgroups of $O^+(3)$.

* **1.9.12.** FUNDAMENTAL DOMAINS. Let $G \subset \mathrm{Is}(E)$ be a subgroup of the group of isometries of an affine Euclidean plane; we assume all orbits of G are discrete subgroups of E. For a fixed $a \in E$, show that G and the set P defined by

$$P = \big\{\, x \in E \mid d(x,a) \le d(x,g(a)) \text{ for any } g \in G \,\big\}$$

satisfy the axioms of a crystallographic group (1.7.3).

1.9.13. VALENCIES OF A TILING. Consider a plane tiling by convex tiles satisfying 1.7.3. Show that the number of points in a tile P which belong to at least three different tiles is finite, greater than or equal to three, and does not depend on P. Let this number be r, let the corresponding points, ordered along the boundary of P, be $m_1, \ldots, m_r$, and call α_i the number of tiles containing m_i. Show that the sequence $(\alpha_i)_{i=1,\ldots,r}$ is independent, up to a circular permutation and a reversal, of the tile P; we call it the sequence of *valencies* of the tiling. Show that one always has

$$\frac{r}{2} - 1 = \sum_{i=1}^{r} \frac{1}{\alpha_i}$$

(if necessary, check 12.7.2 and 12.7.4). Deduce that only 23 sequences of valencies are possible *a priori*. Draw tilings whose sequences of valencies are $(3,3,3,3,3,3)$, $(3,3,3,3,6)$, $(3,3,3,4,4)$, $(3,3,4,3,4)$, $(4,4,4,4)$, $(3,6,3,6)$, $(3,4,6,4)$, $(6,6,6)$, $(4,8,8)$, $(3,12,12)$ and $(4,6,12)$. Finally, prove that the other sequences cannot be realized by any tiling. If necessary, see [GR–SH1], [GR–SH2] and the references therein.

* **1.9.14.** Does any triangle tile the plane? Any convex quadrilateral? Any quadrilateral? For pentagons, see [SC2].

1.9.15. Consider, in S_3, the two transpositions $s = (213)$ and $t = (132)$. Show that $t \circ s$ has order 3. Deduce that if S_3 acts on a group G by homomorphisms and some element g of G is invariant under s (i.e., $s(g) = g$) and anti-invariant under t (i.e., $t(g) = g^{-1}$) we have $g = g^{-1}$. For an application of this, see 9.5.4.9.

* **1.9.16.** Show that with the six tiles in figure 1.9.16.1 it is possible to tile the plane. Show also that any tiling with these tiles can never be periodic, i.e., the group of isometries which leaves the tiling globally invariant cannot contain a non-trivial translation. If you have trouble, consult [RN1], or make a sufficient number of photocopies of the tiles and start working with them. See figure 1.9.16.2 for another necessarily non-periodical tiling, due to Penrose.

Figure 1.9.16.1

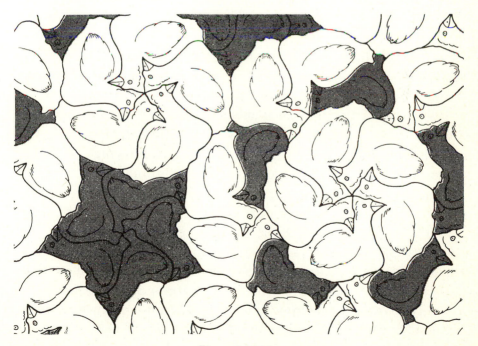

Figure 1.9.16.2

Chapter 2

Affine spaces

In this chapter we study affine spaces, the most frequent geometric spaces in this book. An affine space is nothing more than a vector space whose origin we try to forget about, by adding translations to the linear maps. It follows that the elementary properties of affine spaces, of their morphisms and of their subspaces are all properties from linear algebra, more or less disguised. As a consequence, most demonstrations are automatic: to prove a result about an affine space, the idea is generally to use some appropriate translation to reduce the statement to a vector form. This often means vectorializing the space at the appropriate point.

From section 2.5 on, geometry again occupies the foreground, starting with the theorems of Thales, Pappus and Desargues. The only harder, but very elegant, result is the fundamental theorem of affine geometry, section 2.6, which states that a set-theoretical map between affine spaces which takes collinear points into collinear points is, in essence, an affine map.

The end of the chapter is devoted to finite-dimensional real affine spaces. Among the results about such spaces that will be needed in the rest of the book is the following: a finite-dimensional real affine space (as well as its affine group $GA(X)$) has a canonical topology, and also a Lebesgue measure, unique up to multiplication by a scalar. Using this measure we associate to each compact subset K of an affine space X a point, called its center, which is invariant under automorphisms of X that leave K globally invariant. Along the same lines, we show that the subgroup of $GA(X)$ formed by automorphisms that leave globally invariant a compact set K with non-empty interior is compact; conversely, every compact subgroup of $GA(X)$ is contained in the stabilizer of a point of X.

For more details on the proofs in sections 2.1 through 2.6, the best is to consult [FL].

All vector spaces considered in this chapter are over a commutative field. When we consider together two vector or affine spaces, they will always be spaces over the same field, with the single and important exception of section 2.6.

It is an interesting exercise to study carefully where exactly the commutativity of the base field is used, and, when appropriate, to modify the statements of the results so as to render them valid for skew fields as well.

2.1. Definitions

2.1.1. DEFINITION. *An affine space over a field K is a faithful and transitive group action $(X, \vec{X}, \Phi)$, where $\vec{X}$ is a vector space over K considered with its additive group structure. The vector space $\vec{X}$ is said to underlie the affine space X.*

2.1.2. NOTATION. We will generally *denote* affine spaces by X. We put

$$\Phi(\vec{\xi}, x) = x + \vec{\xi}.$$

The maps $\Phi(\vec{\xi})$ are called *translations* of X; more precisely, $\Phi(\vec{\xi})$ is called a translation by the vector $\vec{\xi}$, and will be denoted by $t_{\vec{\xi}}$.

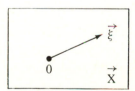

Figure 2.1.2

2.1.3. By 1.4.4.1, the action $(X, \vec{X}, \Phi)$ is simply transitive, so there exists a map $\Theta : X \times X \to \vec{X}$ such that $y = \Phi(\Theta(x, y), x)$ for all $x, y \in X$. We set $\overrightarrow{xy} = \Theta(x, y)$, and sometimes say that $\overrightarrow{xy}$ is the *free vector* associated with the *bound vector* $(x, \overrightarrow{xy})$, or with the pair (x, y). We also write

2.1.4 $$\overrightarrow{xy} = y - x;$$

a certain justification for this notation is given in 3.1.7. The fact that Φ is an $\vec{X}$-action can be translated as follows:

$$(x + \vec{\xi}) + \vec{\eta} = x + (\vec{\xi} + \vec{\eta});$$

we can simply write $x + \vec{\xi} + \vec{\eta}$. In particular, Θ satisfies

2.1.5 $\begin{cases} \Theta_x : X \ni y \mapsto \Theta(x, y) \in \vec{X} \text{ is a bijection} & \text{for all } x \in X, \\ \Theta(x, y) + \Theta(y, z) = \Theta(x, z) & \text{for all } x, y, z \in X, \end{cases}$

since we have $\Theta_x^{-1}(\vec{\xi}) = x + \vec{\xi}$. The identity $\Theta(x,y) + \Theta(y,z) = \Theta(x,z)$ is known as *Chasles' relation*.

2.1.6. ALTERNATIVE DEFINITION. *Let X be a non-empty set, $\vec{X}$ a vector space over K, and $\Theta : X \times X \to \vec{X}$ a map satisfying the conditions in 2.1.5. Then X is an affine space under the action $\Phi(\vec{\xi})(x) = \Theta_x^{-1}(\vec{\xi})$. This definition is indeed equivalent to 2.1.1, for we have $\Theta(x,x) = 0$, $\Theta(y,x) = -\Theta(x,y)$, $\Phi(-\vec{\xi}) \circ \Phi(\vec{\xi}) = \mathrm{Id}_X$, and thus $\Phi(\vec{\eta}) \circ \Phi(\vec{\xi}) = \Phi(\vec{\eta} + \vec{\xi})$.*

The bijections $\Theta_x : X \to \vec{X}$ also satisfy the following relations:

2.1.7 $\Theta_y^{-1} \circ \Theta_x = \Phi(\overrightarrow{xy})$, $\Theta_y \circ \Theta_x^{-1} : z \mapsto z + \overrightarrow{yx}$ for all $x, y \in X$.

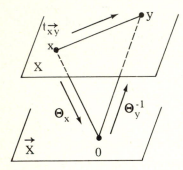 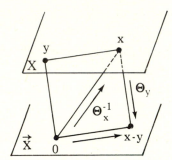

Figure 2.1.7

2.1.8. DEFINITION. *The dimension of X, denoted by $\dim X$, is defined as the dimension of $\vec{X}$. If $\dim X = 0$, X is called a point; if $\dim X = 1$, X is an affine line, and if $\dim X = 2$, X is called an affine plane.*

2.1.9. DEFINITION. *If $a \in X$, we denote by X_a the set X endowed with the vector space structure such that $\Theta_a : X \to \vec{X}$ is a vector space isomorphism. We say that X_a is the vectorialization of X at a.*

2.2. Examples. Affine frames

2.2.1. Take $X = \vec{X}$ and $\Phi(\vec{\xi})(\vec{\eta}) = \vec{\xi} + \vec{\eta}$ for all $\vec{\xi}, \vec{\eta} \in \vec{X}$. In this way we obtain a natural affine structure on any vector space.

2.2.2. If $(X, \vec{X}, \Phi)$, $(X', \vec{X}', \Phi')$ are two affine spaces, then so is $(X \times X', \vec{X} \times \vec{X}', \Phi \times \Phi')$, where

$$(\Phi \times \Phi')(\vec{\xi}, \vec{\xi}')(x, x') = \big(\Phi(\vec{\xi})(x), \Phi'(\vec{\xi}')(x')\big).$$

2.2.3. Let E be a vector space, F a vector subspace of E, and X a *translate* of F, i.e., a coset of F under the additive group structure of E. Then X has a natural affine space structure (X, F, Φ), where $\Phi(f)(x) = x + f$ for all $x \in X$ and $f \in F$. It is important to observe that here the set X does *not* possess a distinguished point, contrary to the trivial example in 2.2.1, where the origin plays a special role. In particular, if F is a hyperplane of E, giving a point in X is equivalent to giving a complementary subspace to F in E, and, in the absence any structure other than that of vector space, there is no such thing as a "natural" complementary subspace. (See 2.2.5 for an example that elaborates on this remark.) This "homogeneity" is one of the motivations for the concept of an affine space.

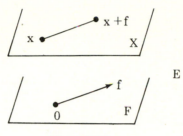

Figure 2.2.3

An important case of a space X as above is the set of solutions of a system of linear equations $\sum_j a_{ij} x_j = b_i$ $(i = 1, \dots, k)$. Indeed, this set is just the inverse image $f^{-1}(\vec{b})$ of the point $\vec{b} = (b_1, \dots, b_k)$ under the appropriate map f, and is thus a translate of the kernel $f^{-1}(0) = F$.

2.2.4. Let $(X, \vec{X}, \Phi)$ be an affine space and $\vec{S} \subset \vec{X}$ be a vector subspace. We have an equivalence relation defined on X by

$$x \sim x' \iff \overrightarrow{xx'} \in \vec{S}.$$

Then $(X/\sim, \vec{X}/\vec{S}, \Phi)$ has a natural affine structure, where Φ is defined by passing to the quotients. We shall see in 2.4.2 what the equivalence classes are.

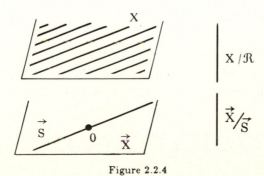

Figure 2.2.4

2.2.5. Let E be a vector space, and $H \subset E$ a hyperplane. Set

$$E_H = \{\, W \mid W \text{ is a complementary subspace to } H \text{ in } E \,\}.$$

Observe that each W is a one-dimensional vector subspace, and that, with the notation of 1.2.5, we can write $E_H = G_{E,1} \backslash G_{H,1}$. Then we have the following

2.2.6. PROPOSITION. *The set E_H has a natural affine structure, the underlying vector space being $\vec{E}_H = L(E/H; H)$ (cf. 0.2).*

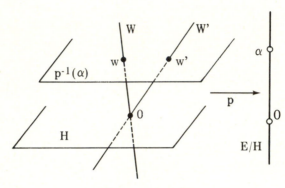

Figure 2.2.6

Proof. Observe that $\dim E/H = 1$, whence

$$\dim \vec{E}_H = \dim H = \dim E - 1.$$

We define the affine structure via the map $\Theta : E_H \times E_H \to L(E/H; H)$. Let $W, W' \in E_H$, $\alpha \in E/H$. If we denote by $p : E \to E/H$ the canonical projection, $p^{-1}(\alpha)$ is an affine hyperplane of E (cf. 2.2.3); we know that p, restricted to an arbitrary complementary subspace to H, is an isomorphism, so we can put

$$\Theta(W, W')(\alpha) = (p|_{W'})^{-1}(\alpha) - (p|_W)^{-1}(\alpha) \in H.$$

See figure 2.2.6, where $w' = (p|_{W'})^{-1}(\alpha)$, $w = (p|_W)^{-1}(\alpha)$. It is immediately seen that Θ satisfies axioms 2.1.5.   □

The example above will be essential in section 5.1. It is natural to view this example and the one in 2.2.3 together (figures 2.2.3 and 2.2.6); the reader can demonstrate the following proposition, looking up the notion of isomorphism between affine spaces in section 2.3 if necessary.

2.2.7. PROPOSITION. *Let E be a vector space, F a hyperplane of E and X a translate of X. Then the map $W \to W \cap X$ gives a natural affine space isomorphism between E_H and X.*   □

2.2.8. In order to be able to calculate in affine spaces, we introduce something similar to vector space bases. By convention,

> we define affine frames for finite-dimensional affine spaces only.

2.2.9. DEFINITION. *An (affine) frame for the affine space X consists of $d+1$ points $\{x_i\}_{i=0,1,\ldots,d}$ in X such that $\{\overrightarrow{x_0 x_i}\}_{i=1,\ldots,d}$ is a basis for $\vec{X}$ (thus we necessarily have $d = \dim X$). The coordinates of $x \in X$ in this frame are the scalars $\{\lambda_i\}_{i=1,\ldots,d}$ such that $\overrightarrow{x_0 x} = \sum_i \lambda_i \overrightarrow{x_0 x_i}$, i.e., the coordinates of $\overrightarrow{x_0 x}$ in the basis $\{\overrightarrow{x_0 x_i}\}_{i=1,\ldots,d}$ of $\vec{X}$. We can write $x = (\lambda_1, \ldots, \lambda_d)$.*

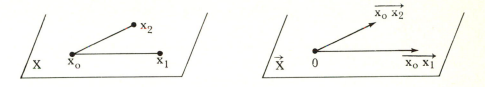

Figure 2.2.9

2.2.10. REMARK. It amounts to the same thing to say that $\{x_i\}_{i=1,\ldots,d}$ is a basis for the vectorialization X_{x_0} (cf. 2.1.9), or that $\{x_0, x_1, \ldots, x_d\}$ is a simplex of X (cf. 2.4.7).

2.3. Affine morphisms

Let $(X, \vec{X}, \Theta)$ and $(X', \vec{X}', \Theta')$ be two affine spaces (over the same field, cf. 0.2), and $f : X \to X'$ a map (in the set-theoretical sense).

2.3.1. PROPOSITION AND DEFINITION. *The following conditions are equivalent:*

  i) $f \in L\big(X_a; X'_{f}(a)\big)$ *for some $a \in X$;*
  ii) $f \in L\big(X_a; X'_{f}(a)\big)$ *for all $a \in X$;*
  iii) $\Theta_{f(a)} \circ f \circ \Theta_a^{-1} \in L(\vec{X}; \vec{X}')$ *for some $a \in X$;*
  iv) $\Theta_{f(a)} \circ f \circ \Theta_a^{-1} \in L(\vec{X}; \vec{X}')$ *for all $a \in X$.*

Moreover, if these conditions are satisfied, $\Theta_{f(a)} \circ f \circ \Theta_a^{-1} \in L(\vec{X}; \vec{X}')$ depends only on f, and will be denoted by $\vec{f}$ or $L(f)$. We have $\vec{f} \circ \Theta = \Theta \circ (f \times f)$. A map satisfying the properties above is called an (affine) morphism, or an affine map, between X and X'. The set of such maps is denoted by $A(X; X')$.

We say that f is an (affine) isomorphism if f is a morphism and is bijective; f is an automorphism if f is an isomorphism from X into itself. $\square$

$$X_a \xrightarrow{f} X'_{f(a)}$$

$$\begin{array}{ccc} X & \xrightarrow{f} & X' \\ \scriptstyle\theta_a\downarrow & & \downarrow\scriptstyle\theta'_{f(a)} \\ \vec{X} & \xrightarrow{\vec{f}} & \vec{X}' \end{array} \qquad \begin{array}{ccc} X \times X & \xrightarrow{f\times f} & X' \times X' \\ \scriptstyle\theta\downarrow & & \downarrow\scriptstyle\theta' \\ \vec{X} & \xrightarrow{\vec{f}} & \vec{X}' \end{array}$$

2.3.2. If $f : X \to X'$ is an affine map, we have, for all $x, y \in X$:

$$\overrightarrow{f(x)f(y)} = \vec{f}(\overrightarrow{xy}), \qquad f(y) = f(x) + \vec{f}(\overrightarrow{xy}).$$

The conclusion is that, heuristically, f consists of a translation and a linear map.

2.3.3. EXAMPLES.

2.3.3.1. If $X = X' = \mathbf{R}$, we recover the well-known maps $x \to ax + b$, for $a, b \in \mathbf{R}$.

2.3.3.2. Every constant map f is affine, and satisfies $\vec{f} = 0$. Conversely, if f is affine and $\vec{f} = 0$, then f is constant.

2.3.3.3. The identity map $f = \mathrm{Id}_X$ is affine, and $\vec{f} = \mathrm{Id}_{\vec{X}}$. For the converse, see below.

2.3.3.4. If $f \in A(X; X')$ and $g \in A(X'; X'')$ we have $g \circ f \in A(X; X'')$, and $\overrightarrow{g \circ f} = \vec{g} \circ \vec{f}$. In particular, the set

$$\mathrm{GA}(X) = A(X; X) \cap S_X$$

is a group under composition of maps, and is called the *affine group* of X. We have a homomorphism

$$L : \mathrm{GA}(X) \ni f \to \vec{f} \in \mathrm{GL}(\vec{X}),$$

where $\mathrm{GL}(\vec{X}) = L(\vec{X}; \vec{X}) \cap S_{\vec{X}}$ is the linear group of $\vec{X}$. The kernel of L (defined as $L^{-1}(\mathrm{Id}_{\vec{X}})$) is evidently $\vec{X}$, the set of translations of X (cf. 2.1.2). It is a normal subgroup of $\mathrm{GA}(X)$, and will most often be *denoted by* $T(X) = \mathrm{Ker}\, L$.

2.3.3.5. The affine group acts simply transitively on affine frames. This can be seen, for instance, by observing that for two affine frames $\{x_i\}_{i=0,\dots,d}$ and $\{x'_i\}_{i=0,\dots,d}$ of X and X', respectively, there exists a unique $f \in A(X; X')$ such that $f(x_i) = x'_i$ for $i = 0, \dots, d$.

2.3.3.6. In particular, X is a homogeneous space for $\mathrm{GA}(X)$ (which is a bigger group than $\vec{X} = T(X)$ used in the definition, 2.1.1). Denote by $\mathrm{GA}_a(X)$ the stabilizer of $a \in X$ in $\mathrm{GA}(X)$. Then (cf. 1.5.5)

$$X \cong \mathrm{GA}(X)/\mathrm{GA}_a(X)$$

as sets, and the restriction $L : \mathrm{GA}_a(X) \to \mathrm{GL}(X)$ is a group isomorphism. To recover $\mathrm{GA}(X)$ from $T(X)$ and $\mathrm{GA}_a(X)$, we introduce the following

2.3.3.7. Definition. *A group G is called the* semidirect product *of two of its subgroups H and K if $G = H \cdot K = \{\, hk \mid h \in H, k \in K \,\}$, $H \cap K = \{e\}$ and H is normal.*

It follows that any g can be written as hk in a unique way, that $G = K \cdot H$ and that g can be written as $k'h$ in a unique way, but in general $k \neq k'$.

2.3.3.8. Proposition. *If $a \in X$, we have a semidirect product*

$$\mathrm{GA}(X) = T(X)\,\mathrm{GA}_a(X).$$

Proof. The requisite map from $\mathrm{GA}(X)$ into $T(X) \times \mathrm{GA}_a(X)$ is evidently $f \mapsto (t_{\overrightarrow{f(a)a}}, t_{\overrightarrow{f(a)a}} \circ f)$. $\qquad\square$

2.3.3.9. For every $a \in X$, $\lambda \in K^* = K \setminus 0$, we call the map

$$H_{a.\lambda} : x \to a + \lambda \overrightarrow{ax}$$

the *homothety of center a and ratio λ*. Homotheties are affine maps, and we have $\vec{H}_{a.\lambda} = \lambda\,\mathrm{Id}_{\vec{X}}$ (use 2.3.2). Here the commutativity of K is essential.

Figure 2.3.3.9.1

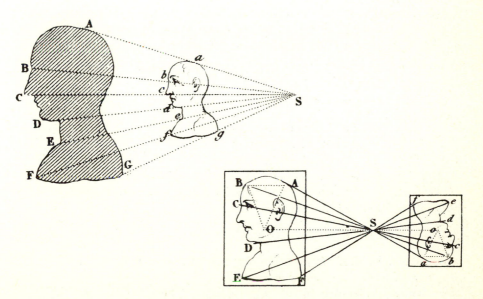

Figure 2.3.3.9.2

The converse is also true:

2.3.3.10. Proposition. *Let* $f \in GA(X)$ *be such that* $\vec{f} = \lambda \operatorname{Id}_{\vec{X}}$, *where* $\lambda \in K \setminus 1$. *Then there exists a unique* a *such that* $f = H_{a,\lambda}$.

Proof. For this, as for most not too subtle results about affine spaces, it is enough to vectorialize. Here we take an arbitrary $b \in X$ and work in the vector space X_b. Finding a such that $f(a) = a$, is, by 2.3.2, equivalent to solving $a = f(b) + \lambda(a - b)$. There is a unique solution

$$a = \frac{1}{1 - \lambda}\big(f(b) - \lambda b\big). \qquad \Box$$

We also have, for every $a \in X$ and $\lambda, \mu \in K$,

2.3.3.11 $$H_{a,\lambda} \circ H_{a,\mu} = H_{a,\lambda\mu}.$$

When $\lambda = 1$, proposition 2.3.3.10 doesn't work, but then we know that f is a translation (cf. 2.3.3.4). In order to collect together all morphisms f such that $\vec{f} = \lambda \operatorname{Id}_{\vec{X}}$ ($\lambda \in K^*$), we work as follows:

2.3.3.12. Proposition and definition. *The center of* $GL(\vec{X})$ *is* $K^* \operatorname{Id}_{\vec{X}}$. *The inverse image under* L *(cf. 2.3.3.4) of* $K^* \operatorname{Id}_{\vec{X}}$ *is a normal subgroup of* $GA(X)$, *denoted by* $\operatorname{Dil}(X)$. *As a set,* $\operatorname{Dil}(X)$ *is the union of* $T(X)$ *and the set of* $H_{a,\lambda}$ *for* $a \in X$ *and* $\lambda \in K^* \setminus 1$; *furthermore,* $T(X)$ *is a normal subgroup of* $\operatorname{Dil}(X)$. *The elements of* $\operatorname{Dil}(X)$ *are called dilatations of* X. $\qquad \Box$

The proof of the first assertion is exercise 2.8.4 (see 8.2.16 if necessary). In 2.5.6 we shall see a nice geometric characterization of dilatations.

2.3.4. Does $A(X; X')$ have a natural algebraic structure, similar to the vector space structure of $L(\vec{X}; \vec{X}')$? The interested reader can refer to [FL] to see how $A(X; X')$ is naturally made into an affine space of dimension

$$\dim A(X; X') = (\dim X + 1) \dim X'.$$

See 2.3.9 for a heuristic derivation of this value.

2.3.5. Let E, E' be vector spaces, H, H' hyperplanes of E, E', respectively. Consider the associated affine spaces $X = E_H$, $X' = E'_{H'}$ (see 2.2.5). Take $f \in L(E; E')$; is there a derived morphism $\bar{f} : X \to X'$? A necessary condition is that, for any one-dimensional vector subspace W of E complementary to H, the image $f(W)$ still be one-dimensional and transversal to H'. This is guaranteed by introducing the set

2.3.6 $$L_{H,H'}(E; E') = \big\{ f \in L(E; E') \mid f(H) \subset H' \text{ and } \underline{f} \text{ is injective} \big\},$$

where $\underline{f} \in L(E/H; E'/H')$ is the quotient map, well-defined because $f(H) \subset H'$. For such an f we indeed have $f(W) \in E'_H$ for any $W \in E_H$; the map thus defined is denoted by $\bar{f} : E_H \to E'_{H'}$. The following proposition is straightforward:

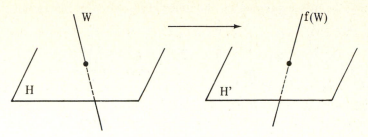

Figure 2.3.5

2.3.7. PROPOSITION. *For $f \in L_{H.H'}(E; E')$ we have $\bar{f} \in A(X; X')$. The map $L(\bar{f}) = \vec{\bar{f}} : L\big(L(E/H; H)\big); L(E'/H'; H')\big)$ is identical with*

$$\eta \mapsto f \circ \eta \circ \underline{f}^{-1}.$$

Moreover, $\bar{f} = \bar{g}$ if and only if there exists $k \in K^*$ such that $g = kf$. □

$$
\begin{array}{ccc}
E/H & \xrightarrow{\;f\;} & E'/H' \\
\big\downarrow{\scriptstyle \eta} & & \big\downarrow \\
H & \xrightarrow{\;f|_H\;} & H'
\end{array}
$$

For a converse of 2.3.7, see 3.2.1 and 5.1.3. The motivation for 2.3.7 comes in chapter 5.

2.3.8. USE OF COORDINATES. Let $f \in A(X; X')$, and fix two affine frames $\{x_i\}_{i=0.1.....n}$ and $\{x'_j\}_{j=0.1.....p}$ for X and X'. Put $\vec{e}_i = \overrightarrow{x_0 x_i}$, $\vec{e}'_j = \overrightarrow{x'_0 x'_j}$, and introduce the matrix associated with $\vec{f}$:

$$
M(\vec{f}) = \begin{pmatrix} \alpha_{11} & \cdots & \alpha_{1n} \\ \vdots & \ddots & \vdots \\ \alpha_{p1} & \cdots & \alpha_{pn} \end{pmatrix}.
$$

From 2.3.2 we have

$$f(x) = f(x_0) + \vec{f}(\overrightarrow{x_0 x}) = f(x_0) + \vec{f}\Big(\sum_i \lambda_i e_i\Big),$$

where $x = (\lambda_1, \ldots, \lambda_n)$. We can write $\overrightarrow{x'_0 f'(x_0)} = \sum_j a_j e'_j$, whence $\lambda'_j = a_j + \sum_k \alpha_{jk} \lambda_k$, where $f(x) = (\lambda'_1, \ldots, \lambda'_p)$. In other words, we have

$$f(x) = \Big(a_1 + \sum_i \alpha_{1i} \lambda_i, \ldots, a_p + \sum_i \alpha_{pi} \lambda_i\Big).$$

This can be also expressed by the following product of matrices:

2.3.9
$$
\begin{pmatrix} 1 \\ \lambda'_1 \\ \vdots \\ \lambda'_p \end{pmatrix} = M(f) \begin{pmatrix} 1 \\ \lambda_1 \\ \vdots \\ \lambda_p \end{pmatrix}, \quad \text{where} \quad M(f) = \begin{pmatrix} 1 & 0 & \cdots & 0 \\ a_1 & & & \\ \vdots & & M(\vec{f}) & \\ a_p & & & \end{pmatrix}.
$$

This allows us to perform any calculations in practice, for example, computing the composition $g \circ f$ of $f \in A(X; X')$ with $g \in A(X'; X'')$. The true reason why this works will be given in 3.2.5. Finally, it is clear that the natural dimension of $A(X; X')$ is the number of parameters involved in $M(f)$, which is equal to $p + pn = p(n+1) = \dim X'(\dim X + 1)$ (cf. 2.3.4).

2.4. Affine subspaces

Affine subspaces are the first objects introduced in geometry: lines in the plane, lines and planes in space. From the mathematical point of view, however, it is preferable to introduce morphisms before subspaces.

2.4.1. DEFINITION AND PROPOSITION. *Let X be an affine space and Y a non-empty subset of X. The following conditions are equivalent:*

  i) *There exists $a \in Y$ such that $\Theta_a(Y)$ is a vector subspace of $\vec{X}$;*
 ii) *For any $a \in Y$, $\Theta_a(Y)$ is a vector subspace of $\vec{X}$;*
iii) *There exists $a \in Y$ such that Y is a vector subspace of X_a;*
 iv) *For any $a \in Y$, Y is a vector subspace of X_a;*
  v) *There exist $a \in X$ and a vector subspace $\vec{V} \subset \vec{X}$ such that $Y = a + \vec{V}$.*

If Y satisfies the conditions above we call it an (affine) subspace of X. The vector space $\Theta(Y \times Y) = \Theta_a(Y) = \vec{V}$ ($a \in Y$ arbitrary) is called the direction of Y and is denoted by $\vec{Y}$; this is consistent because $(Y, \vec{Y}, \Theta \mid Y \times Y)$ is an affine space. The injection $i : Y \to X$ belongs to $A(Y; X)$. The dimension of Y is its dimension as an affine space. □

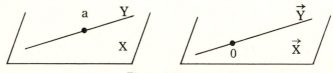

Figure 2.4.1

Condition (e) displays very well the real nature of affine subspaces—they are vector spaces, up to translations (cf. 2.3.2).

2.4.2. EXAMPLES.

2.4.2.1. Zero-dimensional subspaces of X called *points* of X. One- and two-dimensional subspaces are *(affine) lines* and *planes*, respectively. Affine subspaces whose direction is a vector hyperplane of $\vec{X}$ are also called *(affine) hyperplanes* of X.

2.4.2.2. Let $f \in A(X; X')$, let Y be a subspace of X and Y' a subspace of X'. Then $f(Y)$ is a subspace of X' and $f^{-1}(Y')$ is a subspace of X (if it is non-empty). Furthermore, $\overrightarrow{f(Y)} = \vec{f}(\vec{Y})$ and $\overrightarrow{f^{-1}(Y')} = (\vec{f})^{-1}(\vec{Y}')$.

2.4.2.3. There is a one-to-one correspondence between the affine subspaces Y of E_H (see 2.2.5) and the vector subspaces $\hat{Y}$ of E, where $\hat{Y}$ is the vector subspace of E spanned by $\bigcup_{W \in Y} W$. Notice that $\dim \hat{Y} = \dim Y - 1$, and $\vec{Y} = L(E/H; \hat{Y} \cap H)$.

Similarly, in example 2.2.3, there is a one-to-one correspondence between subspaces Y of X and vector subspaces $\hat{Y}$, where $\hat{Y}$ is the vector subspace of E spanned by Y. We then have $\vec{Y} = \hat{Y} \cap F$.

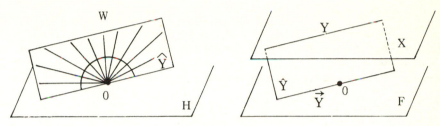

Figure 2.4.2.3

2.4.2.4. Let $(Y_i)_{i \in I}$ be an arbitrary family of subspaces of X. Then the intersection $\bigcap_i Y_i$ is either empty or a subspace; in the latter case we have

$$\overrightarrow{\bigcap_i Y_i} = \bigcap_i \vec{Y}_i.$$

The following result is a classical consequence of this:

2.4.2.5. Proposition. *Let $S \subset X$ be an arbitrary non-empty subset of an affine space X. There exists a smallest subspace containing S, called the subspace spanned by S, and denoted by $\langle S \rangle$. It is equal to the intersection of all subspaces containing S.* □

2.4.2.6. We conclude that (if $\dim X = d < \infty$) a $(d+1)$-point subset $\{x_i\}_{i=0.1.....d}$ of X is an affine frame if and only if $\langle x_0, x_1, \ldots, x_d \rangle = X$.

2.4.3. DEFINITION AND PROPOSITION. *The points $\{e_i\}_{i=0.....k}$ of X are said to be (affine) independent if $\dim \langle e_0, \ldots, e_k \rangle = k$. This happens if and only if $e_i \notin \langle e_1, \ldots, \hat{e}_i, \ldots, e_k \rangle$ (cf. 0.1) for every $i = 0, \ldots, k$.* □

2.4.4. For instance, two points x, y are independent if and only if $x \neq y$. In this case there is a unique line containing them, namely, $\langle x, y \rangle$. We thus recover one of Euclid's axioms: two distinct points determine a unique line.

2.4.5. PROPOSITION. *Let X be an affine space over a field of characteristic $\neq 2$. In order that a non-empty subset Y of X be a subspace it is necessary and sufficient that $\langle x, y \rangle \subset Y$ for every $x, y \in Y$.* □

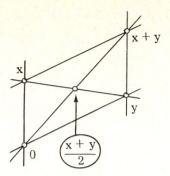

Figure 2.4.5

This, too, is an ancient axiom: a plan in space is a subset which contains the line joining any two of its points. The proof consists of vectorializing X at $0 \in Y$, and calculating in X_0. For x and y in Y, we have $\langle 0, x \rangle = Kx \subset Y$ and $x + y \in \langle 0, (x + y)/2 \rangle \subset Y$. This doesn't work in characteristic 2, and indeed the proposition is not valid in this case.

2.4.6. NOTE. Let a, b, c be three collinear points (i.e., lying on the same line). Then it makes sense to talk about the ratio $\overrightarrow{ac}/\overrightarrow{ab} \in K$.

2.4.7. SIMPLICES. A *simplex* in an affine space X of finite dimension d is a set of $d + 1$ independent points of X. A two-dimensional simplex is called a *triangle*, and a three-dimensional simplex is a *tetrahedron*. The *sides* of a triangle $\{x, y, z\}$ are the lines $\langle x, y \rangle$, $\langle y, z \rangle$, $\langle y, t \rangle$ and $\langle z, x \rangle$. The *edges* of a tetrahedron $\{x, y, z, t\}$ are the lines $\langle x, y \rangle$, $\langle x, z \rangle$, $\langle x, t \rangle$, $\langle y, z \rangle$, $\langle y, t \rangle$ and $\langle z, t \rangle$, and its *faces* are the planes $\langle x, y, z \rangle$, $\langle y, z, t \rangle$, $\langle z, t, x \rangle$ and $\langle t, x, y \rangle$. Sometimes a set of three independent points in a space X of arbitrary dimension is also called a *triangle*.

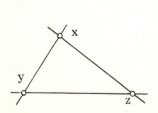
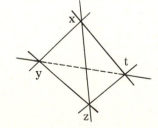

Figure 2.4.7

2.4.8. ANALYTIC GEOMETRY: EQUATIONS DEFINING SUBSPACES. Here, as in 2.3.8, we develop tools to perform calculations related with subspaces. Recall (cf. 2.2.8) that dim $X = d < \infty$. Recall also the following facts from linear algebra:

2.4.8.1. Proposition. *Let E be a vector space of dimension d; its dual E^* has the same dimension. Let F be an arbitrary subset of E, and $[F]$ the vector* *subspace of E spanned by F. Denote by*

$$F^\perp = \{f \in E^* \mid f(F) = 0\}$$

the orthogonal of F in E^*. We have $F^\perp = [F]^\perp$ and $\dim[F] + \dim F^\perp = d$. If S is a subset of E^*, its orthogonal in E is also denoted by $S^\perp$:

$$S^\perp = \{x \in E \mid f(x) = 0 \text{ for any } f \in S\}.$$

Again, $[S]^\perp = S^\perp$ and $\dim S^\perp + \dim[S] = d$. If F, F' are vector subspaces of E and S is a vector subspace of E^*, we always have

$$F^{\perp\perp} = F, \quad S^{\perp\perp} = S, \quad (F \cap F')^\perp = F^\perp + F'^\perp, \quad (F + F')^\perp = F^\perp \cap F'^\perp. \quad \Box$$

For infinite-dimensional vector spaces, refer to [BI5, section II.2].

2.4.8.2. Corollary. *Let $H \subset E$ be a hyperplane. There exists $f \in E^* \setminus 0$ such that $H = f^{-1}(0)$, and if $H = f^{-1}(0) = g^{-1}(0)$ for $f, g \in E^*$, we must have $g = \lambda f$ for some $\lambda \in K^*$. A subset $F \subset E$ is a subspace of dimension p if and only if there exist linearly independent functionals $f_1, \ldots, f_{d-p} \in E^*$ such that $F = \bigcap_{i=1}^{d-p} f_i^{-1}(0)$.* $\Box$

We can now move on to the affine case. The thing to watch out for is that the kernel of non-zero constant affine forms is empty, contrary to the case of linear forms, whose kernel always contains zero.

2.4.8.3. Definition. *An affine form over an affine space X is an element $f \in A(X; K)$, where K has its natural affine structure over itself (cf. 2.2.1). We consider on $A(X; K)$ its standard vector space structure.*

2.4.8.4. For example, if $\{x_i\}_{i=0,\ldots,d}$ is a frame in X, every affine form can be written in a unique way as

$$x = (\lambda_1, \ldots, \lambda_d) \to f(x) = a + \sum_i \alpha_i \lambda_i$$

(cf. 2.3.8). A constant is also an affine form, and its kernel (i.e., the inverse image of 0) is empty unless the constant is zero.

2.4.8.5. Proposition. *Let X be a d-dimensional affine space and V a non-empty subset of X. Then V is a p-dimensional affine subspace of X if and only if there exists a $(d-p)$-dimensional vector subspace $V' \subset A(X; K)$ not containing the constant affine form 1 and such that*

$$V = V'^\perp = \{x \in X \mid f(x) = 0 \text{ for any } f \in V'\}.$$

Proof. The idea is to vectorialize X at $a \in V$, and work in X_a. One direction is trivial: if V is an affine space, take $V' = V^\perp$, and $1 \notin V'$ because $1(x) = 1 \neq 0$ for any x. For the converse, we only have to prove that $V'^\perp \neq \emptyset$, and then apply 2.4.8.1 with $E = X_a$. We proceed by induction on $k = \dim V'$; the case $k = 0$ is trivial. Let $\{f_1, \ldots, f_k\}$ be a basis of V', and put $V_1' = Kf_1 + \cdots + Kf_{k-1}$. By the induction assumption, $V_1 = V_1'^\perp \neq \emptyset$, and we just have to show that $f_k|_{V_1}$ is not a non-zero constant. Assume f_k is constant on V_1; then $\overrightarrow{f_k(x_0)f_k(x)} = \vec{f_k}(\overrightarrow{x_0 x}) = 0$ for any $x_0, x \in V_1$. We can write $\vec{V}_1' = K\vec{f_1} + \cdots + K\vec{f}_{k-1}$, whence

$$\vec{f_k} = \sum_{i=1}^{k-1} \lambda_i \vec{f_i},$$

which implies

$$f_k(x) - f_k(x_0) + \sum_{i=1}^{k-1} \lambda_i f_i(x)$$

for any $x \in X$, since $f_i(x_0) = 0$, and finally $f_k(x_0) = 0$.   □

2.4.8.6. Examples. The most important example is when $\dim H = p = 1$ in 2.4.8.5 (hyperplanes). Then the proposition says that there exists an affine form $f \in A(X; K)$ such that $H = f^{-1}(0)$; conversely, if f is a non-constant affine form, $f^{-1}(0)$ is a hyperplane, and finally

$$f^{-1}(0) = g^{-1}(0)$$

is equivalent to $f = kg$, for $k \in K^*$.

The commonest practical cases are $K = \mathbf{R}$, $d = 2$ or 3, which correspond to our everyday physical space. If $\dim X = 3$, for example, the equation of a plane Y in an arbitrary frame will be

$$Y = \big\{ (x, y, z) \mid ax + by + cz + d = 0 \big\},$$

where a, b, c do not all three vanish, and are otherwise arbitrary. In the same space X, a line D is written

$$D = \big\{ (x, y, z) \mid ax + by + cz + d = a'x + b'y + c'z + d' = 0 \big\},$$

where the triples (a, b, c) and (a', b', c') must be non-zero and not proportional to one another.

2.4.8.7. Parametric representations. A k-dimensional subspace Y of X can be determined by giving a frame $\{x_i\}_{i=0,\ldots,k}$ of Y. We have

$$Y = \left\{ x_0 + \sum_{i=1}^{k} \lambda_i \overrightarrow{x_0 x_i} \,\middle|\, \lambda_i \in K \text{ for all } i = 1, \ldots, k \right\}.$$

Thus, a line in $\mathbf{R}^3$ will be

$$\big\{ (a + \lambda b, a' + \lambda b', a'' + \lambda b'') \mid \lambda \in \mathbf{R} \big\},$$

and a plane will be

$$\big\{ (a + \lambda b + \mu c, a' + \lambda b' + \mu c', a'' + \lambda b'' + \mu c'') \mid \lambda, \mu \in \mathbf{R} \big\}.$$

2.4.8.8. An important tool for working in the set of (vector, affine, projective) subspaces is the use of "grassmannian coordinates": see [HO–PE, chapter VII].

2.4.9. PARALLELISM AND INTERSECTION OF SUBSPACES

2.4.9.1. Definition. *Two subspaces S, T of the affine space X are called parallel if $\vec{S} = \vec{T}$, and we write $S \parallel T$. They are called weakly parallel if $\vec{S} \subset \vec{T}$, and we write $S \triangleleft T$.*

We see that $\parallel$ is an equivalence relation, and $\triangleleft$ is a partial order. We will later recover the equivalence between these definitions and Euclid's definitions for the case of lines in a plane and lines and planes in (three-dimensional) space, but we must first study the intersection properties of subspaces.

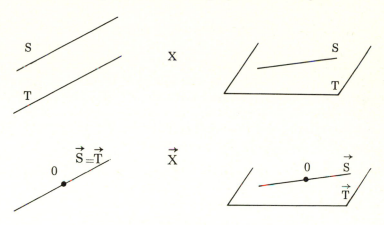

Figure 2.4.9.1

2.4.9.2. Proposition. *If $S \parallel T$, either $S = T$ or $S \cap T = \emptyset$. If $S \triangleleft T$, either $S \subset T$ or $S \cap T = \emptyset$. $S \triangleleft T$ if and only if there exists a subspace S' of T such that $S \parallel S'$. Let $x \in X$ and $S \subset X$ be a subspace; there exists a unique subspace T such that $S \parallel T$ and $x \in T$. Let S, T be subspaces of X. If $S \cap T = \emptyset$,*

$$\dim S + \dim T < \dim X + \dim(\vec{S} + \vec{T})$$

and

$$\dim\langle S \cup T \rangle = \dim(\vec{S} + \vec{T}) + 1 = \dim S + \dim T + 1 - \dim(\vec{S} \cap \vec{T}).$$

If $S \cap T \neq \emptyset$,

$$\dim\langle S \cup T \rangle = \dim S + \dim T - \dim(S \cap T).$$

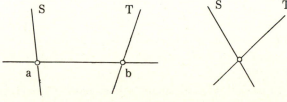

Figure 2.4.9.2

Proof. This proposition follows immediately from Lemma 2.4.9.3 and the relation

$$\dim \vec{S} + \dim \vec{T} = \dim(\vec{S} + \vec{T}) + \dim(\vec{S} \cap \vec{T}),$$

which holds for any subspaces $\vec{S}, \vec{T}$ of a vector space.   □

2.4.9.3. Lemma. *Let $a \in S$ and $b \in T$. Then $S \cap T \neq \emptyset$ is equivalent to $\overrightarrow{ab} \in \vec{S} + \vec{T}$.*

Proof. The lemma is shown by vectorializing at a (cf. 2.1.9). Notice also that

$$\overrightarrow{\langle S \cup T \rangle} = \vec{S} + \vec{T} + K \cdot \overrightarrow{ab}.$$ □

2.4.9.4. Corollary. *If $\vec{S}$ and $\vec{T}$ are complementary (i.e., $\vec{X} = \vec{S} \oplus \vec{T}$), then $S \cap T$ has exactly one point, and S and T are also called complementary.*   □

2.4.9.5. Examples. If X is a plane, $D \subset X$ a line and $x \in D$, there exists a unique line D' of X containing x and such that $D \cap D' = \emptyset$, namely, the line parallel to D. We thus recover Euclid's famous fifth postulate, where the notion of parallelism is formulated as a property of non-intersection.

For $d = 3$, D a line and P a plane of X, either $D \subset P$, or $D \cap P$ consists of one point, or $D \cap P = \emptyset$, in which case $D \triangleleft P$.

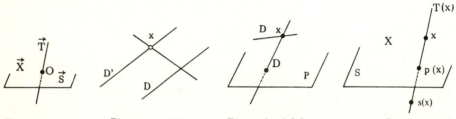

Figure 2.4.9.4   Figure 2.4.9.5.1   Figure 2.4.9.5.2   Figure 2.4.9.6

2.4.9.6. Projections and reflections. Let X be an affine space, S a subspace, and $\vec{T}$ a complement of $\vec{S}$ in $\vec{X}$. By 2.4.9.2 and 2.4.9.4 there exists, for every $x \in X$, a unique subspace $T(x)$ with direction $\vec{T}$ and such that $S \cap T(x)$ has a single point, which we *denote* by $p(x)$. We thus obtain a map $p : X \to S$, called the *projection* from X onto S, parallel to T. We have $p \in A(X; S)$.

Similarly, the *reflection* s through S and parallel to T is the map $s : X \to X$ defined by

$$\overrightarrow{xp(x)} = \overrightarrow{p(x)s(x)}$$

for all $x \in X$. We have $s^2 = s \circ s = \mathrm{Id}_X$ and $s \in A(X; X)$. In the nomenclature of 3.4.2, we can say that $s(x)$ is defined by the property that $p(x)$ is the midpoint of x and $s(x)$. See exercise 2.8.5 for maps with the properties $s^2 = \mathrm{Id}_X$ and $p^2 = p$.

2.4.9.7. See [FL, 68] for the relationship between subspaces and fixed points of endomorphisms of X (i.e., elements of $A(X; X)$).

2.5. Geometry at last: Thales, Pappus, Desargues

2.5.1. PROPOSITION (THALES' THEOREM). *Let H, H', H'' be three distinct parallel hyperplanes in an affine space X, and $(D_i)_{i \in I}$ a family of lines of X, none of which is weakly parallel to H. Then the points $d_i = H \cap D_i$, $d_{i'} = H' \cap D_i$, $d_i'' = H'' \cap D_i$ $(i \in I)$, well-defined by 2.4.9.4 and 2.4.6, satisfy the following condition: The scalar $\overrightarrow{d_i d_i''}/\overrightarrow{d_i d_i'}$ does not depend on $i \in I$, but only on H, H', H''. Conversely, if for some i we have $d''' \in \langle d_i, d_i' \rangle$ and if $\overrightarrow{d_i d_i'''}/\overrightarrow{d_i d_i'}$ is equal to this common value, then $d_i''' = d_i'' = H'' \cap D_i$.*

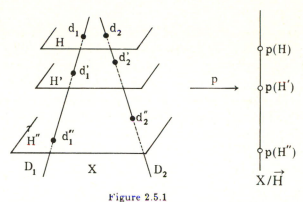

Figure 2.5.1

Proof. We introduce the quotient space $X/\vec{H}$ (cf. 2.2.4) and the canonical projection $p : X \to X/\vec{H}$, which is easily seen to be a morphism. In particular, 2.3.2 shows that, for all i,

$$\frac{\overrightarrow{d_i d_i''}}{\overrightarrow{d_i d_i'}} = \frac{\overrightarrow{p(H)p(H'')}}{\overrightarrow{p(H)p(H')}}.$$

The converse comes from the fact that there exists a unique point c on a line $\langle a, b \rangle$ $(b \notin a)$ such that $\overrightarrow{ac}/\overrightarrow{ab}$ is a fixed $k \in K$.   □

For another proof, see 6.5.4.

All the rest of this section is a consequence of the following elementary remark:

2.5.2. LEMMA. *Let $a, b \in X$ be distinct points and f a dilatation different from the identity. Set $a' = f(a)$ and $D = \langle a, b \rangle$, and let D' be the parallel to D through a' (cf. 2.4.9.2). If $f \in T(X)$, the point $f(b)$ is the intersection of D' with the parallel to $\langle a, a' \rangle$ through b. If f is a homothety of center 0, we have $f(b) = D' \cap \langle 0, b \rangle$.*

Proof. The case of a translation is the famous "parallelogram rule": $\overrightarrow{ab} = \overrightarrow{cd}$ implies $\overrightarrow{ac} = \overrightarrow{bd}$, a consequence of the commutativity of $\vec{X}$. If $f \notin T(X)$, we consider the plane spanned by $0, a, b$ and apply Thales' theorem, with the

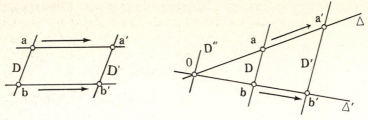

Figure 2.5.2

lines D, D', D'', where D'' is the parallel to D through 0, playing the role of H, H', H''.   □

2.5.3. PROPOSITION (THEOREM OF PAPPUS). *Let X be an affine plane, D, D' two distinct lines of X. Let the points $x, y, z \in D$ and $x', y', z' \in D'$ be all distinct and not contained in $D \cap D'$. If $\langle x, y' \rangle \parallel \langle x', y \rangle$ and $\langle y, z' \rangle \parallel \langle y', z \rangle$ then $\langle x, z' \rangle \parallel \langle x', z \rangle$.*

Proof. If D, D' are not parallel, $D \cap D'$ contains a single point 0, by 2.4.9.5. Let f (resp. g) be the homothety of center 0 taking x into y (resp. y into z). By 2.5.2 and our assumptions, $x' = f(y')$, and $g(z') = y'$. But $g \circ f = f \circ g$ (use 2.3.3.11 and the commutativity of K), so, setting $h = g \circ f$, we have $z = h(x)$ and $x' = h(z')$, whence $\langle x, z' \rangle \parallel \langle x', z \rangle$ by 2.5.2.

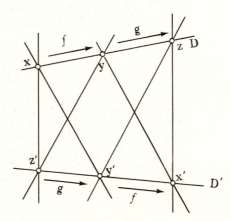

Figure 2.5.3

If $D \parallel D'$, the same proof works if we replace homotheties by translations.
                                                                              □

This is only one possible (affine) version of the Theorem of Pappus; for another one, see 5.4.2. For a converse, see 2.8.6.

2.5.4. PROPOSITION (DESARGUES'S THEOREM). *Let $\langle a, b, c \rangle, \langle a', b', c' \rangle$ be triangles in an affine space (cf. 2.4.7). Assume they have no vertices in common and that their sides are parallel: $\langle a, b \rangle \parallel \langle a', b' \rangle$, $\langle b, c \rangle \parallel \langle b', c' \rangle$,*

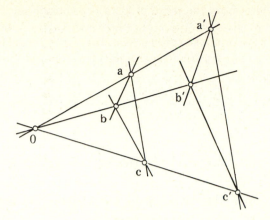

Figure 2.5.4

$\langle c, a \rangle \parallel \langle c', a' \rangle$. *Then the three lines* $\langle a, a' \rangle$, $\langle b, b' \rangle$, $\langle c, c' \rangle$ *are either parallel or concurrent.*

Proof. First, $\langle a, b \rangle \parallel \langle a', b' \rangle$ implies that the four points a, b, a', b' lie on the same plane. By 2.4.9.5, either $\langle a, a' \rangle \parallel \langle b, b' \rangle$ or $\langle a, a' \rangle$ and $\langle b, b' \rangle$ intersect in a single point 0. We discuss the second case. Let f be the homothety of center 0 taking a into a'. By 2.5.2, $f(b) = b'$. Putting $c'' = f(c)$ and applying 2.5.2 twice more, we get $\langle b, c \rangle \parallel \langle b', c'' \rangle$ and $\langle a, c \rangle \parallel \langle a', c'' \rangle$, whence $c' = c''$ by 2.4.9.2. $\square$

For other versions of Desargues, see 5.4.7.

2.5.5. NOTE. In axiomatic theories of affine and projective geometry (2.6.7), the theorems of Pappus and Desargues play an important role, and are sometimes taken as axioms. In particular, Desargues's and Pappus' theorems are connected with the associativity and commutativity, respectively, of the base field. The interested reader can refer to 2.8.9, [AN, 73–75], and [DI, 158–160]. See also 2.6.7 and 4.8.

2.5.6. COROLLARY (CHARACTERIZATION OF DILATATIONS). *Let* $f \in \mathcal{S}_X$ *be a bijection of an affine space* X *of dimension* ≥ 2. *Then* $f \in \mathrm{Dil}(X)$ *if and only if, for any line* D *of* X, $f(D)$ *is a line of* X *parallel to* D.

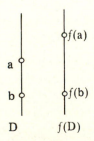

Figure 2.5.6

Proof. This follows from the proof of 2.5.4, by considering the dilatation f' defined by the points $a, b, f(a), f(b)$, where $a, b \in X$ are arbitrary distinct points. The condition of the corollary implies $f = f'$ outside the line $\langle a, b \rangle$. Since $\dim X \geq 2$, moving a and b around proves equality everywhere. The converse is trivial. $\square$

2.5.7. See sections 5.4 and 5.5 for practical applications of the preceding results.

2.6. The fundamental theorem of affine geometry

This is the only delicate result of this chapter. In spite of its simple statement, its proof is long and involved.

2.6.1. INTRODUCTION. We have seen (2.4.2.2) that if $f \in A(X; X')$ the image of every line D of X is a line (or possibly a point) of X'. The converse is false. First of all, in the case $\dim X = \dim X' = 1$, any bijection satisfies this property. Next, take $X = X' = \mathbf{C}^2$ and $f(z_1, z_2) = (\bar{z}_1, \bar{z}_2)$, where $^-$ denotes complex conjugation. It is clear that f takes lines into lines: If $D = x + \mathbf{C} \cdot z$ (see 2.4.8.7), then $f(D) = \bar{x} + \overline{\mathbf{C} \cdot z}$ is still a line.

Generally speaking, the fundamental theorem of affine geometry says that the two counterexamples just mentioned are essentially the only ones possible. Observe that complex conjugation is a (field) automorphism of $\mathbf{C}$. Before stating the theorem, we introduce the following

2.6.2. DEFINITION. *Let V, V' be vector spaces over K, K', respectively. A map $f : V \rightarrow V'$ is called semilinear if there exists an automorphism $\sigma : K \rightarrow K'$ such that*

$$f(\lambda x + \mu y) = \sigma(\lambda) f(x) + \sigma(\mu) f(y)$$

for all $x, y \in V$, $\lambda, \mu \in K$. If X, X' are affine spaces over K, K', a map $f : X \rightarrow X'$ is called semiaffine if there exists $a \in X$ such that $\vec{f} : X_a \rightarrow X'_{f(a)}$ is semilinear.

We have mentioned one example of a semiaffine map, taking $(z_1, z_2) \in \mathbf{C}^2$ into $(\bar{z}_1, \bar{z}_2)$. The reader can, as an exercise, extend most things in section 2.3 to the semiaffine case, and show that semiaffine maps do indeed take collinear points into collinear points.

2.6.3. FUNDAMENTAL THEOREM OF AFFINE GEOMETRY. *Let X, X' be affine spaces of same finite dimension $d \geq 2$. Let $f : X \rightarrow X'$ be a bijection which takes any three collinear points $a, b, c \in X$ into collinear points $f(a), f(b), f(c) \in X'$. Then f is semiaffine.*

There are sharper results in [FL], for example. See also 5.4.8 and 5.4.9. The following proposition gives an idea of how powerful this theorem is:

2.6.4. PROPOSITION. *The only automorphism of* **R** *is the identity. The only continuous automorphisms of* **C** *are the identity and* $z \to \bar{z}$ (*observe that* **C** admits other automorphisms). ☐

See, for example, [FL, 88], exercise 2.8.10 and [PO, 48]. See also 8.12.11 for the automorphisms of the skew field of quaternions.

2.6.5. COROLLARY. *Let* X, X' *be two real affine spaces of same finite dimension* $d \geq 2$. *If* $f : X \to X'$ *is bijective and takes collinear points into collinear points,* f *is an affine map.* ☐

2.6.6. PROOF OF 2.6.3. As we have mentioned, this is a longish proof; we restrict ourselves to the main ideas. See [FL, 83 ff.] for more details.

2.6.6.1. Step 1. *If the points* $\{x_i\}_{i=0,...,k}$ *are independent in* X, *their images are independent in* X'.

Proof. We complete $\{x_i\}_{i=0,...,k}$ into a frame $\{x_i\}_{i=0,...,d}$ of X. Assume the points $\{f(x_i)\}_{i=0,...,d}$ are not independent; then $\langle f(x_0), \ldots, f(x_d) \rangle \neq X'$. But the proof of 2.4.5 and the assumptions of 2.6.3 imply that

$$f(X) \subset \langle f(x_0), \ldots, f(x_d) \rangle \neq X',$$

contradicting the fact that f is surjective. ☐

2.6.6.2. Step 2. *For any line* D *of* X, $f(D)$ *is a line of* X'. *If* D *and* D' *are parallel, so are* $f(D)$ *and* $f(D')$.

Proof. Let $D = \langle a, b \rangle$, and put

$$D' = \langle f(a), f(b) \rangle.$$

Let $c' \in D'$ and $x \in X$ be such that $f(x) = c'$ (using the surjectivity of f). Then $x \in \langle a, b \rangle$ by 2.6.6.1.

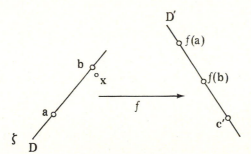

Figure 2.6.6.2.1

Now let $D \parallel D'$ be lines of X, and P the plane containing them (we can obviously suppose $D \neq D'$, else there is nothing to show). Put $P' = \langle f(P) \rangle$; we have $f(D), f(D') \subset f(P) \subset P'$. By 2.4.9.5 we have $f(D) \parallel f(D')$ if $f(D) \cap f(D') = \emptyset$. On the other hand, if $\gamma \in f(D) \cap f(D')$, we can find by the first part of 2.6.6.2 points $c \in D$ and $c' \in D'$ such that $f(c) = f(c') = \gamma$. But $D \cap D' = \emptyset$ and f is injective, a contradiction. ☐

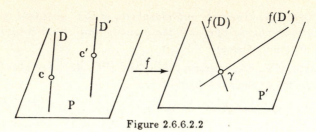

Figure 2.6.6.2.2

2.6.6.3. Step 3. *The map f is additive.*

This means that f, considered as a map between the vectorializations X_0 and $X'_{0'}$, where $0 \in X$ and $0' = f(0) \in X'$, satisfies $f(x + y) = f(x) + f(y)$ for all $x, y \in X_0$.

Proof. If $x, y \in X_0$ are linearly independent, $x + y$ can be geometrically constructed by using parallel lines (figure 2.6.6.3). By 2.6.6.1, $f(x)$ and $f(y)$ are also linearly independent, so the assumptions on f and repeated application of 2.6.6.2 show that $f(x) + f(y) = f(x + y)$. If y and x are not independent, additivity follows from the next step. □

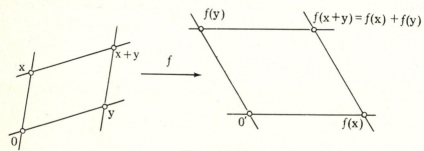

Figure 2.6.6.3

2.6.6.4. Step 4. *Construction of the isomorphism $\sigma : K \to K'$.*

Fix $0 \in X$, $x \in X \setminus 0$ and let $D = K \cdot x$ be the line spanned by x. Similarly, put $0' = f(0)$, $D' = f(D) = K' \cdot f(x)$. The map $\sigma : K \to K'$ is defined by diagram 2.6.6.4.2: the restriction $f|_D : D \to D'$ is indeed a bijection, by 2.6.6.2, and the bijections $K \to D$, $K' \to D'$ are the maps $\lambda \to \lambda x$ and $\lambda' \to \lambda' f(x)$. There remains to check that σ is a field homomorphism. The essential remark is again that $(\lambda + \mu)x$ and $(\lambda\mu)x$ can be geometrically constructed from λx and μx, using parallel lines (figures 2.6.6.4.1 and 2.6.6.4.3), as long as we have a point $y \in X \setminus D$. This is this case exactly because $\dim X = d \geq 2$ by hypothesis. The figures are justified in the following way: figure 2.6.6.4.1, by the parallelogram rule (proof of 2.5.2), and figure 2.6.6.4.3, by the proof of 2.5.3 and 2.3.3.11. □

2.6.6.5. Step 5. *The map f is semiaffine.*

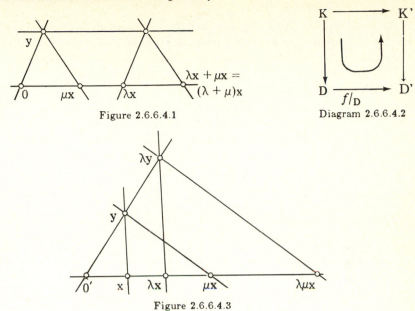

Figure 2.6.6.4.1   Diagram 2.6.6.4.2

$$\lambda x + \mu x = (\lambda + \mu)x$$

Figure 2.6.6.4.3

The previous step showed that f, restricted to an arbitrary affine line of X through 0, is a semiaffine map, but *a priori* the isomorphism $\sigma : D \to D'$ could depend on D. That it doesn't is shown by figures 2.6.6.5, where D and D_1 are two lines through 0. □

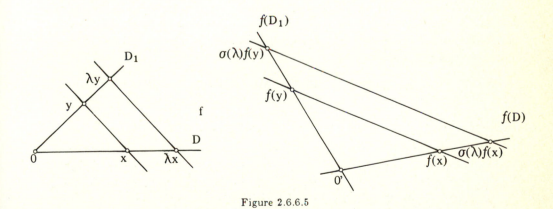

Figure 2.6.6.5

2.6.7. AXIOMATIZATION OF AFFINE GEOMETRY. In a sense, the fundamental theorem of affine geometry says that an affine space is essentially determined by its lines and their set-theoretical intersection properties. In other words, it is reasonable for one to expect to be able to reconstruct the full affine structure of a space $(X, \vec{X}, \Phi)$ starting from X, its lines and their intersection properties. Such an approach, which amounts to finding the weakest possible axioms that must be imposed on X and its subspaces in order to recover $(X, \vec{X}, \Phi)$, is called an *axiomatization* of affine geometry. Euclid's *Elements*, for example, propose such an axiomatization, though he adds also a metric structure, among others. There is an immense amount of literature about such constructions, which are still the object of research, generally together with axiomatizations of projective geometry. Open problems arise particularly in dimension two, because then Desargues's theorem does not follow from intersection axioms for subspaces. Roughly speaking, there exist more affine or projective planes in the axiomatic sense than in the sense of 2.1.1 or 4.1.1, for example, the projective plane of the division ring of octonions: see 4.8.3, [PO, 285], [BES, chapter 3].

Here are a few general references on axiomatic geometry: [AN, chapter 2] is a very readable and fundamental book; [BR] is very complete; [V–Y] is an old reference, useful for appreciating "modern geometry". Recent references for projective planes include [DI] and [H–P]. For a simple example, see [FL, 319].

2.7. Finite-dimensional real affine spaces

> In this whole section all vector and affine spaces are real (i.e., over **R**) and finite-dimensional. Their dimension is generally denoted d.

Real affine spaces of dimension two and three are the objects of our everyday experience (without the notion of distance) and of classical geometry, as studied by the Greeks. Furthermore, finite-dimensional real affine spaces, sometimes endowed with an additional Euclidean structure, will be the main subject of this book.

2.7.1. THE CANONICAL TOPOLOGY. We recall that a real (or complex) vector space has a canonical topological structure (see, for instance, [CH1, 19]). Consequently, the linear group $\mathrm{GL}(E)$ of a vector space E of dimension d also has a canonical topology, induced by $L(E; E)$, which is a vector space of dimension d^2.

2.7.1.1. Proposition. *Let X be an affine space, $a \in X$. The topology induced on X by the canonical topology of the vectorialization X_a does not depend on a, and is called the canonical topology on X.*

Proof. Observe that translations of a finite-dimensional vector space are homeomorphisms, and apply 2.1.8.   □

2.7.1.2. For example, every morphism $f \in A(X; X')$ is automatically continuous; every affine subspace Y is closed, and its complement $X \setminus Y$, if not empty, is dense.

2.7.1.3. If we don't want to resort to the general result 2.3.4, we can proceed in the following way to find a canonical topology for $\mathrm{GA}(X)$: Take $a \in X$, write X as the semidirect product $T(X) \, \mathrm{GA}_a(X)$ (2.3.3.8). We have isomorphisms $T(X) \cong \vec{X}$ and $\mathrm{GA}_a(X) \cong \mathrm{GL}(\vec{X})$, so we can successively consider canonical topologies on $T(X)$, $\mathrm{GA}_a(X)$, $T(X) \, \mathrm{GA}_a(X)$ and $\mathrm{GA}(X)$.

2.7.1.4. Proposition. *The topology thus defined on* $\mathrm{GA}(X)$ *depends only on* X, *and is called the canonical topology.*

Proof. If $f = s \circ g = s' \circ g'$ with $g \in \mathrm{GA}_a(X)$ and $g' \in \mathrm{GA}_b(X)$, we necessarily have $s' = t_{\overrightarrow{f(a)a}}$ and $g' = s'^{-1} \circ f$. Thus s' and g' are clearly continuous at the point $(s, g) \cong f$.   □

Thus $\mathrm{GA}_a(X)$ and $T(X)$ are closed in $\mathrm{GA}(X)$. The compact subsets of $\mathrm{GA}(X)$ will be studied in 2.7.5.

2.7.2. ORIENTATION. We recall briefly the notion of orientation for vector spaces. There are two equivalent definitions, one algebraic, using determinants, and the other geometric, using homotopy between bases.

2.7.2.1. For the first definition, remark that the vector space $A^d E^*$ of alternating d-linear forms over a d-dimensional vector space E is a real one-dimensional vector space; thus $A^d E^* \setminus 0$ has exactly two connected components.

2.7.2.2. Definition. *An orientation for* E *is the choice of one of the two components of* $A^d E^* \setminus 0$. *Let this component be* $\mathcal{O}$. *A form* $\omega \in A^d E^*$ *is called positive if* $\omega \in \mathcal{O}$; *a basis* $\mathcal{B} = \{e_i\}_{i=1,\ldots,d}$ *of* E *is called positively oriented if* $\omega(e_1, \ldots, e_d) > 0$ *for any* $\omega \in \mathcal{O}$.

The last sentence makes sense because $\omega, \omega' \in \mathcal{O}$ implies that $\omega' = k\omega$ for some $k \in \mathbf{R}_+^*$.

2.7.2.3. Observe that if $\omega \in \mathcal{O}$ and $f \in \mathrm{GL}(E)$, the pullback $f^*\omega$ is in $\mathcal{O}$ if and only if $\det f > 0$. Thus it makes sense to introduce the following notation (useful even if E is not oriented):

2.7.2.4. Notation. *We put* $\mathrm{GL}^+(E) = \{ f \in \mathrm{GL}(E) \mid \det f > 0 \}$, *and* $\mathrm{GL}^-(E) = \{ f \in \mathrm{GL}(E) \mid \det f < 0 \}$.

2.7.2.5. With this notation we have, for an arbitrary basis $\mathcal{O}$ of E: If $\omega \in \mathcal{O}$ and $f \in \mathrm{GL}^+(E)$, then $f^*\omega \in \mathcal{O}$. If $\mathcal{B}$ is a positively oriented basis for E, the basis $f(\mathcal{B})$ is positively oriented if and only if $f \in \mathrm{GL}^+(E)$. For this reason we call elements of $\mathrm{GL}^+(E)$ *orientation-preserving*, and will do so even if E is not oriented. On the other hand, if E, E' are real vector spaces of same finite dimension, the following definition only makes sense if E, E' are oriented: $f \in \mathrm{Isom}^+(E; E')$ if and only if $f^*\omega' \in \mathcal{O}$ for any $\omega' \in \mathcal{O}'$, where $\mathcal{O}, \mathcal{O}'$ are the orientations of E, E'.

2.7.2.6. The second way of defining orientation is geometric, or even mechanical; it essentially says that two bases have the same orientation if they can be deformed into one another:

2.7.2.7. Definition. *Two bases* $\mathcal{B}, \mathcal{B}'$ *of a vector space* E *are called homotopic if there exists a continuous map* $F : [0, 1] \to E^d$ *such that* $F(0) = \mathcal{B}$,

$F(1) = \mathcal{B}'$ *and* $F(t)$ *is a basis for all* $t \in [0, 1]$. *An orientation for* E *is an equivalence class of homotopic bases.*

2.7.2.8. The only non-trivial result in this section consists in proving that 2.7.2.2 and 2.7.2.7 are equivalent definitions. This is done by showing that there are exactly two equivalence classes in the space of bases, under the homotopy equivalence relation, and that if we choose an orientation for E in the sense of 2.7.2.2 the set of positively oriented bases is exactly one of these equivalence classes. Observe first that the choice of a fixed basis $\mathcal{B}_0$ of E determines a map $\phi : \mathrm{GL}(E) \to E^d$,

$$\phi(f) = f(\mathcal{B}_0),$$

whose image is the set of bases of E. This map is continuous, and, in fact, a homeomorphism onto its image. Thus the desired equivalence follows from the

2.7.2.9. Proposition. *The space* $\mathrm{GL}(E)$ *has exactly two (path)-connected components, namely* $\mathrm{GL}^+(E)$ *and* $\mathrm{GL}^-(E)$.

Proof. There are at least two components since det : $\mathrm{GL}(E) \to \mathbf{R}^*$ is a surjective continuous function. The path-connectedness of $\mathrm{GL}(E)$ will be proved in 8.4.3.   □

2.7.2.10. Definition. *An orientation for an affine space* X *is an orientation for the underlying vector space* $\vec{X}$. *We call* $\{x_i\}_{i=0,1,\ldots,d}$ *a positively oriented frame of the oriented affine space* X *if* $\{\overrightarrow{x_0 x_i}\}_{i=1,\ldots,d}$ *is a positively oriented basis for* $\vec{X}$. *We put*

$$\mathrm{GA}^+(X) = \{f \in \mathrm{GA}(X) \mid \vec{f} \in \mathrm{GL}^+(\vec{X})\},$$

$$\mathrm{GA}^-(X) = \{f \in \mathrm{GA}(X) \mid \vec{f} \in \mathrm{GL}^-(\vec{X})\}.$$

2.7.2.11. One can also define $\mathrm{Isom}^+(X; X')$ if X, X' are two oriented affine spaces. One can talk about homotopic affine frames (whether or not X is oriented). In this way we obtain an equivalent notion of orientation; to see this we just prove that $\mathrm{GA}^+(X)$ is still path-connected.

For $a \in X$, the group $\mathrm{GA}^+(X)$ is the semidirect product $T(X)\, \mathrm{GA}^+(X)$ (cf. 2.7.1.3), where $T(X)$ is homeomorphic to $\vec{X}$, which is path-connected, and $\mathrm{GA}_a^+(X)$ is homeomorphic to $\mathrm{GL}^+(\vec{X})$, which is also path-connected by 2.7.2.9.

2.7.3. HYPERPLANES AND HALF-SPACES

2.7.3.1. Let H be a hyperplane of an affine space X. By 2.4.8.6, there exists an affine form $f \in A(X; \mathbf{R})$ on X such that $H = f^{-1}(0)$. It is thus natural to consider the subsets $f^{-1}(\mathbf{R}_+)$ and $f^{-1}(\mathbf{R}_-)$, as well as $f^{-1}(\mathbf{R}_+^*)$ and $f^{-1}(\mathbf{R}_-^*)$. The first two are closed subsets of X, and the second two open subsets, whose closures are the first two.

2.7.3.2. Proposition. *The sets $f^{-1}(\mathbf{R}_+)$ and $f^{-1}(\mathbf{R}_-)$ (resp. $f^{-1}(\mathbf{R}_+^*)$ and $f^{-1}(\mathbf{R}_-^*)$) depend only on the hyperplane H; they are called the closed (resp. open) half-spaces bounded by H. Every half-space is path-connected. The set $X \setminus H$ has exactly two connected components, $f^{-1}(\mathbf{R}_+^*)$ and $f^{-1}(\mathbf{R}_-^*)$.*

Proof. The first assertion follows form the beginning of 2.4.8.6. The second is true because half-spaces are convex sets; and the last, because $X \setminus H$ must have at least two connected components, since affine forms are continuous (2.7.1.2) and $\mathbf{R}^*$ has two connected components. $\qquad \square$

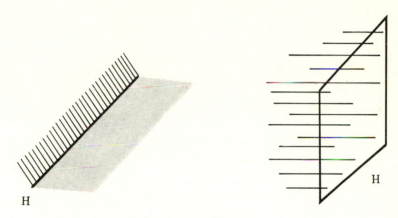

Figure 2.7.3.2

Half-spaces play a fundamental role in chapters 11 and 12.

2.7.4. LEBESGUE MEASURE ON AN AFFINE SPACE

2.7.4.1. We shall make free use of classical results from integration theory (see 0.6).

2.7.4.2. Let E be a vector space and $f : \mathbf{R}^d \to E$ an isomorphism, obtained by choosing a basis of E, for example. Let μ_0 denote the Lebesgue measure on $\mathbf{R}^d$ and $f(\mu_0)$ the image measure on E. If $g : \mathbf{R}^d \to E$ is another isomorphism, we want to compare $g(\mu_0)$ and $f(\mu_0)$. The change of variable theorem (cf. [GT, 33]) shows that

$$g(\mu_0) = \left|\det(g^{-1} \circ f)\right| f(\mu_0),$$

where $\left|\det(g^{-1} \circ f)\right|$ is a strictly positive real number. Thus a vector space E does not have a canonical measure, but has a family of proportional natural measures $f(\mu_0), f \in \mathrm{Isom}(\mathbf{R}^d; E)$, called *Lebesgue measures* on E. (This is the same as saying that E has a canonical measure up to scalar multiplication.)

The passage to the affine case is allowed by the observation that translations of $\mathbf{R}^d$ preserve μ_0. Thus

2.7.4.3. Proposition. *A Lebesgue measure on an affine space X is the image, under any bijection $\Theta_a^{-1} : \vec{X} \to X$, $a \in X$, of a Lebesgue measure*

of $\vec{X}$. *Two Lebesgue measures on X are proportional. If μ is a Lebesgue measure on X, we denote by $\vec{\mu}$ the Lebesgue measure on $\vec{X}$ which gives rise to it. For every Lebesgue measure μ on X and every $f \in \mathrm{GA}(X)$ we have $f(\mu) = |\det \vec{f}|\mu$.*   □

We shall see in section 9.12 that a Euclidean affine space possesses a canonical Lebesgue measure.

2.7.4.4. Notes. Lebesgue measures are the only measures on an affine space invariant under all translations. The idea in proving this is to regularize the measure in question, so as to obtain a continuous invariant measure, which must be proportional to a Lebesgue measure. The proportionality factor, being invariant under translation, must be a constant.

2.7.5. Center of a compact set. Compact subgroups of $\mathrm{GA}(X)$

In this section we use Lebesgue measures to associate to a compact subset K of an affine space X a well-defined point, fixed by every $f \in \mathrm{GA}(X)$ such that $f(K) = K$. This latter property will be essential in 2.7.5.10.

2.7.5.1. We first fix a Lebesgue measure μ on X, and suppose that K has non-empty interior $\overset{\circ}{K} \neq \emptyset$. In particular,

$$\mu(K) = \int_X \chi_K \mu > 0,$$

where χ_K is the characteristic function of K (see 0.6). Recall that μ gives rise to a measure $\vec{\mu}$ with values in a vector space E by putting, for any $f : X \to E$ and $\phi \in E^*$,

$$\phi\left(\int_X f\mu\right) = \int_X \phi(f)\vec{\mu}.$$

This is the same as choosing arbitrary coordinates, in which $f = (f_1, \ldots, f_d)$, and writing

$$\int_X f\mu = \left(\int_X f_1\mu, \ldots, \int_X f_d\mu\right).$$

Given the compact set K and an arbitrary point $a \in X$, we set

$$I_\mu(a) = \int_{x \in X} \chi_K \overrightarrow{ax}\,\vec{\mu}.$$

If b is another point,

$$I_\mu(b) = \int \chi_K \overrightarrow{bx}\,\vec{\mu} = \int \chi_K (\overrightarrow{ba} + \overrightarrow{ax})\vec{\mu} = \mu(K)\overrightarrow{ba} + I_\mu(a).$$

We thus get the following

2.7.5.2. Proposition. *If K is a compact set in X with $\overset{\circ}{K} \neq \emptyset$, the point*

$$a = \left(\mu(K)\right)^{-1} I_\mu(a)$$

does not depend on $a \in X$ or on the choice of μ.   □

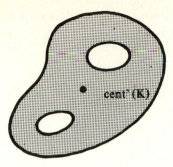

Figure 2.7.5.3

This point, called the *centroid* of K, shall be denoted $\mathrm{cent}'(K)$, or $\mathrm{cent}'_X(K)$ if necessary.

2.7.5.3. Remark. If X is a Euclidean affine space of dimension 2 or 3, and if K is considered as a homogeneous (i.e., constant-density) plate or solid, $\mathrm{cent}'(K)$ is exactly the center of mass of K, as defined in mechanics.

2.7.5.4. In the general case of an arbitrary compact K, possibly with empty interior, we generalize the definition above by using the following trick, borrowed from chapter 11: Take the convex hull $\mathcal{E}(K)$ of K and the affine subspace $\langle K \rangle$ spanned by K. Then $\mathcal{E}(K)$ has non-empty interior in $\langle K \rangle$, by 11.2.7, which allows us to state the following

2.7.5.5. Definition. *The center of a compact subset K of an affine space is the point*

$$\mathrm{cent}(K) = \mathrm{cent}'_{\langle K \rangle}\big(\mathcal{E}(K)\big).$$

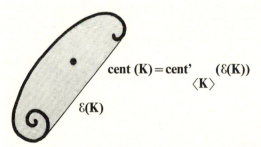

Figure 2.7.5.5

2.7.5.6. Remark. This is no longer the center of mass considered in mechanics (if K is, say, a curved homogeneous piece of wire in the plane or in space), cf. 3.4.2 and 3.7.14. Nor does it coincide with the other fixed points which we shall find in 9.8.6.

2.7.5.7. Application. *Let K be a compact subset of an affine space X. If $f \in \mathrm{GA}(X)$ satisfies $f(K) = K$, then $f\big(\mathrm{cent}(K)\big) = \mathrm{cent}(K)$. In other* *words,*

$$\mathrm{GA}_K(X) = \{f \in \mathrm{GA}(X) \mid f(K) = K\} \subset \mathrm{GA}_{\mathrm{cent}(K)}(X).$$

Furthermore, if $K \neq \emptyset$ *we have* $|\det \vec{f}| = 1$ *for any* $f \in \mathrm{GA}_K(X)$.

Proof. The last assertion is a consequence of the end of 2.7.4.3:

$$f(\mu)(K) = \mu(f^{-1}(K)) = \mu(K) = |\det \vec{f}|\mu(K).$$

To show the first we can, in view of 2.7.5.4 and 2.7.4.3, restrict ourselves to the case $\overset{\circ}{K} \neq \emptyset$. Then

$$f\big(\mathrm{cent}(K)\big) = f\big(a + (\mu(K))^{-1}I(a)\big) = f(a) + \vec{f}\big((\mu(K))^{-1}I(a)\big)$$

$$= f(a) + (\mu(K))^{-1}\vec{f}\big(I(a)\big),$$

where $a \in X$ is arbitrary. Using 2.7.5.1 we get

$$\vec{f}\big(I(a)\big) = \vec{f}\left(\int_{x \in X} \chi_K \overrightarrow{a\vec{x}} \mu\right) = \int_{x \in K} \chi_K \vec{f}(\overrightarrow{a\vec{x}})\mu =$$

$$= \int_{x \in K} \chi_K \overrightarrow{f(a)f(x)}\mu = I\big(f(a)\big).$$

Thus

$$f\big(\mathrm{cent}(K)\big) = f(a) + (\mu(K))^{-1}I\big(f(a)\big) = \mathrm{cent}(K). \qquad \square$$

2.7.5.8. Notes. If X is affine Euclidean, one can show that every isometry $f \in \mathrm{Is}(X)$ such that $f(K) = K$ leaves invariant a point of X which does not depend on f. This can be done without resorting to integration theory (see 9.8.6). On the other hand, given a compact set $K \subset X$ with $\overset{\circ}{K} \neq \emptyset$, there exists a Euclidean structure on X which is invariant under $\mathrm{GA}_K(X)$, so in particular $\mathrm{GA}_K(X) \subset \mathrm{Is}(X)$ for this structure (see 2.7.5.10 and 8.2.5).

Finally, observe that in general $\mathrm{GA}_{\mathcal{E}(K)} \neq \mathrm{GA}_K(X)$, as shown by figure 2.7.5.8.

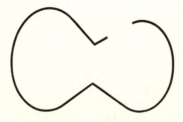

Figure 2.7.5.8

2.7.5.9. Proposition. *Let* K *be a compact subset of* X *with non-empty interior. The subgroup* $\mathrm{GA}_K(X) \subset \mathrm{GA}(X)$ *is compact in the canonical topology* (2.7.1.3).

The condition $\mathring{K} \neq \emptyset$ is necessary; if $K = \{a\}$, for $a \in X$, we have $\mathrm{GA}_a(X)$ homeomorphic to $\mathrm{GL}(\vec{X})$, which is not compact.

Proof. Use 2.7.5.7 to vectorialize X at $\mathrm{cent}(K)$, reducing the problem to the vector case. It can be shown that K contains the origin, i.e., $\mathrm{cent}(K) \in K$ (2.8.11), but instead we observe that if $f(K) = K$ for $f \in \mathrm{GL}(X)$, we also have

$$f(-K) = -K,$$

where $-K = \{-x \mid x \in K\}$, and 0 is an interior point of $\mathcal{E}\big(K \cap (-K)\big)$ (figure 2.7.5.9).

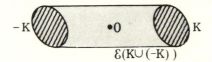

$$\mathcal{E}(K \cup (-K))$$

Figure 2.7.5.9

Now consider any Euclidean structure on X, and take $\epsilon > 0$ such that $B(0, \epsilon) \subset K$. Put $M = \sup_{x \in K} \|x\|$. Then $f(K) = K$ implies

$$\|f(x)\| \leq M \text{ for any } x \text{ such that } \|x\| \leq \epsilon;$$

thus $\|f\| \leq \epsilon^{-1} M$ in the usual norm of $L(X; X)$. Thus $\mathrm{GL}_K(X)$ is a bounded closed set of $L(X; X)$, and hence compact. $\qquad\square$

The converse is also true:

2.7.5.10. Proposition. *Let G be a compact subgroup of $\mathrm{GA}(X)$. Then there exists $x \in X$ such that $G \subset \mathrm{GA}_x(X)$. Moreover, there exists a Euclidean structure on X left invariant by G (i.e., such that $G \subset \mathrm{Is}(X)$ for this structure).*

Proof. Let $a \in X$ be arbitrary, and $G(a)$ its orbit under G. The orbit is a compact subset of X, so we can take $x = \mathrm{cent}\big(G(a)\big)$. The "moreover" part will be show in three different ways (see 8.2.5 and 9.8.6). $\qquad\square$

2.7.5.11. Corollary. *All maximal compact subgroups of $\mathrm{GA}(X)$ are conjugate (in $\mathrm{GA}(X)$, under inner automorphisms).*

Proof. Consider the Euclidean structure on X given by a quadratic form q on $\vec{X}$, and let $\mathrm{Is}_a(X, q)$ be the group of isometries of X (for this Euclidean structure) leaving a fixed. This group is compact, so 2.7.5.10 implies that every maximal compact subgroup of $\mathrm{GA}(X)$ is of the form $\mathrm{Is}_a(X, q)$. The corollary follows from 8.1.6 and formula 1.5.3. $\qquad\square$

2.7.5.12. Notes. Corollary 2.7.5.11 is included here as an aside; it is a particular case of a fundamental general theorem which says that all maximal compact subgroups of a Lie group are conjugate. See, for example, [HN, 218 and note on page 240].

A result as general as 2.7.5.10 cannot be extended even to very well-behaved spaces. For example, take X as the sphere S^n and $G = \mathrm{Is}(X)$

(cf. 18.5). Then G has no fixed point. For a general result in this direction, see 9.8.6.5.

2.7.6. EQUIAFFINE GEOMETRY. At the end of 2.7.5.7 we hit upon the subgroup
$$\mathrm{SA}(X) = \{\, f \in \mathrm{GA}(X) \mid |\det \vec{f}| = 1 \,\},$$
sometimes called the *unimodular* or *special* affine group. Its elements leave invariant any Lebesgue measure on X, but $\mathrm{SA}(X)$ is much larger than $\mathrm{Isom}(X)$ (for an arbitrary Euclidean structure on X), and is in fact not compact. However, one can construct a fairly rich geometry dealing with properties invariant under $\mathrm{SA}(X)$: see [SK, vol. 2, p. 1-50 to 1-56], [FT1, 40–52], [BLA2], [DE4, volume IV, p. 367, exercise 12]. For example, in 2.8.12 we have notions of length and curvature of curves of X that are invariant under $\mathrm{SA}(X)$.

2.7.7. DIFFERENTIAL CALCULUS OF AFFINE SPACES. The notion of differentiability can be easily extended from $\mathbf{R}^n$ to maps $f : U \to X'$ from an open set U of an affine space X into another affine space X'. One way to do this is to remark that different vectorializations of X and X' differ only by translations, which are differentiable infinitely often; another way is to observe that the definition of differentiability uses the difference
$$f(x) - f(a) - g(x - a), \qquad g \in L(\vec{X}; \vec{X}'),$$
which makes sense (in $\vec{X}'$) since
$$x - a = \overrightarrow{ax} \in \vec{X} \text{ and } f(x) - f(a) = \overrightarrow{f(a)f(x)} \in \vec{X}'.$$
The only thing to watch out for is that the derivative f' of f is in $L(\vec{X}; \vec{X}')$. A general reference for this subject is [CH1].

2.8. Exercises

* **2.8.1.** THEOREM OF CEVA. Let a, b, c be a triangle in an affine plane, and let $a' \in \langle b, c \rangle, b' \in \langle c, a \rangle, c' \in \langle a, b \rangle$ be three points on the sides of this triangle. Prove that the three lines $\langle a, a' \rangle, \langle b, b' \rangle, \langle c, c' \rangle$ are concurrent (or parallel) if and only if we have
$$\frac{\overrightarrow{a'b}}{\overrightarrow{a'c}} \cdot \frac{\overrightarrow{b'c}}{\overrightarrow{b'a}} \cdot \frac{\overrightarrow{c'a}}{\overrightarrow{c'b}} = -1.$$

* **2.8.2.** THEOREM OF MENELAUS. With the same data as in the previous exercise, show that the three points a', b', c' are collinear if and only if
$$\frac{\overrightarrow{a'b}}{\overrightarrow{a'c}} \cdot \frac{\overrightarrow{b'c}}{\overrightarrow{b'a}} \cdot \frac{\overrightarrow{c'a}}{\overrightarrow{c'b}} = 1.$$

Deduce the theorems of Pappus, Pascal (16.8.5) and Desargues from the theorems of Ceva and Menelaus. Generalize.

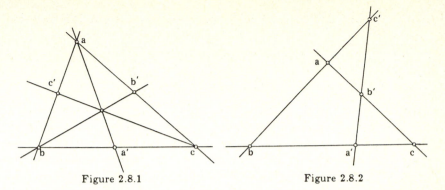

Figure 2.8.1   Figure 2.8.2

2.8.3. Let X be an affine space of dimension n over a finite field with k elements. Compute $\#X$, $\#\,\mathrm{GA}(X)$, and the number of p-dimensional subspaces of X.

2.8.4. Show that the center of $\mathrm{GL}(X)$, where X is a vector space, consists only of $K^* \cdot \mathrm{Id}_X$.

2.8.5. For X an affine space, study the maps $f \in \mathrm{GA}(X)$ such that $f^2 = \mathrm{Id}_X$. Study the maps f such that $f^2 = f$.

2.8.6. State and prove the converses of 2.5.3, 2.5.4.

2.8.7. Let X be a complex affine space (i.e., over **C**). Show that X has a canonical topology. If H is a hyperplane, is $X \backslash H$ connected in this topology?

* **2.8.8.** Show that if X is a finite-dimensional real affine space and Y is a subspace the complement $X \setminus Y$ is connected if $\dim Y \le \dim X - 2$. Is $X \setminus Y$ simply connected? (See 18.2.2.)

2.8.9. PAPPUS AND COMMUTATIVITY. Show that if an affine plane X, over a (possibly non-commutative) field, satisfies theorem 2.5.3 for all sextuples of points in X, then the base field is commutative.

2.8.10. Prove proposition 2.6.4.

* **2.8.11.** Let K be a compact subset of a finite-dimensional real affine space such that $\overset{\circ}{K} \ne \emptyset$; let H be a hyperplane of X and X', X'' its two closed half-spaces. Assume also that H is such that $\overset{\circ}{K'} \ne \emptyset$ and $\overset{\circ}{K''} \ne \emptyset$, where $K' = K \cap X'$ and $K'' = K \cap X''$. Show that $\mathrm{cent}'(K)$ is the weighted average of $\mathrm{cent}'(K')$ and $\mathrm{cent}'(K'')$ with weights $\mu(K')$ and $\mu(K'')$, or, in the terminology of chapter 3, the barycenter of the family $\big\{(\mathrm{cent}'(K'), \mu(K')), (\mathrm{cent}'(K''), \mu(K''))\big\}$. Deduce that if in addition K is convex, then $\mathrm{cent}(K) \in \overset{\circ}{K}$. (See figure 2.8.11.)

* **2.8.12.** EQUIAFFINE LENGTH AND CURVATURE. Let X be a real affine plane, and fix a basis for $\vec{X}$. We can then define the determinant $\det(\vec{u}, \vec{v})$ of any two vectors in $\vec{X}$ relative to this basis by writing the 2×2 matrix whose columns are $\vec{u}$, $\vec{v}$ in this basis. Let $c : [a, b] \to X$ be a differentiable curve of

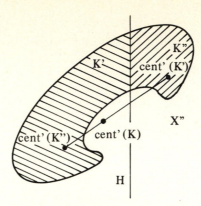

Figure 2.8.11

class C^3 in X. The *equiaffine length* of c (in X, relative to this basis) is the real number

$$\int_a^b \left(\det\left(\bar{c}'(t), \bar{c}''(t)\right)\right)^{1/3} dt.$$

Show that c has same equiaffine length as $f \circ c$ for any $f \in SA(X)$ (see 2.7.6). Show that we can reparametrize c by its equiaffine length if for every t we have $\det\left(\bar{c}'(t), \bar{c}''(t)\right) \neq 0$. The *equiaffine curvature* is the number $K = \det(\bar{c}'', \bar{c}''')$ when c is parametrized by its equiaffine length; show that this curvature, too, is invariant under $SA(X)$. Find the equiaffine length and curvature for an ellipse, a parabola or a hyperbola in X (always fixing a basis). In the same way that a curve in the Euclidean plane is determined up to an isometry by giving the curvature as a function of the arclength (see for example [CAR, 33], or [SK, vol. 2, p. 1–1]), show that a curve in X is determined, up to an element of $SA(X)$, by giving the equiaffine curvature as a function of its equiaffine arclength.

2.8.13. Find a polygon in an affine space, given the midpoints of the sides (watch out for the parity of the number of sides). Compare with 10.11.4 and 16.3.10.3.

Chapter 3
Barycenters; the universal space

In this rather technical chapter we construct for each affine space X a "universal" vector space $\hat{X}$ where X embeds as an affine hyperplane not containing the origin (section 3.1). In this model $\vec{X}$ is a vector hyperplane of $\hat{X}$—namely, the direction of X in $\hat{X}$. The construction of this universal space may at first appear to come out of the blue, but we hope the reader will soon be convinced of its usefulness. Indeed, we apply it to the study of many classical techniques and constructions: barycenters (section 3.4), barycentric coordinates (section 3.6), homogeneization of polynomials (section 3.3), completion of affine spaces into projective spaces (chapter 5) and the study of affine quadrics (chapter 15).

> All fields considered are commutative.

3.1. The universal space

In an affine space X there is no intrinsic calculation with vectors, since X has no distinguished point (see introduction to chapter 2 and 2.1.9). It is, however, possible to calculate with vector fields on X, since the set of maps from an arbitrary set into a vector space has a natural vector space structure.

3.1.1. DEFINITION. *A vector field on an affine space* $(X, \vec{X})$ *is a map* $f : X \to \vec{X}$. *The set of vector fields on* $(X, \vec{X})$ *has a canonical vector space structure and will be denoted by* $\mathcal{V}(X)$.

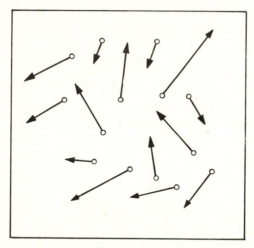

Figure 3.1.1

3.1.2. EXAMPLES. Vector fields are generally drawn as arrows joining x to $x + f(x)$, as in figure 3.1.1. We shall never consider wild vector fields as in this figure, but only the following two examples:

3.1.2.1. The *constant field* $f_{\vec{\xi}}$, where $\vec{\xi} \in \vec{X}$, is defined by $f(x) = \vec{\xi}$ for all $x \in X$ (figure 3.1.2.1). The *trajectories* of this field are lines parallel to $\vec{\xi}$. Geometrically speaking, the map $x \mapsto x + f(x)$ is the translation $t_{\vec{\xi}}$.

3.1.2.2. The *central field* $f_{(k,a)}$ associated with the pair $(k, a) \in K^* \times X$ is defined by $f_{(k,a)}(x) = k\overrightarrow{xa}$ for any $x \in X$ (figure 3.1.2.2). The map $x \mapsto x + f(x)$ is the homothety $H_{a,1-k}$, and the trajectories are open half-lines starting at the origin. See also 3.7.6.

3.1.3. NOTATION. The following two subsets of $\mathcal{V}(X)$ will be used:

$$\mathcal{C}(X) = \{\, f_{\vec{\xi}} \mid \vec{\xi} \in \vec{X} \,\}, \qquad \mathcal{C}'(X) = \{\, f_{(k,a)} \mid (k,a) \in K^* \times X \,\}.$$

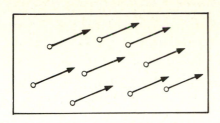

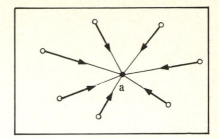

Figure 3.1.2.1   Figure 3.1.2.2

3.1.4. Observe that $\vec{\xi} \mapsto f_{\vec{\xi}}$ defines a linear isomorphism between $\vec{X}$ and $C(X)$, which is a vector subspace of $\mathcal{V}(X)$. Analogously, $(k, a) \mapsto f_{(k,a)}$ is a bijection between $K^* \times X$ and $C'(X)$, this latter not being a vector subspace of $\mathcal{V}(X)$. Observe also that, as elements of $\mathcal{V}(X)$, the vector fields $\lambda f_{(k,a)}$ and $f_{(\lambda k,a)}$ are identical, and finally that $C(X) \cap C'(X) = \emptyset$.

3.1.5. THEOREM. *The union $D(X) = C(X) \cup C'(X)$ is a vector subspace of $\mathcal{V}(X)$. The map $\phi : D(X) \to K$ defined by $\phi(f_{(k,a)}) = k$ for $(k, a) \in K^* \times X$ and $\phi(f_{\vec{\xi}}) = 0$ for $\vec{\xi} \in \vec{X}$ is a linear form on $D(X)$. The map $x \mapsto f_{(1,x)}$ from X into $D(X)$ induces an affine isomorphism between X and the affine hyperplane $\phi^{-1}(1)$ of $D(X)$ (cf. 2.2.3).*

Proof. The proof is trivial, but the theorem is important. We first check that ϕ is linear:

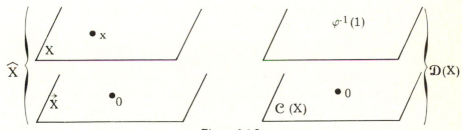

Figure 3.1.5

3.1.6. The vector field $f_{(k,a)} + f_{(k'.a')}$ takes x to $k\overrightarrow{xa} + k'\overrightarrow{xa'}$; thus, for $k + k' = 0$ we have $k\overrightarrow{xa} + k'\overrightarrow{xa'} = k(\overrightarrow{xa} - \overrightarrow{xa'}) = k\overrightarrow{a'a}$, so $f_{(k,a)} + f_{(k'.a')} = f_{k\overrightarrow{a'a}}$. On the other hand, if $k + k' \neq 0$, there exists b such that

$$\overrightarrow{ab} = \frac{k'}{k + k'}\overrightarrow{aa'},$$

so that $k\overrightarrow{xa} + k'\overrightarrow{xa'} = (k + k')\overrightarrow{xb}$, and $f_{(k,a)} + f_{(k'.a')} = f_{(k+k'.b)}$. (The justification for this b comes from looking at the values of the two identical vector fields $f_{(k,a)} + f_{(k'.a')}$ and $f_{(k+k'.b)}$, where b is as yet unknown, at the point $x = a$.)

Similarly, one has $(f_{(k,a)} + f_{\bar{x}})(x) = k\overrightarrow{xa} + \vec{\xi} = k\overrightarrow{xa'}$, where a' is such that $\overrightarrow{aa'} = k^{-1}\vec{\xi}$. Thus $f_{(k,a)} + f_{\vec{\xi}} = f_{(k,a')}$. The linearity of ϕ follows from these calculations. To show that $\Sigma : x \mapsto f_{(1,x)}$ is an affine morphism, fix $a \in X$. By 2.3.1, Σ is affine if and only if there exists a linear map $\vec{\Sigma} : \vec{X} \to \mathcal{C}(X)$ such that $\overrightarrow{\Sigma(a)\Sigma(b)} = \vec{\Sigma}(b) - \vec{\Sigma}(a)$ (as vectors in $\mathcal{D}(X)$). But, by definition, $\overrightarrow{\Sigma(a)\Sigma(b)} = \Sigma(b) - \Sigma(a)$ takes x into $\overrightarrow{xb} - \overrightarrow{xa} = \overrightarrow{ab}$, so that $\overrightarrow{\Sigma(b)\Sigma(a)} = f_{\overrightarrow{ab}}$, and the linear map $\vec{\Sigma}$ given by $\vec{\Sigma}(\vec{\xi}) = f_{\vec{\xi}}$ satisfies the desired condition. $\square$

We now forget about vector fields, and synthesize our results in the following (seemingly artificial) corollary:

3.1.7. COROLLARY. *Let $(X, \vec{X})$ be an affine space, and let $\hat{X}$ be the disjoint union $\vec{X} \cup (K^* \times X)$. We identify X and $1 \times X \subset \hat{X}$. Then $\hat{X}$ is a vector space with the multiplication given by*

$$k(h, x) = (kh, x) = (kh)x, \qquad 0(kx) = 0$$

(in particular, $(k, x) = kx$), and the addition given by

$$\begin{cases} kx + k'x' = (k + k')x'' & \text{if } k + k' \neq 0, \text{ where } x'' = \dfrac{k'}{k + k'}x, \\ kx + k'x' = k\overrightarrow{x'x} & \text{if } k + k' = 0, \\ kx + \vec{\xi} = k(x + k^{-1}\vec{\xi}) \end{cases}$$

(and the original multiplication and addition on $\vec{X} \subset X$). The set $\vec{X}$ is a hyperplane of $\hat{X}$. The map $M : \hat{X} \to X$ defined by $M(kx) = k$, $M(\vec{\xi}) = 0$ for $\vec{\xi} \in \vec{X}$ is a linear form on $\hat{X}$. The affine hyperplane $M^{-1}(1)$ is affine isomorphic to X, and does not contain 0. For any $a \in X$, we have a direct sum decomposition $\hat{X} = \vec{X} \oplus Ka$. The restriction to X of the projection $q : \hat{X} \to \vec{X}$ onto the first factor of this direct sum is identical with Θ_a (cf. 2.1.5—recall that we identified X and $1 \times X \subset \hat{X}$). $\square$

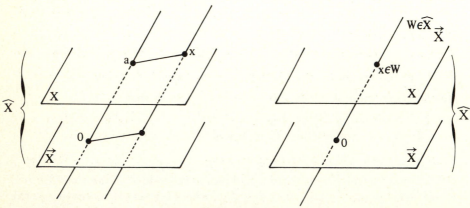

Figure 3.1.7

3.1.8. REMARKS. Corollary 3.1.7 completely justifies the notation $\vec{xy} = y - x$ introduced in 2.1.3.1. Watch out for the dangers arising from the use of certain identifications; if there is any confusion, it's a good idea to start over, carefully distinguishing all objects. A typical example is afforded by the affine space $X = \vec{X}$ (2.2.1): in $\hat{X}$, the notation $\vec{X}$ means both the affine subspace

$$\vec{X} = X = M^{-1}(1)$$

and the vector subspace $\vec{X}$ (figure 3.1.8). Thus it's a bad idea to write $X = \vec{X}$; we have to distinguish between the two when considering the associated universal space.

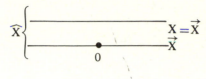

Figure 3.1.8

3.1.9. THE SNAKE THAT BITES ITS TAIL. Starting from X, we form the universal vector space $\hat{X}$, which contains $\hat{X}$ as a hyperplane. According to 2.2.5, we consider the affine space $\hat{X}_{\vec{X}}$; then 2.2.7 and 3.1.7 show that X is naturally isomorphic to $\hat{X}_{\vec{X}}$ via the two isomorphisms $X \cong M^{-1}(1)$ and $M^{-1}(1) \cong \hat{X}_{\vec{X}}$.

3.2. Morphisms and the universal space

Let X and X' be affine spaces. We search for the relationship between $A(X; X')$ and $L(\hat{X}; \hat{X}')$. Going from $L(\hat{X}; \hat{X}')$ to $A(X, X')$ is easy, just use 3.1.7 and 2.3.7. The converse is given by the following proposition, whose proof is straightforward:

3.2.1. PROPOSITION. *Let X, X' be two affine spaces, and f a morphism in $A(X; X')$. We define $\hat{f} : \hat{X} \to \hat{X}'$ by $\hat{f}|_{X} = f$, $\hat{f}(ks) = k\hat{f}(x)$ (using the identifications $X \subset \hat{X}$, $X' \subset \hat{X}'$). Then $\hat{f} \in L(\hat{X}; \hat{X}')$, and in fact $\hat{f} \in L_{\vec{X}, \vec{X}'}(\hat{X}; \hat{X}')$. Using the notation of 2.3.6 and the identification in 3.1.8, we have $\overline{\hat{f}} = f$. If $g \in A(X', X'')$, where X'' is a third affine space, then $\widehat{g \circ f} = \hat{g} \circ \hat{f}$.* □

3.2.2. This justifies calling $\hat{X}$ a "universal" space, and shows that $X \to \hat{X}$ is, in a sense to be made precise later, a functorial correspondence.

3.2.3. COROLLARY. *For every affine space X, the affine group $GA(X)$ of X is naturally isomorphic to the quotient group $GL_{\vec{X}}(\hat{X})/K^* \cdot Id_{\vec{X}}$ of $GL_{\vec{X}}(\hat{X}) = \{ f \in GL(\hat{X}) \mid f(\hat{X}) = \vec{X} \}$ by the center $K^* \cdot Id_{\vec{X}}$ of $GL(X)$ (see 2.3.3.12).* □

3.2.4. EXAMPLE. With the definitions of 2.3.3.12, we have $f \in \mathrm{Dil}(X)$ if and only if $f|_{\vec{X}} = \mathrm{Id}_{\vec{X}}$.

3.2.5. ELABORATING ON 2.3.8. We use the same notation and same data. From the affine frame $\{x_i\}_{i=0,\ldots,n}$ for X we can form two bases for X. The first is $\{x_0, \overrightarrow{x_0 x_1}, \ldots, \overrightarrow{x_0 x_n}\}$, where $x_0 \in X \subset \hat{X}$ and $\overrightarrow{x_0 x_i} \in \vec{X} \subset \hat{X}$. The second is $\{x_0, x_1, \ldots, x_n\} \subset X \subset \hat{X}$, and will be studied more closely in 3.6. For now we form the two bases $\{x_0, \overrightarrow{x_0 x_1}, \ldots, \overrightarrow{x_0 x_n}\}$ and $\{x_0', \overrightarrow{x_0' x_1'}, \ldots, \overrightarrow{x_0' x_n'}\}$ of X and X', respectively, and interpret 2.3.8 in this light. The matrix $M(f)$ of 2.3.9 is simply the matrix of $\hat{f}$ (3.2.1) in these bases, for if $x = (\lambda_1, \ldots, \lambda_n)$ in X, the coordinates of x in the basis $\{x_0, \overrightarrow{x_0 x_1}, \ldots, \overrightarrow{x_0 x_n}\}$ are clearly $(1, \lambda_1, \ldots, \lambda_n)$, and the elements of $\vec{X} \subset \hat{X}$ are exactly those whose first coordinate is zero. This proves 2.3.9.

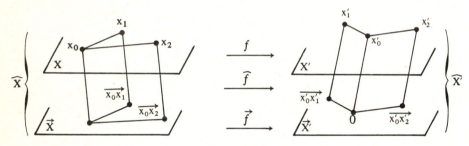

Figure 3.2.5

3.3. Polynomials over an affine space

> In this section fields have characteristic zero.

We recall the definition and some properties of polynomials and polynomial maps on vector spaces. A very good reference is [CH1, 79].

3.3.1. DEFINITIONS. *Let V, W be vector spaces. A map $f : V \to W$ is called a homogeneous polynomial map of degree k if there exists a symmetric k-linear map $\phi : V^k \to W$ such that $f = \phi \circ \Delta$, where $\Delta : V \to V^k$ is the diagonal map $V(x) = (x, \ldots, x)$. We denote by $P_k^\bullet(V; W)$ the set of homogeneous polynomial maps of degree k from V into W. If $W = K$, the base field of V, we call f a homogeneous polynomial of degree k, and put $P_k^\bullet(V) = P_k^\bullet(V; W)$. If f is a homogeneous polynomial map, the map ϕ such that $f = \phi \circ \Delta$ is uniquely determined by f. A map $f : V \to W$ is called a polynomial map of degree less than or equal to k if there exist $f_i \in P_i^\bullet(V; W)$ $(i = 0, \ldots, k)$ such that $f = \sum_{i=0}^k f_i$ (by convention, $P_0^\bullet(V; W)$ is the set of constant maps from V into W). If $W = K$ we say that f is a polynomial of degree less than or equal to k. If f is a polynomial map of degree less than or equal to k, f is uniquely written as a sum $\sum_i f_i$, with $f_i \in P_i^\bullet(V; W)$. The set* *of polynomial maps of degree less than or equal to k is denoted by $P_k(V;W)$, and we put $P_k(V) = P_k(V;K)$.*

3.3.2. REMARKS. There are explicit formulas to calculate $\phi : V^k \to W$ from a polynomial map $f : V \to W$ (see [CH1, 85, formula 6.3.5]). When $k = 2$, we are treading familiar ground: the elements f of $P_2^\bullet(V)$ are called *quadratic forms*, and the map ϕ such that $f = \phi \circ \Delta$ is the *polar form* of f, equal to

3.3.2.1
$$\phi(x,y) = \frac{1}{2}\big(f(x+y) - f(x) - f(y)\big)$$

for all $x, y \in V$. It was in order to guarantee the uniqueness of ϕ that we assumed the characteristic of the base field to be zero. For quadratic forms we see from 3.3.2.1 that it is enough to have characteristic $\neq 2$; and in general the characteristic just has to be strictly larger than the degree of the polynomials.

The vector space $P_1^\bullet(V)$ is simply the dual V^* of V, the set of all linear forms on V. More generally, $P_k^\bullet(V;W)$ has a natural vector space structure, being a set of maps into a vector space.

The expression "homogeneous of degree k" comes from the property $f(\lambda x) = \lambda^k f(x)$ for any $f \in P_k^\bullet(V)$ and any $x \in V$. This property by itself is not sufficient to guarantee that f is a polynomial, unless we assume f is differentiable enough times (3.7.12).

3.3.3. COORDINATES. Suppose V has finite dimension n, and let $\{e_i\}_{1,\ldots,n}$ be a basis for V and $x = (\lambda_1, \ldots, \lambda_n)$ the coordinates of x in this basis. A map $f \in P_k^\bullet(V;W)$ can be written in two ways, both useful. The first is

$$f(\lambda_1, \ldots, \lambda_n) = \sum_{\alpha_1 + \cdots + \alpha_n = k} a_{\alpha_1,\ldots,\alpha_n} \lambda_1^{\alpha_1} \cdots \lambda_n^{\alpha_n},$$

where the sum is taken over all integers ≥ 0 which add up to k, and the $a_{\alpha_1,\ldots,\alpha_n}$ are fixed vectors in W (scalars if $W = K$). The second is

$$f(\lambda_1, \ldots, \lambda_n) = \sum_{i_1 \leq \cdots \leq i_k} a_{i_1,\ldots,i_k} \lambda_{i_1} \cdots \lambda_{i_k},$$

where the $a_{i_1,\ldots,i_k}$ are fixed in W, and the sum is taken over sequences of (possibly repeated) increasing integers $i_1, \ldots, i_k$ between 1 and n. In both cases we have written multiplication by scalars on the right, so as to recover the habitual notation for polynomials when $W = K$.

3.3.4. The quickest way to define polynomials over an affine space X is to use the associated universal space $\hat{X}$. A more elementary definition is given in 3.7.11 (cf. 3.3.14).

3.3.5. DEFINITION. *Let X be an affine space, W a vector space. A polynomial map of degree k on X is a map $f : X \to W$ such that there exists $\hat{f} \in P_k^\bullet(\hat{X})$ satisfying $\hat{f}|_X = f$. The set of such maps is a vector space in a natural way; it is denoted by $P_k(X;W)$. If $W = K$, f is called a polynomial of degree k, and we write $P_k(X) = P_k(X;W)$. The symbol of f, denoted by $\vec{f}$, is the homogeneous polynomial map $\vec{f} = \hat{f}|_{\vec{X}}$.*

3.3.6. EXAMPLES. $P_0(X;W)$ is just the set of constant maps; $P_1(X;W) = A(X;W)$, and in particular $P_1(X)$ is the set of affine forms of X (verify this directly or use 3.3.14). This justifies calling the elements of $P_2(X)$ affine quadratic forms.

3.3.7. Definition 3.3.5 is simple and quick, but often useless in practice. Besides, one would like to study the restriction map $P_k^\bullet(\hat{X}) \to P_k(X)$. It is surjective by definition, but is it injective? One also wants to perform explicit calculations in finite dimension and study the relation between $P_k(X)$ and $P_k(X_a)$, where X_a is the vectorialization of X at a. To do all this, we fix $a \in X$, and consider the direct sum $\hat{X} = \vec{X} \oplus Ka$ (cf. 3.1.6); we work in $P_k(X)$ for simplicity. Let $\hat{\phi} : \hat{X}^k \to K$ such that $\hat{f} = \hat{\phi} \circ \Delta$. We have
3.3.8

$$\hat{\phi}(\vec{\xi}_1 + t_1 a, \ldots, \vec{\xi}_k + t_k a) = \sum_{i=0}^{k} \sum_{l_1 < \cdots < l_i} \prod_{s \neq l_1, \ldots, l_i} t_s \hat{\phi}(\vec{\xi}_{l_1}, \ldots, \vec{\xi}_{l_i}, a, \ldots, a),$$

due to the symmetry of ϕ. For $i = 0, 1, \ldots, k$, we define $\phi_i : X_a^i \to K$ by

3.3.9 $$\phi_i(\vec{\xi}_1 + a, \ldots, \vec{\xi}_i + a) = \binom{k}{i} \hat{\phi}(\vec{\xi}_1, \ldots, \vec{\xi}_i, a, \ldots, a).$$

Each ϕ_i is a symmetric i-linear map, so the map $f_i = \phi_i \circ \Delta : X_a \to K$ belongs to $P_i^\bullet(X_a)$, and we have, by 3.3.8,

3.3.10 $$f = \sum_{i=0}^{k} f_i.$$

Thus any $f \in P_k(X)$ is a polynomial of degree $\leq k$ over X_a for any $a \in X$. The converse is also true, and 3.3.8 allows us to obtain $\hat{f}$ from f in a unique way. Indeed, if $a \in X$ and $f = \sum_{i=0}^{k} f_i$ with $f_i \in P_i^\bullet(X_a)$ for $i = 0, 1, \ldots, k$ and $f_i = \phi_i \circ \Delta$ with $\phi_i : X_a^i \to K$, we can write
3.3.11

$$\hat{\phi}(\vec{\xi}_1 + t_1 a, \ldots, \vec{\xi}_k + t_k a) = \sum_{i=0}^{k} \binom{k}{i}^{-1} \sum_{l_1 < \cdots < l_i} \prod_{s \neq l_1, \ldots, l_i} t_s \phi_i(\vec{\xi}_{l_1} + a, \ldots, \vec{\xi}_{l_i} + a).$$

Clearly $\hat{f}|_X = f$, for $\hat{f} = \hat{\phi} \circ \Delta$, and $\hat{\phi}$ is symmetric and k-linear. Moreover, for $t \neq 0$, we have

3.3.12
$$f(\vec{\xi} + ta) = \sum_{t=0}^{k} t^{k-i} f_i(\vec{\xi} + a) = t^k \sum_{i=0}^{k} f_i(t^{-1}\vec{\xi} + a)$$

$$= t^k \sum_{i=0}^{k} f_i(t^{-1}(\vec{\xi} + a)) = t^k f(t^{-1}(\vec{\xi} + a)).$$

Identifying X_a and $\vec{X}$ by means of Θ_a gives

3.3.13 $$\hat{f}(\vec{\xi}, t) = t^k f(t^{-1}\vec{\xi})$$

for any $x \in X$, $t \in K^*$ and $\hat{f}|_X = f$.

Formula 3.3.13 expresses the fact that $\hat{f}$ is obtained from f by introducing the homogenizing variable t, which in practice means using 3.3.12 for the actual calculations. The following proposition is a consequence of the remarks above:

3.3.14. PROPOSITION. *The restriction map*

$$P_k^\bullet(\hat{X}) \ni \hat{f} \mapsto \hat{f}|_X \in P_k(X)$$

is bijective; the inverse map is given by 3.3.13. The elements of $P_k(X)$ can be characterized as the maps $f : X \to K$ such that $f \in P_k(X_a)$ for any $a \in X$ (or for some $a \in X$). Under the identification afforded by Θ_a, the symbol $\vec{f}$ is simply the degree-k part of f. $\qquad\square$

3.3.15. EXAMPLE. Let X have finite dimension n and let $\{x_i\}_{i=0.1.....n}$ be a frame for X, with associated coordinates $(\lambda_i)_{i=0,.....n}$. Then $f \in P_2(X)$ if and only if there exist $a_{ij}, b_j, c \in K$ $(i, j = 1, \ldots, n)$ such that

$$f(\lambda_1, \ldots, \lambda_n) = \sum_{i,j} a_{ij}\lambda_i\lambda_j + \sum_i b_i\lambda_i + c,$$

and then $\hat{f}$ is given by

$$\hat{f}(\lambda_1, \ldots, \lambda_n, t) = \sum_{i,j} a_{ij}\lambda_i\lambda_j + \sum_i b_i\lambda_i t + ct^2.$$

3.4. Barycenters

Embedding an affine space X into the vector space $\hat{X}$ allows us to form linear combinations like $\sum_i \lambda_i x_i$, where $x_i \in X$ and $\lambda_i \in K$. *A priori*, the result is just in $\hat{X}$, but it will in fact be in X if, in the notation of 3.1.6,

$$M\left(\sum_i \lambda_i x_i\right) = \sum_i \lambda_i M(x_i) = \sum_i \lambda_i = 1$$

This gives rise to the following

3.4.1. DEFINITION. *Let $\{x_i\}_{i \in I}$ be a family of points in an affine space X, and $\{\lambda_i\}_{i \in I}$ a family of scalars such that $\lambda_i = 0$ except for finitely many indices and that $\sum_{i \in I} \lambda_i = 1$. The barycenter of the points x_i with masses λ_i is the point x of X given by*

$$x = \sum_{i \in I} \lambda_i x_i.$$

3.4.2. EXAMPLES. If $\lambda_1 = 1$, $\lambda_i = 0$ for $i \neq 1$, then $x = x_1$. If $I = \{1, 2\}$ and $\lambda_1 = \lambda_2 = \frac{1}{2}$ (thus the characteristic of K is necessarily $\neq 2$), we find $x = (x_1 + x_2)/2$; this point is called the *midpoint* of x_1 and x_2. More generally, if $I = \{1, \ldots n\}$ and $\lambda_i = 1/n$ for all $1 \leq i \leq n$ (thus $n \neq 0$ in K), the point

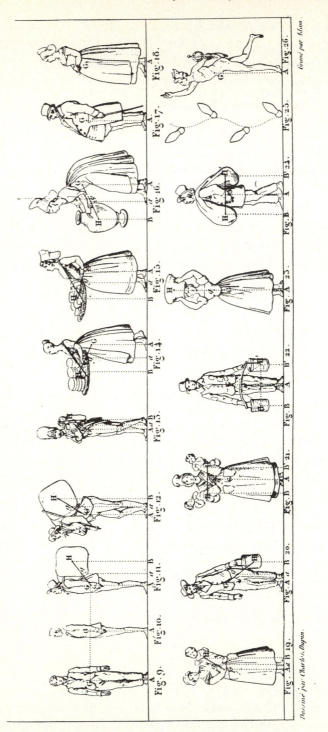

Figure 3.4.1 (Source: [DP])

$x = (x_1 + \cdots + x_n)/n$ is called the *center of mass*, or *equibarycenter*, of the x_i. Observe that if X is an affine Euclidean plane, the equibarycenter of a triangle $\{x_1, x_2, x_3\}$ (cf. 2.4.7) made up of three points x_1, x_2, x_3 coincides with the center of mass (in the mechanics sense, that is, the centroid) of a triangular homogeneous plate defined by the convex hull of these three points (figure 3.4.2.1); but the same is not true any longer if we consider the center of mass of four points

$$\frac{x_1 + x_2 + x_3 + x_4}{4}$$

and the centroid of the homogeneous plate defined by them (cf. figure 3.4.2.2, exercise 3.7.14).

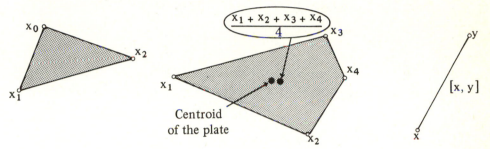

Figure 3.4.2.1   Figure 3.4.2.2   Figure 3.4.3

3.4.3. A FUNDAMENTAL EXAMPLE is when $K = \mathbf{R}$, X is a real affine space and x, y are two points of X. The *segment with endpoints* x, y is the subset $[x, y] = \left\{ \lambda x + (1 - \lambda)y \mid \lambda \in [0, 1] \right\}$ of X. This is the foundation of the notion of convexity (chapters 11 and 12).

3.4.4. If $\sum_i \lambda_i \neq 1$, the sum $\sum_i \lambda_i x_i$ will lie in $\hat{X} \setminus X$. We can still work with it, though, if we pull it back to X by dividing it by the scalar $\sum_i \lambda_i$, as long as this is not zero.

3.4.5. DEFINITION. *An element of $\hat{X} = \vec{X} \cap (K^* \times X)$ is called a* punctual mass, *and written as* (λ, x), *where $x \in \vec{X}$ if $\lambda = 0$ and $x \in X$ otherwise. A punctual mass (λ, x) is also referred to as the* point x with mass λ. *If $\left\{ (\lambda_i, x_i) \right\}_{i=1,\dots,n}$ is a family of punctual masses, the* barycenter *of this family, also called the* barycenter *of the x_i with masses λ_i, is the point $(\sum_i \lambda_i, x)$ of $\hat{X}$, where x is the vector $\sum_{\lambda_i} x_i \in \vec{X}$ if $\sum_i \lambda_i = 0$ and the point*

$$\frac{\sum_i \lambda_i x_i}{\sum_i \lambda_i} \in X$$

otherwise.

3.4.6. REMARKS.

3.4.6.1. This definition coincides with the mechanics notion of center of mass for punctual masses when $\sum_i \lambda_i \neq 0$. For example, the center of mass of three points x_i, each with mass 1, is the point $(x_1 + x_2 + x_3)/3$, but it must be ascribed the mass 3, obtained by adding the masses of the x_i.

3.4.6.2. If the family consists of two punctual masses $\{(1, x), (-1, y)\}$ we recover the notation $\overrightarrow{yx} = x - y$; cf. 2.1.4 and 3.1.7.

3.4.6.3. An equivalent, more classical definition employs the sums $\sum_i \lambda_i \overrightarrow{yx_i}$, concealing the use of vector fields. The barycenter of a family $\{(\lambda_i, x_i)\}$, $\sum_i \lambda_i \neq 0$ is then the point x satisfying

3.4.6.4
$$\sum_i \lambda_i \overrightarrow{yx_i} = \left(\sum_i \lambda_i \right) \overrightarrow{yx}$$

for all $y \in X$, and in particular

3.4.6.5
$$\sum_i \lambda_i \overrightarrow{xx_i} = 0.$$

3.4.6.6. The explicit calculation of the barycenter x of a family $\{(\lambda_i, x_i)\}$ $(\sum_i \lambda_i \neq 0)$, is performed by vectorializing X at an arbitrary point $a \in X$:

$$x = a + \frac{\sum_i \lambda_i \overrightarrow{ax_i}}{\sum_i \lambda_i}.$$

In particular, if X is finite-dimensional and the x_i have coordinates $(x_{i1}, \ldots, x_{in})$ in a fixed affine frame, the coordinates of their barycenter in the same frame will be

$$\left(\frac{\sum_i \lambda_i x_{i1}}{\sum_i \lambda_i}, \ldots, \frac{\sum_i \lambda_i x_{in}}{\sum_i \lambda_i} \right).$$

3.4.7. The associativity of addition in X has several consequences for the barycenters, giving rise to properties with an intuitive mechanical interpretation and nice geometrical applications.

3.4.8. PROPOSITION (ASSOCIATIVITY OF BARYCENTERS OF PUNCTUAL MASSES). *Let I be a finite set, $I = I_1 \cup \cdots \cup I_p$, a partition of I, $\{(\lambda_i, x_i)\}_{i \in I}$ a family of punctual masses. The barycenter of this family coincides with the barycenter of $\{(\mu_l, \xi_l)\}_{l=1,\ldots,p}$, where, for each l, the punctual mass (μ_l, ξ_l) is the barycenter of $\{(\lambda_i, x_i)\}_{i \in I_l}$.* □

3.4.9. PROPOSITION (NON-ASSOCIATIVITY OF BARYCENTERS IN X). *Let I be a finite set, $I = I_1 \cup \cdots \cup I_p$, a partition of I, $\{x_i\}_{i \in I}$ a family of points in X and $\{\lambda_i\}_{i \in I_l}$ families of scalars such that $\sum_{i \in I_l} \lambda_i = 1$ for all $l = 1, \ldots, p$. Let $\alpha_1, \ldots, \alpha_p \in K$ be such that $\sum_l \alpha_l = 1$. If ξ_l is the barycenter of the x_i with masses λ_i $(i \in I_l)$ and ξ is the barycenter of the ξ_l $(l = 1, \ldots, p)$ with masses α_l, then ξ is the barycenter of the x_i $(i \in I)$ with masses $\alpha_l \lambda_i$ $(i \in I_l)$.* □

3.4.10. CLASSICAL GEOMETRICAL CONSEQUENCES. We start with three points x_1, x_2, x_3 of X, and put $I = \{1, 2, 3\} = \{1\} \cup \{2, 3\}$, $\lambda_1 = 1$, $\lambda_2 = \lambda_3 = \frac{1}{2}$, $\alpha_1 = \frac{1}{3}$, $\alpha_2 = \frac{2}{3}$ (in particular, the characteristic of K is not 2 or

3). Then $\xi_1 = x_1$, and $\xi_2 = (x_2 + x_3)/2$ is the midpoint of $\{x_2, x_3\}$. Thus the barycenter x of x_1, x_2, x_3 with masses $\{1/3, 1/3, 1/3\}$, i.e., their center of mass, is the barycenter of $x_1, (x_2 + x_3)/2$ with masses 1 and 2, respectively. In geometrical terms, the three medians

$$\left\langle x_1, \frac{x_2 + x_3}{2} \right\rangle, \quad \left\langle x_2, \frac{x_3 + x_1}{2} \right\rangle, \quad \left\langle x_3, \frac{x_1 + x_2}{2} \right\rangle$$

·must be concurrent; their common point ξ is the barycenter of the triangle $\{x_1, x_2, x_3\}$, and it divides each median in segments whose ratio is $\frac{1}{2}$.

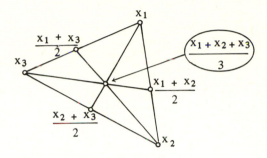

Fig 3.4.10.1

Considering now four points (again char $K \neq 2, 3$), we obtain seven concurrent lines: three joining the midpoints of opposite sides (including diagonals), and four joining a vertex to the center of mass of the other three (figure 3.4.10.2).

3.5. Relationship with affine maps and subspaces

3.5.1. PROPOSITION. *Let X, X' be two affine spaces. A necessary and sufficient condition for a map $f : X \to X'$ to be affine is that f preserve barycenters, i.e., that for all finite families $\{x_i\}_{i \in I} \subset X$ and $\{\lambda_i\}_{i \in I} \subset K$ with $\sum_i \lambda_i = 1$ we have*

$$f\left(\sum_i \lambda_i x_i \right) = \sum_i \lambda_i f(x_i).$$

Proof. Necessity follows from 3.2.1 and 3.4.1. To prove sufficiency, we have to show that $f : X_a \to X'_{f(a)}$ is a linear map. But if $x, x' \in X$ and $\lambda, \mu \in K$, the point $x'' = \lambda x + \mu x'$ in X_a is none other than the barycenter of $\{(a, 1 - \lambda - \mu), (x, \lambda), (x', \mu)\}$, and similarly for $f(a)$, $f(x)$ and $f(x')$. $\square$

The next proposition can be shown by the same argument:

3.5.2. PROPOSITION. *Let X be an affine space. A subset $S \subset X$ is a subspace of X if and only if it contains the barycenter of any family of points in*

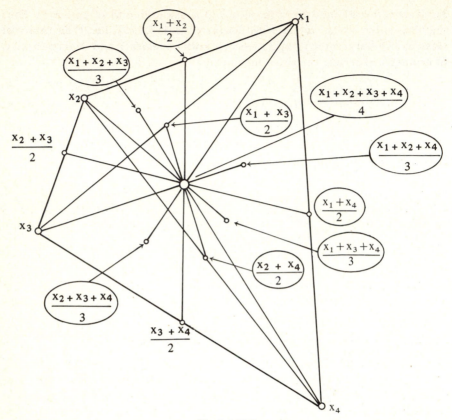

Fig 3.4.10.2

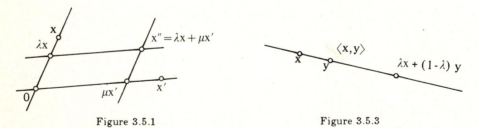

Figure 3.5.1   Figure 3.5.3

S, *i.e., if for all finite families* $\{\lambda_i\}_{i\in I} \subset K$ *and* $\{x_i\}_{i\in I} \subset S$ *with* $\sum_i \lambda_i = 1$
we have

$$\sum_i \lambda_i x_i \in S. \qquad \square$$

3.5.3. COROLLARY. *If S is a subset of X, the subspace $\langle S \rangle$ spanned by X
is the set of barycenters of points of S (for all possible choices of masses).* $\square$

Thus the line $D = \langle x, y \rangle$ spanned by two distinct points x, y on it is simply

$$D = \langle x, y \rangle = \{\, \lambda x + (1 - \lambda) y \mid \lambda \in K \,\}.$$

3.5.4. PROPOSITION. *Let X be an affine space over a field of characteristic zero. For any finite subset $F \subset X$ there exists a point $x \in X$ such that $\mathrm{GA}_F(X) \subset \mathrm{GA}_x(X)$. In other words, any $f \in \mathrm{GA}(X)$ such that $f(F) = F$ fixes x.*

Proof. If $F = \{x_1, \ldots, x_n\}$, take $x = (x_1 + \cdots + x_n)/n$ and apply 3.5.1. $\square$

This can be extended to the case of F compact: see 2.7.5.7 and 9.8.6. Exercise 3.7.3 provides a nice corollary.

3.6. Barycentric coordinates

3.6.1. We will exploit here the idea, introduced in 3.2.5, of associating to an affine frame $\{x_i\}_{i=0,1,\ldots,n}$ of the affine space X the corresponding basis $\{x_i\}_{i=0,1,\ldots,n}$ of $\hat{X}$.

3.6.2. PROPOSITION. *Let $\{x_i\}_{i=0,1,\ldots,n}$ be a frame for an affine space X. For any $x \in X$ there exist $\lambda_i \in K$ $(i = 0, 1, \ldots, n)$ such that $\sum_i \lambda_i = 1$ and $x = \sum_i \lambda_i x_i$. The scalars λ_i are uniquely defined by this property and are called the barycentric coordinates of x in the frame $\{x_i\}_{i=0,1,\ldots,n}$.*

Proof. As an element of $\hat{X}$, x can be written in a unique way as $\sum_i \lambda_i x_i$. By 3.1.6, $\sum_i \lambda_i = 1$ because $x \in X \subset \hat{X}$. $\square$

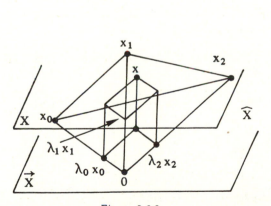

Figure 3.6.2

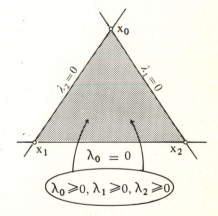

Figure 3.6.3

3.6.3. For example, if $n = 2$, the points with coordinate $\lambda_0 = 0$ in the frame $\{x_0, x_1, x_2\}$ are those which belong to the side $\langle x_1, x_2 \rangle$ of the triangle $\{x_0, x_1, x_2\}$. (See figure 3.6.3 for the case $K = \mathbf{R}$; the shaded part corresponds to points with coordinates $\lambda_i \geq 0$ for $i = 0, 1, 2$.) This remark will be relevant to the study of convexity, see 11.1.8.4.

3.6.4. Barycentric coordinates will be used in 10.6.8.

3.6.5. BARYCENTRIC SUBDIVISION. The *barycentric subdivision* of a simplex $\{x_0, \ldots, x_d\}$ in a real affine space of dimension d is defined by induction on d. For $d = 1$, it is just the set consisting of the two simplices $\{x_0, (x_0 + x_1)/2\}$ and $\{(x_0 + x_1)/2, x_1\}$. For $d = 2$, it is the set of six triangles having one vertex at $(x_0 + x_1 + x_2)/3$ and opposite side equal to each simplex in the barycentric subdivisions of the faces of $\{x_1, x_2, x_3\}$ (figure 3.6.5.1). The generalization is obvious.

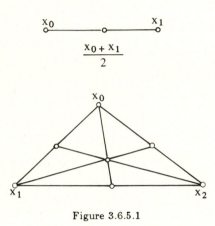

Figure 3.6.5.1

Figure 3.6.5.2 shows the barycentric subdivision of the barycentric subdivision (i.e., the *second barycentric subdivision*) of $\{x_0, x_1, x_2\}$. It consists of $6 \times 6 = 36$ triangles. See exercise 3.7.8 about iterated subdivisions.

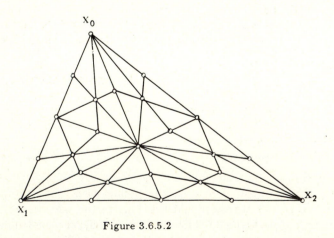

Figure 3.6.5.2

This application of barycenters is important in algebraic topology: see, for example, [GG2, 60–68], [H–Y, 206–209], or [CA, 82–86].

3.7. Exercises

3.7.1. Is Proposition 3.5.4 still valid in prime characteristic?

3.7.2. Extend the geometric results in 3.4.10 to sets of five and six points.

3.7.3. Let X be an affine space over a field of characteristic zero. Show that every finite subgroup of $GA(X)$ has a fixed point.

3.7.4. Let X be a finite-dimensional real affine space, K a compact subset of X with non-empty interior. Suppose $K = \bigcup_{i=1}^{n} K_i$, where $\mathring{K}_i \neq \emptyset$ and $K_i \cap K_j$ has measure zero for all $i \neq j$. Put $\xi_i = \operatorname{cent}'(K_i)$, $\xi = \operatorname{cent}'(K)$ (cf. 2.7.5.2). Prove that ξ is the barycenter of the ξ_i with masses $\mu(K_i)$.

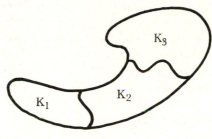

Figure 3.7.4

3.7.5. Let X be an affine space. Give an intrinsic definition (i.e., one that uses neither a vectorialization X_a nor the universal space $\hat{X}$) for $P_1(X)$ and $P_2(X)$. Find a formula for the symbol $\vec{f}$ of $f \in P_1(X)$ and of $f \in P_2(X)$.

3.7.6. Why did we take the vector field $x \mapsto k\overrightarrow{xa}$, instead of $x \mapsto k\overrightarrow{ax}$, in definition 3.1.2.2?

3.7.7. Let X be an affine space, $a, b \in X$, $f \in P_k(X)$. Put $f = \sum_{i=0}^{k} f_i$ in X_a, with $f_i \in P_i^{\bullet}(X_a)$, and $f = \sum_{i=0}^{k} g_i$ in X_b, with $g_i \in P_i^{\bullet}(X_b)$. Fixing an affine frame for X, write an explicit expression for the g_i as a function of the f_i and of $\xi = \overrightarrow{ab}$, first for $k = 1$, then for $k = 2$.

* 3.7.8. Let Σ be a simplex in a Euclidean affine space of dimension n; let d be its diameter (cf. 0.3). Show that all simplices of the barycentric subdivision of Σ have diameter less than or equal to $nd/(n+1)$; deduce that when we iterate the process of barycentric subdivision, the diameter of all simplices tends towards zero.

3.7.9. Generalize formula 3.3.2.1 for $k = 3$.

3.7.10. Let $\dim E = n$. Show that, for all $k \geq 0$,

$$\dim P_k^{\bullet}(E) = \binom{n+k-1}{k}, \qquad \dim P_k(E) = \binom{n+k}{k}.$$

* 3.7.11. Give a direct proof of the fact that if $f : X \to W$ belongs to $P_k(X_a; W)$ for some $a \in X$, then $f \in P_k(X_b; W)$ for any $b \in X$.

3.7.12. Let X be a real vector space and $f : X \to \mathbf{R}$ a C^1 map such that $f(\lambda x) = \lambda^k f(x)$ for all $x \in X$ and $\lambda \in \mathbf{R}$. Show that the derivative f' of f satisfies the *Euler identity* $f'(x)(x) = kf(x)$ for any $x \in X$. Write and prove an analogous formula for the p-th derivative of f, when f is of class C^p and again homogeneous of degree k. Deduce that if f is of class C^k and homogeneous of degree k, it is necessarily a polynomial.

* **3.7.13.** Determine the center of mass of the physical object consisting of three homogeneous pieces of wire of same linear density, lying on the three sides of a triangle. Give a geometrical construction for this point.

* **3.7.14.** Give a geometrical construction for the centroid of a homogeneous plate in the shape of a quadrilateral. Compare this point with the center of mass of the four vertices (figure 3.4.2.2).

* **3.7.15.** Using the notation of 3.3.1, show that for every $f \in P_k^\bullet(V;W)$ we have

$$\phi(v_1,\ldots,v_k) = \frac{1}{k!}\sum_{j=1}^k (-1)^{k-j} \sum_{1\le i_1\le\cdots\le i_j\le k} f(v_{i_1} + \cdots + v_{i_j}).$$

* **3.7.16.** In a real affine space, consider p points $x_{1,1},\ldots,x_{1,p}$ $(p \ge 2)$. For $i = 1, 2, \ldots, p$ denote by $x_{2,i}$ the center of mass of the $(x_{1,j})_{j\ne i}$. Define by recurrence the center of mass $x_{k+1,i}$ of the $(x_{k,j})_{j\ne i}$, for all $k \ge 1$. Prove that every sequence $(x_{k,i})_{k\in\mathbf{N}}$ converges. What can you say about the limit of these sequences for different values of i?

3.7.17. Prove the theorems of Ceva and Menelaus (2.8.1 and 2.8.2) using barycenters.

3.7.18. Show that the midpoints of the diagonals of a *complete quadrilateral* (see figure 3.7.18) are collinear. Compare with 6.4.4 and 17.16.2.1.

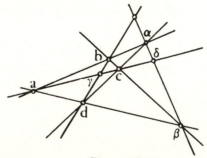

Figure 3.7.18.

Chapter 4
Projective spaces

This chapter is devoted to projective spaces. It starts with an introduction (section 4.0), meant to lessen the shock of a somewhat sudden definition in 4.1. Section 4.1 also contains a number of examples. The aim of the rather long sections 4.2 and 4.3 is to give the reader a concrete idea of what projective spaces are, and to show that projective spaces now play a role that transcends the framework of elementary geometry, where they originated. Section 4.2 describes natural coordinate systems; section 4.3 studies finite-dimensional real and complex projective spaces, including their topological properties. Keeping in mind the relatively elementary level of this book, we have not included proofs of results from algebraic topology, but they should, together with the references mentioned, be enough to convince the reader of the continuing importance of projective spaces in mathematics (though the focus of interest has changed somewhat over time).

The following sections present a classical, and consequently terse, exposition of projective morphisms and subspaces, together with their elementary properties. In the last section (4.8) we say some words about the non-commutative case.

Base fields are assumed commutative, unless we state otherwise.

4.0. Introduction

Here are some notes to motivate the introduction of projective geometry and show that definition 4.1.1 doesn't just come out of the blue.

4.0.1. Affine geometry is not satisfactory under certain points of view: the intersection of subspaces, for example, gives rise to numerous exceptions, as attested by the statements of the results in 2.4.2.4, 2.4.9.2 and 2.4.9.4. One would like to have a geometry where the intersection of two subspaces is always a subspace, and where the relation $\dim\langle S \cup T\rangle = \dim S + \dim T - \dim(S \cap T)$ is always satisfied.

4.0.2. Desargues was the first (and led the field by about two centuries) to construct the projective completion of a real affine plane by adding to this plane points "at infinity", one for each set of parallel lines in the plane (in the completion, these lines all pass through the point they define). This construction will be the object of chapter 5.

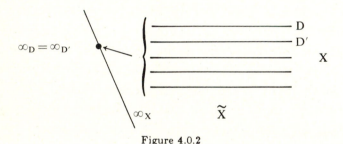

Figure 4.0.2

4.0.3. It is natural in geometry to consider the set of lines passing through a fixed point, or, alternatively, the set $G_{E,1}$ of one-dimensional vector subspaces of a given vector space E. Such a set appears, for example, in the study of the tangent to a curve at a point (cf. 4.3.4). The definition of projective spaces given here (4.1.1) uses precisely these $G_{E,1}$.

4.0.4. Our eyesight is not plane, but conical, as all rays converge to the center of the eye. It also poses the problem of combining two images obtained from different centers; this is connected with the notion of perspective, and plays a role in aerial photography, for example (see section 4.7).

4.0.5. Projective spaces provide a nice and historically relevant way of translating linear algebraic concepts, by associating a quadric to a quadratic form, for instance (see 14.1). Further, real and complex projective spaces play a fundamental role in differential geometry, algebraic topology (cobordism, for instance), and, sure enough, in algebraic geometry. Part of their interest is due to their being the simplest compact manifolds, except for spheres. For a modern discussion of these subjects, see for example [GG2], [HU], [B–H].

4.0.6. Projective spaces arise naturally in quantum mechanics, cf. [C–D–L, 219], for instance.

4.1. Definition and examples

4.1.1. DEFINITION. *Let E be a vector space. The projective space derived from E, denoted by $P(E)$, is the quotient of $E \setminus 0$ by the equivalence relation "$x \sim y$ if and only if $y = \lambda x$ for some $\lambda \in K$". The dimension of $P(E)$ is $\dim E - 1$. The canonical projection is $p : E \setminus 0 \to P(E)$.*

4.1.2. REMARKS. The definition of the dimension will be justified in 4.2.1 and 5.1.3, but it is quite natural, since $P(E)$, being a set of lines of E, should have dimension one lower than E.

Definition 4.1.1 can bother the purists, since it hitches up to a projective space the vector space from which it derives. A possibly more satisfying, but also lengthier, definition is found in [BI5, 337]; see also 4.8.3.

There are axiomatizations of projective geometry, as for affine geometry; see references in 2.6.7.

4.1.3. EXAMPLES.
4.1.3.1. For every integer $n \geq 1$ we set $P^n(K) = P(K^{n+1})$, and call this the *standard* projective space of dimension n over the field K.

4.1.3.2. A projective space is called *real* if $K = \mathbf{R}$, and *complex* if $K = \mathbf{C}$.

4.1.3.3. Zero-dimensional projective spaces are called *points;* one- and two-dimensional projective spaces are called *projective lines* and *planes*, respectively.

4.1.3.4. There is a natural bijection between the projective space $P(E)$ and $G_{E.1}$ (cf. 1.2.5); the two spaces are often identified.

4.1.3.5. By 2.4.8.1 and 2.4.8.6, the hyperplanes of a vector space E are in one-to-one correspondence with the points of $P(E^*)$, where E^* is the dual of E. We denote by $\mathcal{K}(E)$ the set of hyperplanes of E; thus there is a bijection $\mathcal{K}(E) \mapsto P(E^*)$, which serves to identify the two spaces.

4.1.3.6. Let $\mathcal{P}_k(X)$ be the vector space of polynomials of degree k over an affine space X (3.3.5), and let N be the map

$$N : \mathcal{P}_k(X) \ni f \to f^{-1}(0) \subset X.$$

Since $(\lambda f)^{-1}(0) = f^{-1}(0)$ for all $\lambda \in K^*$, the map N can be factored by the equivalence relation $\sim$ of 4.1.1, giving rise to a map $\underline{N}$ from $P(\mathcal{P}_k(X))$ to the set of subsets of X:

$$\mathcal{P}_k(X) \setminus 0 \xrightarrow{\;\;p\;\;} P(\mathcal{P}_k(X))$$
$$N \searrow \qquad\qquad \swarrow \underline{N}$$
$$\text{subsets of } X$$

We shall see that in certain cases $\underline{N}$ is injective (see 14.1.6). For $k = 2$, the image of $\underline{N}$ is, by definition, the set of quadrics of X.

Similarly, if V is a vector space, by remarking that $f(\lambda x) = \lambda^k f(x)$ for $f \in P_k^{\bullet}(V)$ (see 3.3.2) we are led to consider the diagram

$$
\begin{array}{ccc}
P_k^{\bullet}(V) \setminus 0 & \xrightarrow{\ p\ } & P(P_k^{\bullet}(V)) \\
\downarrow{\scriptstyle N} & \searrow{\scriptstyle \underline{N}} & \downarrow{\scriptstyle \underline{N}} \\
\text{subsets of } V & \longrightarrow & \text{subsets of } P(V)
\end{array}
$$

since each $f^{-1}(0)$ is, in fact, a cone in V. We shall meet these two diagrams again in chapters 14 and 15.

4.1.3.7. Let K be a finite field with k elements, and $P(E)$ a projective space of dimension n over K. Then $\#P(E) = (k^{n+1} - 1)/(k - 1)$ because $\#K^{n+1} = k^{n+1}$ and each line has $k-1$ elements, after deleting the origin. For example, every projective line over $\mathbf{Z}_2$ has three points, and every projective plane seven points.

4.2. Structure of projective spaces: charts

4.2.1. Let $P(E)$ be a projective space. If $H \subset E$ is a hyperplane, one can say, in the notation of 2.2.5, that $E_H = G_{E,1} \setminus G_{H,1} = P(E) \setminus P(H)$ is a subset of $P(E)$, and constitutes, in fact, most of $P(E)$. But $E_H \subset P(E)$ has a natural affine structure, so that, except for the points of $P(H)$, the points of $P(E)$ form an affine space. This makes it much easier to perform calculations, use frames, and so on. We call the bijection $P(E) \setminus P(H) \to E_H$ a *chart* for $P(E)$.

In order to parametrize the whole of $P(E)$, each point must be in the domain of a chart. Suppose that $P(E)$ has dimension n, and take a family $\{H_i\}_{i=0,\ldots,n}$ of hyperplanes of E such that $\bigcap_i H_i = 0$. Then $\bigcap_i E_{H_i} = \emptyset$, so $P(E) = \bigcup_i (P(E) \setminus P(H_i))$ is indeed the union of the domains of the corresponding charts. We say we have an *atlas* for $P(E)$.

4.2.2. In order to be able to use an atlas, we must know how to pass from one chart to the other. In other words, we must calculate the broken arrow in the diagram below:

$$
\begin{array}{c}
P(E) \setminus \big(P(H_i) \cup P(H_j)\big) \\
\swarrow \qquad \searrow \\
E_{H_i} \ \text{--} \ \text{--} \to E_{H_j}
\end{array}
$$

Here $i \neq j$ and the broken arrow indicates that the map is not defined on the whole of E_{H_i}. We shall work out this calculation in 4.2.4.

4.2.3. A simpler idea for calculating in $P(E)$ when $\dim E < \infty$ is to take a basis $\{e_i\}_{i=0,1,\ldots,n}$ of E. Every $m \in P(E)$ is of the form $m = p(x) = p(x_0,\ldots,x_n)$, where $x = (x_0,\ldots,x_n)$ in the basis being considered. We say that $(x_0, x_1, \ldots, x_n)$ is a set of *homogeneous coordinates* of x (with respect to the basis $\{e_i\}_{i=0,1,...,n}$). The word "homogeneous" comes from the fact that the sets of homogeneous coordinates of a point $x \in P(E)$ are all of the form $(\lambda x_0, \ldots, \lambda x_n)$, for all $\lambda \in K^*$ and fixed $x_0, \ldots, x_n$.

4.2.4. We now blend the two viewpoints presented in 4.2.2 and 4.2.3. Take a basis $\{e_i\}_{i=0,1,...,n}$ of E, and define the hyperpanes H_i as $H_i = x_i^{-1}(0)$. The points of $P(E) \setminus P(H_i)$ are those whose homogeneous coordinates $(x_0, x_1, \ldots, x_n)$ satisfy $x_i \neq 0$. On the other hand, by 2.2.7, E_{H_i} is isomorphic to the affine hyperplane $x_i^{-1}(1) = H_i + e_i$, parallel to H_i, and which has an affine frame $\{e_i\} \cup \{e_i + e_j\}_{j \neq i}$. In this frame we can express the chart $P(E) \setminus P(H_i) \to E_{H_i}$ as the bijection $\pi_i : P(E) \setminus P(H_i) \to K^n$ defined by

4.2.4.1

$$\pi_i : P(E) \setminus P(H_i) \ni p(x_0, x_1, \ldots, x_n) \mapsto \left(\frac{x_0}{x_i}, \ldots, \frac{\widehat{x_i}}{x_i}, \ldots, \frac{x_n}{x_i} \right) \in K^n.$$

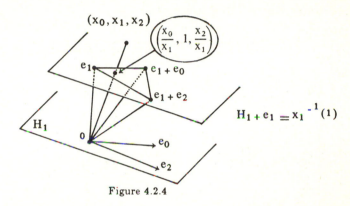

Figure 4.2.4

The calculation of $\pi_j \circ \pi_i^{-1}$ is now immediate. We have, for $j > i$,

4.2.4.2
$$\pi_i^{-1}(v_1, \ldots, v_n) = p(v_1, \ldots, v_{i-1}, 1, v_i, \ldots, v_n),$$

and

4.2.4.3
$$\pi_j \circ \pi_i^{-1}(v_1, \ldots, v_n) = \left(\frac{v_1}{v_{j-1}}, \ldots, \frac{v_{i-1}}{v_{j-1}}, \frac{1}{v_{j-1}}, \frac{v_i}{v_{j-1}}, \ldots, \frac{\widehat{v_{j-1}}}{v_{j-1}}, \ldots, \frac{v_n}{v_{j-1}} \right),$$

where $\pi_j \circ \pi_i^{-1}$ is defined on $K^n \setminus \pi_i(P(H_j)) = v_j^{-1}(0)$.

4.2.5. EXAMPLE: $E = K^2$. This is an important case because $P(K^2) = P^1(K)$ is the simplest non-trivial projective space over K, the projective line. We naturally take $\{e_0, e_1\}$ to be the canonical basis of E^2; thus 4.2.4.2 reduces to the map $K^* \ni v \mapsto 1/v \in K^*$.

One way to interpret this is by saying that $P^1(K)$ is obtained by gluing together two copies of K by the map $v \mapsto 1/v$ on K^*. Don't be surprised to recognize this as one definition of the Riemann sphere (for $K = \mathbf{C}^2$); we will meet it again several times, in 4.3.6, 10.8, 16.3.9 and 20.6.

Another interpretation is that $P^1(K)$ is the union of K, embedded in $P^1(K)$ via the map $\pi_0^{-1} : v \to p(v, 1)$, with the point $p(1, 0)$; from this point of view, $P^1(K)$ is the completion of K, obtained by adjoining one point at infinity, here $p(1, 0)$. We shall discuss this point of view in more detail in 5.2.3.

4.2.6. NOTES. The maps $\pi_j \circ \pi_i^{-1} : v_i^{-1}(0) \to K^n$ belong to the most regular class of maps, after linear and affine maps: they are rational functions. If $K = \mathbf{R}$ or $\mathbf{C}$ they are continuous, C^∞ (which makes sense since they are defined on an open set of K^n), and even C^ω, i.e., real or complex analytic, as the case may be. Thus all kinds of topological, differential and analytic notions make sense on real and complex projective spaces. In the language of manifolds, this means that real and complex projective spaces are topological, C^∞ and even C^ω manifolds.

Finite-dimensional complex projective spaces are the natural habitat of algebraic geometry. Interesting references are [TM, 190–191], [DE5] and [GR–HA2]. The topology of real and complex projective spaces is taken up again in the next section. Their orientability is treated in 4.9.4.

4.3. Structure of projective spaces: topology and algebraic topology

In this section we deal only with finite-dimensional real or complex projective spaces.

A finite-dimensional vector space E over $\mathbf{R}$ or $\mathbf{C}$ has a canonical topology (cf. 2.7.1). This gives rise to a canonical topology on $P(E)$:

4.3.1. DEFINITION. *The canonical topology of a projective space $P(E)$ is the quotient of the canonical topology of $E \setminus 0$ by the equivalence relation $\sim$ of 4.1.1. Projective spaces will always be considered with their canonical topology.*

4.3.2. LEMMA. *If H is a hyperplane of E, the bijection $E_H \to P(E) \setminus P(H)$ (cf. 2.2.5) is a homeomorphism from E_H (cf. 2.7.1.1) into $P(E) \setminus P(H)$ with the induced topology.*

Proof. In coordinates, this bijection is the map π_n^{-1} given in 4.2.4.2, which is obviously continuous. Its inverse

$$\pi_n : p(v_1, \ldots, v_n, 1) \mapsto (v_1, \ldots, v_n)$$

is also continuous, by the properties of the quotient topology. $\square$

Thus the canonical topology of $P(E)$ is the same topology it has as a manifold (cf. 4.2.6).

4.3.3. PROPOSITION. *The space $P(E)$ is Hausdorff, path-connected and compact.*

4.3.3.1. From the geometric point of view, the first property is a consequence of 4.3.2 and the fact that affine spaces are Hausdorff, since for any two points $m, n \in P(E)$, there exists a hyperplane H of E such that $m, n \in P(E) \setminus P(H)$ (figure 4.3.3.1).

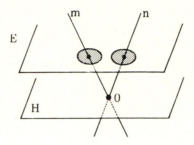

Figure 4.3.3.1

A more algebraic proof goes like this: Let $\Lambda^2 E$ be the second exterior power of E, and introduce the map

$$\alpha : (E \setminus 0) \times (E \setminus 0) \ni (x, y) \mapsto x \wedge y \in \Lambda^2 E.$$

This is a bilinear, hence continuous map, and its kernel $\alpha^{-1}(0)$ is exactly the graph of the equivalence relation $\sim$. Since $\alpha^{-1}(0)$ is the inverse image of 0 by a continuous map, it is a closed set, so the quotient $P(E) = (E \setminus 0)/\sim$ is indeed Hausdorff.

If you don't know any exterior algebra, you can skirt the more esoteric notation in the previous paragraph and define α in coordinates with respect to a fixed basis of E by

$$\alpha\big((x_1, \ldots, x_n), (y_1, \ldots, y_n)\big) = (x_1 y_2 - x_2 y_1, \ldots, x_{n-1} y_n - x_n y_{n-1}).$$

Here the image of α lies in $K^{n(n-1)/2}$ (which is isomorphic to $\Lambda^2 E$), and each component is of the form $x_i y_j - x_j y_i$.

4.3.3.2. To prove the remaining two properties, look at E as a vector space over $\mathbf{R}$ (whether $K = \mathbf{R}$ or $\mathbf{C}$), and consider on E any Euclidean structure. Let $S(E) = \{x \in E \mid \|x\| = 1\}$ be the unit sphere in E. Since $p(x) = p(x/\|x\|)$ for any $x \in E \setminus 0$, we have

$$p\big(S(E)\big) = p(E).$$

Now $S(E)$ is compact (18.2.1) and (unless dim $E = 1$) path-connected, so $P(E)$ is compact (being Hausdorff) and path-connected (since $P(E)$ has only one point when dim $E = 1$). One can also work as in the geometrical proof of Hausdorffness.   □

$$S(E) \quad \hookrightarrow \quad E \setminus 0$$
$$p \searrow \qquad \swarrow p$$
$$P(E)$$

4.3.3.3. Remark. This proof also shows that $P(E)$ is homeomorphic to the topological space $S(E)/\sim$, where $x \sim y$ if and only if $y = \pm x$ for $K = \mathbf{R}$ and $y = ux$ $(u \in \mathbf{U} = \{z \in \mathbf{C} \mid |z| = 1\})$ for $K = \mathbf{C}$.

4.3.4. APPLICATION. Since $P(\vec{X})$ is identified with the space of lines going through a fixed point a of the affine space X, we see from 4.3.3 that the set of secants joining a to other points of a continuous curve C going trough a has at least one accumulation point (figure 4.3.4). This is a good candidate for the tangent to C at a.

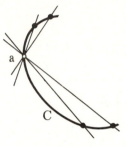

Figure 4.3.4

4.3.5. NOTE. One can define a topology on $P(E)$ for base fields other than $\mathbf{R}$ and $\mathbf{C}$, for example, when K is a locally compact field. In this case 4.3.3 is still valid.

4.3.6. PROPOSITION. *The space $P^1(\mathbf{R})$ is homeomorphic to the sphere S^1, and $P^2(\mathbf{C})$ is homeomorphic to the sphere S^2.*

4.3.6.1. Identify $\mathbf{R}^2$ with $\mathbf{C}$, so that $S^1 = S(\mathbf{R}^2) = S(\mathbf{C})$; we have $P^1(\mathbf{R}) = p(S^1)$. Take the squaring map $c : S^1 \ni z \mapsto z^2 \in S^1$, which is well-defined since the square of a complex number of absolute value 1 also has absolute value 1. Since $p(-z) = p(z)$ and $(-z)^2 = z^2$, we see that c can be factored as $\underline{c} \circ p$, and $\underline{c} : P^1(\mathbf{R}) \to S^1$ is still continuous. It is also bijective and S^1 is compact (cf. 4.3.3), so $\underline{c}$ is homeomorphism.

$$S^1 \xrightarrow{\;p\;} P^1(\mathbf{R})$$
$$c \searrow \qquad \swarrow \underline{c}$$
$$S^1$$

4.3.6.2. Identify $\mathbf{C}^2$ with $\mathbf{R}^4$ endowed with its canonical Euclidean structure, i.e., $\left\| (z, z') \right\|^2 = |z|^2 + |z'|^2$ for $(z, z') \in \mathbf{C}^2 = \mathbf{R}^4$. Consider the sphere $S^3 = S(\mathbf{R}^4) = S(\mathbf{C}^2)$ and the sphere $S^2 = S(\mathbf{R}^3)$, where $\mathbf{R}^3$ is identified with $\mathbf{C} \times \mathbf{R}$. As in 4.3.3.2, we have

$$p(S^3) = P^1(\mathbf{C}) = P(\mathbf{C}^2).$$

Introduce the magic map $H : S^3 \to S^2$ defined by

4.3.6.3 $$H\big((z, z')\big) = (2z\bar{z}', |z|^2 - |z'|^2)$$

(we leave it to the reader to verify that $(z, z') \in S^3$ indeed maps inside S^2). Now for $m, m' \in S^3$ we have $p(m) = p(m')$ if and only if $m' = \lambda m$ for some $\lambda \in \mathbf{C}$, and we then must have $|\lambda| = 1$. But $H\big((\lambda z, \lambda z')\big) = H\big((z, z')\big)$ for $|\lambda| = 1$, so H can be factored as $\underline{H} \circ p$, and $\underline{H} : P^1(\mathbf{C}) \to S^2$ is easily seen to be continuous, bijective, and thus a homeomorphism since $P^1(\mathbf{C})$ is compact.

$$\begin{array}{ccc} S^3 & \xrightarrow{\;p\;} & P^1(\mathbf{C}) \\ {\scriptstyle H}\searrow & & \swarrow{\scriptstyle \underline{H}} \\ & S^2 & \end{array}$$ $\square$

4.3.7. NOTE. The map $H : S^3 \to S^2$ is called the *Hopf fibration*, and is fundamental in geometry. The inverse images $H^{-1}(s)$ of points $s \in S^2$ are all homeomorphic to the circle S^1; this result, a consequence of 4.3.6.2, has already been mentioned in 1.2.9. These circles in S^3 form a structure connected with Clifford parallelism, a topic dealt with in detail in 18.8. In addition, the map $H : S^3 \to S^2$ and its generalizations are essential in algebraic topology; see for example [GG2, 151], [H–W, 387] and [HU, chapter 14].

4.3.8. THE HEURISTICS OF THE HOPF FIBRATION. As in 4.2.5 and 5.2.3, we write $P^1(\mathbf{C}) = \mathbf{C} \cup \infty$, where ∞ is the point $p(1, 0)$ and $\mathbf{C}$ the set of $p(x, 1)$. We suspect that $\mathbf{C} \cup \infty = \mathbf{R}^2 \cup \infty$, the one-point compactification of $\mathbf{C}$, is homeomorphic to S^2. The confirmation comes from the map known as stereographic projection (figure 4.3.8.1; for details see 18.1.4), which here can be written as

4.3.8.1 $$\mathbf{C} \ni x \mapsto \left(\frac{2}{|z|^2 + 1} z, \frac{|z|^2 - 1}{|z|^2 + 1} \right) \in S^2,$$

where we still have $S^2 \subset \mathbf{R}^3 = \mathbf{C} \times \mathbf{R}$.

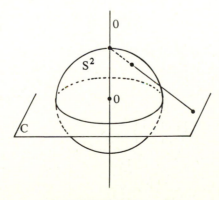

Figure 4.3.8.1

In homogeneous coordinates we have

$$z = p\left(\frac{u}{v}, 1\right) = p(u, v),$$

which leads to defining $z = u/v$. If we replace z by u/v in 4.3.8.1, keeping in mind that $|u|^2 + |v|^2 = 1$ for $(u, v) \in S^3$, we get exactly 4.3.6.3.

4.3.9. The topological characterization of $P^n(\mathbf{R})$ and $P^n(\mathbf{C})$ for $n \geq 2$ is more difficult, and lies in the realm of algebraic topology. Here we limit ourselves to talking about some particular cases and giving references.

4.3.9.1. The first special case is $P^2(\mathbf{R})$. This space is non-orientable (cf. 4.9.4 and 4.9.5), so it cannot be embedded in $\mathbf{R}^3$ (without self-intersections or singularities). This means there is no hope of visualizing $P^2(\mathbf{R})$ as a subset of $\mathbf{R}^3$. There are immersions of $P^2(\mathbf{R})$ in $\mathbf{R}^3$; the images of two of them are given in figures 4.3.9.1.1 and 4.3.9.1.2 (see [H–C, 313–319], for details of these immersions). The surface on the right is called *Boy's surface*.

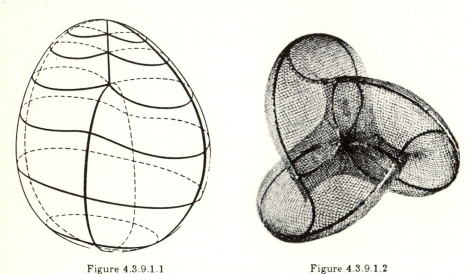

Figure 4.3.9.1.1   Figure 4.3.9.1.2

Source: [H–C]

There are embeddings of $P^2(\mathbf{R})$ in $\mathbf{R}^n$ for $n \geq 4$; a particularly beautiful example in $\mathbf{R}^5$ is the so-called *Veronese surface*, which arises from the map

$$\mathbf{R}^3 \ni (x, y, z) \rightarrow (x^2, y^2, z^2, \sqrt{2}yz, \sqrt{2}zx, \sqrt{2}xy) \in \mathbf{R}^6.$$

The restriction of this map to S^2 has values on the (five-dimensional) affine hyperplane $\sum_{i=1}^{3} u_i = 1$ of $\mathbf{R}^6$. It is an even map, and the quotient $P^2(\mathbf{R}) \rightarrow \mathbf{R}^6$ is an injection, hence a homeomorphism into its image.

It is also possible to see $P^2(\mathbf{R})$ as a Möbius strip whose boundary, a single curve Γ, has been collapsed to a point, or, better yet, identified with the boundary of a disc Γ' (figure 4.3.9.1.3).

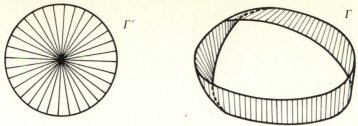

Figure 4.3.9.1.3

4.3.9.2. The projective space $P^3(\mathbf{R})$ is homeomorphic to the group $O^+(3)$ of rotations of $\mathbf{R}^3$, cf. 8.9.3.

4.3.9.3. The spaces $P^n(\mathbf{R})$ are never simply connected; in fact, their fundamental group is isomorphic to $\mathbf{Z}_2$ for $n \geq 2$, since by 4.3.3.3 the map $p : S^n \to P^n(\mathbf{R})$ is a twofold covering, and S^n is simply connected (18.2.2).

4.3.9.4. The spaces $P^n(\mathbf{C})$ are always simply connected. They are not homeomorphic to S^{2n} unless $n = 1$. Real and complex projective spaces can be fully described from the algebraic topological point of view, via their cell decompositions, Betti numbers, cohomology rings over $\mathbf{Z}$ and so on. See [GG2, 90], [SR, 264–265]. The ring structures of the $\mathbf{Z}_2$ cohomology of $P^n(\mathbf{R})$ and the $\mathbf{Z}$ cohomology of $P^n(\mathbf{C})$ are particularly simple: they have a single generator, of degree one for $P^n(\mathbf{R})$ and degree two from $P^n(\mathbf{C})$.

4.4. Projective bases

A coordinate system for an n-dimensional vector space requires n points (a basis); for an n-dimensional affine space, $n+1$ points (an affine frame); and we shall see that a coordinate system for an n-dimensional projective space requires $n+2$ points. One reason is the following: in the notation of 4.2.3, the $n+1$ points $m_i = p(e_i)$ $(i = 0, 1, \ldots, n)$ of $P(E)$ are not enough to determine the homogeneous coordinates of a point (even up to a multiplicative scalar), since the m_i could equally well come from $\lambda_i e_i$, for arbitrary $\lambda_i \in K^*$, not necessarily the same for all i. However, adding a $(p + 2)$-th element resolves the ambiguity:

4.4.1. DEFINITION. *Let $P(E)$ be an n-dimensional projective space. A (projective) base for $P(E)$ is a family $\{n_i\}_{i=0.1.\ldots,n+2}$ of $n + 2$ points of $P(E)$ such that there exists a basis $\{e_i\}_{i=1.\ldots,n+1}$ of E satisfying $m_i = p(e_i)$ for $i = 1, \ldots, n + 1$ and $m_0 = p(e_1 + \ldots + e_{n+1})$.*

This last conditions says that, in the system of homogeneous coordinates defined by the basis $\{e_i\}_{i=1,\ldots,n+1}$, the point m_0 can be written as $(1, \ldots, 1)$. The next result, though elementary, is fundamental for what follows:

4.4.2. LEMMA. *Let m_i be a projective base for $P(E)$. Two bases $\{e_i\}$, $\{e_i'\}$ satisfying definition 4.4.1 are necessarily proportional, i.e., there exists $\lambda \in K^*$ such that $e_i' = \lambda e_i$ for all $i = 1, \ldots, n + 1$.*

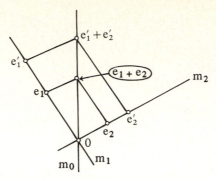

Figure 4.4.2

Proof. Geometrically, this is a consequence of figure 4.4.2 and Thales' theorem (cf. 2.5.1). Alternatively, we can use the assumptions $p(e_i) = p(e'_i)$ and $p(e'_1 + \cdots + e'_n + 1) = p(e_1 + \cdots + e_{n+1})$ to find $\lambda, \lambda_i \in K$ $(i = 1, \ldots, n+1)$ such that $e'_i = \lambda_i e_i$ and

$$e'_1 + \cdots + e'_n + 1 = \lambda(e_1 + \cdots + e_{n+1}).$$

Then

$$\lambda e_1 + \cdots + \lambda_{n+1} = \lambda_1 e_1 + \cdots + \lambda_{n+1} e_{n+1},$$

which implies $\lambda = \lambda_i$ for all $i = 1, \ldots, n+1$ because $\{e_i\}$ is a basis. $\square$

4.4.3. Thus a projective base of $P(E)$ is enough to associate homogeneous coordinates to points of $P(E)$, since it determines a basis for E, well-defined up to a multiplicative scalar. These homogeneous coordinates are called the *projective coordinates* of the point with respect to the projective base.

4.5. Morphisms

4.5.1. Let E, E' be vector spaces, and let $f \in L(E; E')$. We have $f(\lambda x) = \lambda f(x)$ for all $x \in E$, so f is compatible with the equivalence relations defining $P(E)$ and $P(E')$. There is, however, a substantial nuisance: the image $f(E \backslash 0)$ is generally not contained in $E' \backslash 0$, so the quotient map can only be defined from $P(E) \setminus P(f^{-1}(0))$ to $P(E')$. We shall nonetheless talk about such quotient maps as if they were maps from $P(E) \to P(E')$.

4.5.2. DEFINITION. *Let $P(E)$, $P(E')$ be projective spaces. A map $g : P(E) \to P(E')$ is called a morphism if there exists a map $f : L(E; E')$ such that g is obtained from f by passing to the quotients, i.e., $g \circ p = p \circ f$ (in particular, g is generally defined on $P(E) \setminus P(f^{-1}(0))$ only). In this case we put $g = \underline{f}$. We say that f is an isomorphism or a homography if f is a vector space isomorphism; homographies are really defined on the whole of $P(E)$. The set of morphisms from $P(E)$ to $P(E')$ is denoted by $M(P(E); P(E'))$,* *and the set of isomorphisms by* $\mathrm{Isom}\big(P(E); P(E')\big)$.

$$
\begin{array}{ccc}
E \setminus f^{-1}(0) & \xrightarrow{\;f\;} & E' \setminus 0 \\
\Big\downarrow{\scriptstyle p} & & \Big\downarrow{\scriptstyle p} \\
P(E) \setminus P\big(f^{-1}(0)\big) & \xrightarrow{\;\underline{f}\;} & P(E')
\end{array}
$$

4.5.3. QUESTION. The first question is how good is the correspondence $f \mapsto \underline{f}$; the answer is that, for $f, f' \in L(E; E')$, we have

4.5.4 $\qquad\qquad \underline{f} = \underline{f'} \iff f' = \lambda f \quad$ for some $\lambda \in K^*$.

Proof. Observe first that $g = \underline{f}$ determines the kernel $f^{-1}(0)$. Now if x, y are linearly independent vectors not in $f^{-1}(0)$, then $f(x)$ and $f(y)$ are also linearly independent. But $\underline{f}\big(p(x)\big) = \underline{f'}\big(p(x)\big)$, and similarly for $x + y$ and y, so there exist scalars $\lambda(x)$, $\lambda(y)$, $\lambda(x + y)$ such that

$$
f'(x) = \lambda(x) f(x), \quad f'(y) = \lambda(y) f(y), \quad f'(x + y) = \lambda(x + y) f(x + y).
$$

Since $f(x)$ and $f(y)$ are linearly independent we conclude that $\lambda(x) = \lambda(y) = \lambda(x + y)$, showing that $f' = \lambda f$ for some λ. $\qquad\square$

Property 4.5.4 can be interpreted as providing a bijection

4.5.5 $\qquad\qquad M^\bullet\big(P(E); P(E')\big) \cong P\big(L(E; E')\big),$

where $M^\bullet\big(P(E); P(E')\big)$ represents $M\big(P(E); P(E')\big)$ without the trivial morphism arising from the zero map from E into E' (which is, by the way, the only morphism defined nowhere in $P(E)$!)

4.5.6. REMARK. Commutativity of the base field is essential in proving the backward implication in 4.5.4. If the field is not commutative, the map λf obtained from $f \in L(E; E')$ is generally no longer linear.

4.5.7. Bearing in mind the caveat in 4.5.1, we have, for $f \in L(E; E')$ and $g \in L(E'; E'')$:

$$
\underline{g \circ f} = \underline{g} \circ \underline{f}.
$$

4.5.8. Every n-dimensional projective space over K is isomorphic to $P^n(K)$. This, together with continuity, explains why in section 4.3 we only talked about $P^n(\mathbf{R})$ and $P^n(\mathbf{C})$.

The following is a consequence of 4.5.5 and 4.5.7:

4.5.9. PROPOSITION. *The homographies $P(E) \to P(E)$ form a group under composition of maps, called the projective group of E, and denoted by $\mathrm{GP}(E)$ or $\mathrm{PGL}(E)$. There is a group isomorphism $\mathrm{GP}(E) \cong \mathrm{GL}(E)/K^*\,\mathrm{Id}_E$.*

4.5.10. PROPOSITION (FIRST FUNDAMENTAL THEOREM OF PROJECTIVE GEOMETRY). *Let $P(E)$, $P(E')$ be projective spaces of same finite dimension, and $\{m_i\}$, $\{m'_i\}$ projective bases for $P(E)$ and $P(E')$, respectively. There exists a unique homography $g : P(E) \to P(E')$ such that $m'_i = g(m_i)$ for every i.*

Proof. Let $\{e_i\}$, $\{e_i'\}$ be bases associated with $\{m_i\}$, $\{m_i'\}$ as in 4.4.1. Define $f \in \mathrm{Isom}(E; E')$ by $e_i' = f(e_i)$ for all i; then $\underline{f} = g$ has the desired properties. If g and g' both have these properties, the homography $g'^{-1} \circ g : P(E) \to P(E)$ fixes the base $\{m_i\}$ pointwise, and must be of the form $\lambda \, \mathrm{Id}_E$ by lemma 4.4.2. $\qquad\square$

4.5.10.1. Remark. This result is false if the field is non-commutative: see exercise 4.9.8.

4.5.11. COROLLARY. *The group* $\mathrm{GP}(E)$ *acts transitively on* $P(E)$ *and simply transitively on bases of* $P(E)$. $\qquad\square$

4.5.12. The name of the fundamental theorem of projective geometry for the apparently simple proposition 4.5.10 is justified by the difficulty in proving it from axiomatizations of projective spaces. See, for example, [V–Y, volume I, p. 95].

4.5.13. EXPLICIT CALCULATION OF MORPHISMS. Let $\{e_i\}$ and $\{e_i'\}$ be bases for E and E', respectively. Consider $f \in L(E; E')$, and let $M(f) = (a_{ij})$ be the matrix of f with respect to the bases $\{e_i\}$ and $\{e_i'\}$. In terms of the associated homogeneous coordinates in $P(E)$, $P(E')$ (cf. 4.2.3), we have

$$\underline{f}((x_1, \ldots, x_{n+1})) = \left(\sum_i a_{1i} x_i, \ldots, \sum_i a_{n+1,i} x_i \right).$$

Homogeneous coordinates are only defined up to a multiplicative constant; if we want a completely determined formula, we can use the charts in $P(E)$ and $P(E')$ associated with the respective bases. Take, for example, the charts π_{n+1}, $\pi_m' + 1$, where $n = \dim(P(E))$, $m = \dim(P(E'))$. Using 4.2.4.1 and 4.2.4.2 we get the following formula for $\pi_{m+1} \circ \underline{f} \circ \pi_{n+1}^{-1}$:

$$\pi_{m+1} \circ \underline{f} \circ \pi_{n+1}^{-1}((v_1, \ldots, v_n))$$
$$= \left(\frac{\sum_i a_{1i} v_i + a_{1,n+1}}{\sum_i a_{p+1,i} v_i + a_{p+1,n+1}}, \ldots, \frac{\sum_i a_{pi} v_i + a_{p,n+1}}{\sum_i a_{p+1,i} v_i + a_{p+1,n+1}} \right).$$

Naturally, this is only defined where the denominator is non-zero.

4.5.14. PARTICULAR CASE: $n = m = 1$. Take $\underline{f} \in \mathrm{GP}(K^2)$, that is $f \in \mathrm{GL}(K^2)$, and let the matrix of f in the canonical basis of K^2 (cf. 4.2.5) be $M(f) = \left(\begin{smallmatrix} a & c \\ c & d \end{smallmatrix} \right)$. Then $\pi_2 \circ \underline{f} \circ \pi_2^{-1}$ takes the form

$$v \to \frac{av + b}{cv + d}$$

for $v \neq -d/c$, or, in homogeneous coordinates,

$$\underline{f}((v, 1)) = \left(\frac{av + b}{cv + d}, 1 \right)$$

for $v \neq -d/c$. Doing the same thing for other combinations of the two charts π_1, π_2, one obtains the following table:

4.5.15

$$c \neq 0 \begin{cases} (v,1) \mapsto \left(\dfrac{av+b}{cv+d}, 1 \right), & v \neq -\dfrac{d}{c} \\ \left(-\dfrac{d}{c}, 1 \right) \mapsto (1,0) \\ (1,0) \mapsto \left(\dfrac{a}{c}, 1 \right) \end{cases} \qquad c = 0 \begin{cases} (v,1) \mapsto \left(\dfrac{av+b}{d}, 1 \right) \\ (1,0) \mapsto (1,0) \end{cases}.$$

Observe that, for $K = \mathbf{R}$ or $\mathbf{C}$,

$$\frac{a}{c} = \lim_{v \to \infty} \frac{av+b}{cv+d} \qquad \text{and} \qquad \lim_{v \to -d/c} \frac{av+b}{cv+d} = \infty,$$

thus corroborating the observation made at the end of 4.2.5, that $p(1,0)$ plays the role of infinity. We will return to this point in 5.2.5.

4.5.16. STRUCTURE OF HOMOGRAPHIES. Again we assume that $P(E)$ is finite-dimensional. In view of 4.5.4, our objective is to study the structure of a map $f \in \mathrm{GL}(E)$ up to a multiplicative constant. The geometric structure of an element of $\mathrm{GL}(E)$ is described, in the algebraically closed case at least, by its Jordan decomposition; passing to $P(E)$ poses no special difficulties. We do not pursue this question here, but refer the reader to [FL, appendix 1]. It is interesting to compare the modern treatment in this reference with an older exposition not using linear algebra explicitly, for example [V–Y, volume I].

4.5.17. PROPOSITION. *Let $g = \underline{f} \in \mathrm{GP}(E)$ and $m \in P(E)$. Then m is a fixed point of g (i.e., $m = g(m)$) if and only if $p^{-1}(m)$ is an eigenspace of f.* $\square$

The conclusion is that the essential things in the study of morphisms $\underline{f}$ of $P(E)$ are the eigenspaces of f in E, and not its eigenvalues (though these too merit interest, cf. 6.6.3).

4.5.18. COROLLARY. *If $K = \mathbf{C}$, or if $K = \mathbf{R}$ and the dimension of $P(E)$ is even, every homography has at least one fixed point.* $\square$

For example, if $f \in \mathrm{GL}(E)$ has no repeated eigenvalues, $\underline{f}$ has exactly $n+1$ distinct fixed points, where $n = \dim(P(E))$ (exactly because of 4.4.2). In older books such points are called "double".

4.5.19. INVOLUTIONS OF $\mathrm{GP}(E)$. If $f \in \mathrm{GL}(E)$ is involutive (i.e., $f^2 = f \circ f = \mathrm{Id}_E$), so is $\underline{f}$ (in fact, we shall see in 6.4.6 a dazzling geometric construction for $\underline{f}$ in this case). Watch out, however: not every involutive $g \in \mathrm{GP}(E)$ is of the form $\underline{f}$ for f involutive. For example, for $E = \mathbf{R}^2$ and $M(f) = \left(\begin{smallmatrix} 0 & -1 \\ 1 & 0 \end{smallmatrix} \right)$, we have $\underline{f}^2 = \mathrm{Id}_{P^1(\mathbf{R})}$, since $f^2 = -\mathrm{Id}_{\mathbf{R}^2}$. The right condition for $\underline{f}^2 = \mathrm{Id}_{P(E)}$ is then that $f^2 = \lambda \, \mathrm{Id}_E$ with $\lambda \in K^*$; if K is algebraically closed, one can write $\lambda = \mu^2$, and then $g = \underline{f} = \underline{\mu^{-1}f}$, with $(\mu^{-1}f)^2 = \mathrm{Id}_E$.

4.5.20. TOPOLOGY OF GP(E). If $K = \mathbf{R}$ or $\mathbf{C}$ and E is finite-dimensional, the projective group GP(E) is endowed with a canonical topology, the quotient of the topology of GP(E) by the action of $K^* \operatorname{Id}_E$ (cf. 2.7.1). The connectedness of GP(E) depends on the orientability of $P(E)$; see 4.9.4 and 4.9.5, or [FL, 228–230].

4.5.21. REMARK. The isomorphisms $P(E) \to P(E^*)$ lead to a number of geometric results, which will be encountered in their natural context in 14.5, 14.8.12. See also [FL, 260 ff.].

4.6. Subspaces

Observe that if $F \subset E$ is a vector subspace different from $\{0\}$, F is saturated under the equivalence relation $\sim$ of 4.1.1. Thus we can identify $P(F)$ with $p(F) \subset P(E)$; in particular the injection $i : F \to E$ gives rise to the natural injection $\underline{i} : P(F) \to P(E)$, which is a morphism.

4.6.1. DEFINITION. *A subset V of a projective space $P(E)$ is called a (projective) subspace if there exists a vector subspace F of E such that $p(F) = V$. We consider $P(F) = V$ with its natural projective structure. In particular, $\dim V = \dim F - 1$ is called the dimension of the subspace V.*

4.6.1.1. We see that there is a natural bijection between the projective subspaces of $P(E)$ and the vector subspaces of E.

4.6.2. The subspace of dimension -1 is the empty subset. Zero-, one- and two-dimensional subspaces are the *points*, *lines* and *planes* of $P(E)$, respectively. Projective subspaces of $P(E)$ coming from vector hyperplanes of E are called *(projective) hyperplanes* of $P(E)$; their set is *denoted* by $\mathcal{H}\big(P(E)\big)$, and we have (cf. 4.1.3.5) the following bijections:

4.6.3 $$\mathcal{H}\big(P(E)\big) \overset{\text{set}}{\cong} \mathcal{H}(E) \overset{\text{set}}{\cong} P(E^*).$$

4.6.4. Let $\{V_i\}_{i \in I}$ be an arbitrary family of subspaces of a projective space. The intersection $\bigcap_{i \in I} V_i$ is still a subspace. We deduce (as in 2.4.2.5) the following

4.6.5. PROPOSITION. *Let S be an arbitrary subset of a projective space. The subspace $\langle S \rangle$ spanned by S is the smallest subspace containing S. It is the intersection of all subspaces containing S.*   □

4.6.6. DEFINITION. *The points m_i ($i = 1, \ldots, k+1$) are called (projectively) independent if*

$$\dim\big(\langle m_1, \ldots, m_{k+1} \rangle\big) = k.$$

4.6.7. EXAMPLES. A single point is always independent. Two points are independent if and only if they are distinct, in which case they determine a unique line. Three points a, b, c are independent if they are distinct and each does not belong to the line determined by the other two: $a \notin \langle b, c \rangle$,

$b \notin \langle c, a \rangle$, $c \notin \langle a, b \rangle$. In this case they define a unique plane $\langle a, b, c \rangle$. Looking at it the other way around, we can start from the fact that two distinct points determine a unique line and the analogous fact for three points and planes, and take these as the first axioms in establishing an axiomatic theory of projective spaces. See also [V–Y, volume I, 95].

4.6.8. PROPOSITION. *The points m_i $(i = 1, \ldots, k + 1)$ are independent if and only if $m_i \notin \langle m_1, \ldots, \hat{m}_i, \ldots, m_{k+1} \rangle$ for $i = 1, \ldots, k + 1$ (cf. 0.1). In an n-dimensional projective space, $\{m_i\}_{i=0,1,\ldots,n+1}$ forms a projective base if and only if the points $\{m_j\}_{j \neq i}$ are independent for all $i = 0, 1, \ldots, n + 1$.*

Proof. The first assertion is just the projective version of a well-known fact from linear algebra. To prove the second assertion, take $e_i \in E$ such that $p(e_i) = m_i$ $(i = 0, 1, \ldots, m + 1)$; since the points $\{m_i\}_{i=1,\ldots,m+1}$ are independent, $\{e_i\}_{i=1,\ldots,n+1}$ is a basis of E. Put $e_0 = \sum_{i=1}^{n+1} \lambda_i e_i$; the fact that the $\{m_j\}_{j \neq i}$ are independent means that $\lambda_i \neq 0$ $(i = 1, \ldots, n + 1)$, so $\{\lambda_i e_i\}_{i=1,\ldots,n+1}$ is still a basis of E, satisfying 4.4.1. □

4.6.9. COROLLARY. *Let D, D' be projective lines and $\{a, b, c\}$ (resp. $\{a', b', c'\}$) distinct points on D (resp. D'). There exists a unique homography $f : D \to D'$ such that $f(a) = a'$, $f(b) = b'$, and $f(c) = c'$.*

Proof. From 4.6.7 and 4.6.8, $\{a, b, c\}$ and $\{a', b', c'\}$ are bases for D and D', respectively. The corollary follows by applying 4.5.10. □

4.6.10. Axiomatic theories of projective spaces include axioms about the intersection of subspaces; in this case, they consist of the observation in 4.6.1.1 and the classical linear algebra relation

$$\dim(F + G) + \dim(F \cap G) = \dim F + \dim G,$$

which together give the following

4.6.11. PROPOSITION. *Let V, W be subspaces of the same projective space. Then*

$$\dim(\langle V \cup W \rangle) + \dim(V \cap W) = \dim V + \dim W.$$ □

4.6.12. COROLLARY.
  i) *If $\dim V + \dim W \geq \dim(P(E))$, then $V \cap W \neq \emptyset$;*
  ii) *if $\dim(P(E)) = n$, then n hyperplanes of $P(E)$ have at least one point in common;*
  iii) *if H is a hyperplane and m a point outside H, every line D through m intersects H in exactly one point;*
  iv) *two distinct lines in a projective plane intersect in exactly one point.* □

4.6.13. This corollary displays the superiority of projective geometry over affine geometry: intersection properties for subspaces can be stated without special cases; there are no parallel subspaces (cf. 2.4.9). We shall study in chapter 19 a geometry in which there are multiple parallels (19.3.2).

4.6.14. EQUATIONS OF SUBSPACES. Calculation of subspaces is carried out with the help of 2.4.8.1 and 4.6.1.1.

4.6.15. SUBSPACES AND TOPOLOGY. Here $K = \mathbf{R}$ or $\mathbf{C}$ and all projective spaces are finite-dimensional, and have the topology introduced in 4.3.1. Since the complement $E \setminus F$ of a vector subspace $E \subset F$ is dense in F (if $E \neq F$), we get that the complement $P(E) \setminus P(F)$ of a subspace $P(F)$ is dense in $P(E)$. This property is analogous to 2.7.1.2. Similarly, every subspace is closed in $P(E)$. On the other hand, 2.7.3.2 has no analogue here: for any hyperplane $P(H)$ in $P(E)$, the space $P(E) \setminus P(H)$ is path-connected. In fact, $P(E) \setminus P(H)$ is homeomorphic to an affine space (cf. 4.3.3.1); alternatively, one can construct directly an arc joining two given points of $P(E) \setminus P(H)$ (figure 4.6.15).

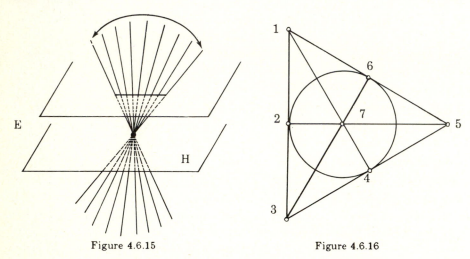

Figure 4.6.15   Figure 4.6.16

4.6.16. FINITE FIELDS. The points, lines and other subspaces of a projective space $P(E)$ over a finite field with k elements form interesting configurations. For example, each line contains the same number of points, and each point lies in the same number of lines. Figure 4.6.16 represents the case $k = 2$, $\dim P(E) = 2$: the space has seven points and seven lines (one of which is represented by a circle). For more details on projective configurations, see [H–C, 94–143], an easy reference with many drawings, and [DI], a more recent reference. See also 4.9.11.

4.7. Perspective; aerial photography

4.7.1. Let H, H' be hyperplanes of a projective space $P(E)$ and m a point not in H or H'. Given an arbitrary $x \in H$, there exists by 4.6.7 and 4.6.12 a unique line $\langle m, x \rangle$, which in turn intersects H' in a single point $g(x) = H' \cap (\langle m, x \rangle)$. This defines a map $g : H \to H'$.

4.7.2. PROPOSITION. *The map g belongs to* $\mathrm{Isom}(H; H')$.

Proof. Let V and V' be the hyperplanes of E such that $H = p(V)$ and $H' = p(V')$; we are looking for $f \in L(V;V')$ such that $\underline{f} = g$. It is enough to take the restriction to V of the linear projection onto $\overline{V'}$ in the direction of the one-dimensional vector subspace $p^{-1}(m)$ (i.e., the projection associated with the direct sum $E = p^{-1}(m) \oplus V'$). $\qquad\qquad\square$

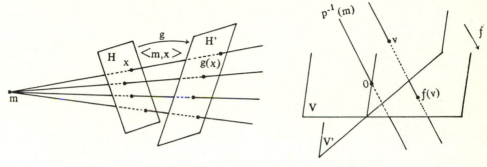

Figure 4.7.1 $\qquad\qquad\qquad\qquad$ Figure 4.7.2

4.7.3. The map g is called the *perspective* of center m taking H onto H'. This is consistent with the everyday usage of the word perspective: the correspondence between two planes established by our vision or a photograph. H contains the plane object to be photographed, and H' contains the film.

In aerial photography, the problem arises of matching partial shots of an area which is too large to fit in a single photo. Assuming the area to be flat, it is possible to use perspectives to make overlapping parts coincide perfectly: the correspondence between matching points in two photographs is a homography, being the composition of the perspectives of the two photos (figure 4.7.3.1), and thus can be composed with another homography to give the identity. By proposition 4.5.10, it is enough to match four points in the two images to obtain a perfect correspondence. See [BUR, 36–51].

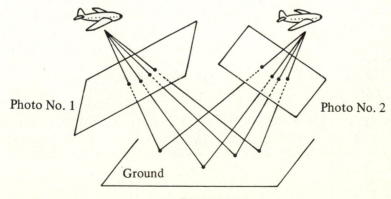

Photo No. 1 $\qquad\qquad\qquad\qquad\qquad\qquad\qquad$ Photo No. 2

Ground

Figure 4.7.3

Figure 4.7.4.1

Figure 4.7.4.2

(Source: IGN, France)

4.7.4. As a practical exercise, the reader can locate four points on figures 4.7.4.1 and 4.7.4.2, transfer them to a blank sheet of paper, and verify that the two sets of points cannot be mapped into one another by an affine map.

4.7.5. AERIAL PHOTOGRAPHY AND CARTOGRAPHY. It is only fair to warn the reader that, when a map is assembled by patching together aerial photographs, the matching of overlapping portions of the patches can never be perfect as indicated in 4.7.3. This is because the surface of the earth, even not considering local elevations, is not plane but spherical (in fact, ellipsoidal, cf. 18.1.5.3). One has to work in three dimensions, with the help of intricate and expensive devices called stereocomparators (see [BUR, 230 and plates 21–23]); only thus can one hope for a theoretically perfect matching. On the subject of cartography, see 18.1.5 and 18.1.8.

4.7.6. Perspective still makes sense in affine spaces, but then it is only defined on a subset of the hyperplane being observed. The perspective g with center m from H onto H' is defined on $H \setminus D$, where D is the intersection of H and the hyperplane parallel to H' and going through m. This affine perspective is often used as an introduction to projective geometry, in particular to the projective completion of an affine space discussed in the next chapter.

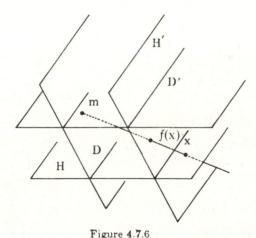

Figure 4.7.6

4.8. The non-commutative case

4.8.1. One can define projective spaces, morphisms and subspaces for skew fields. We have mentioned *en passant* (4.5.6, 4.5.10.1) certain results which do not hold in this case; it is a good exercise for the reader to sift through the chapter to find what results really require commutativity, giving counterexamples when appropriate.

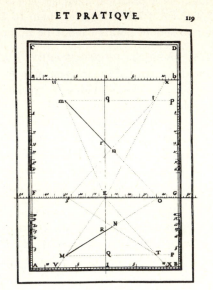

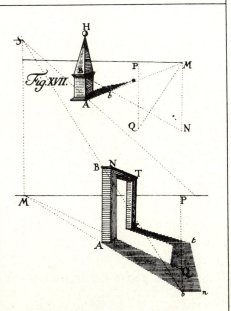

8.45   Estienne   Migon

8.46   Johann   Heinrich   Lambert

Se Vend à Paris chez Jean Rue St. Jean de Beauvais, N.º 10.

M. C. Escher, *Tower of Babel*, 1928. Woodcut, 62 × 38.5 cm.
Escher Foundation–Haags Gemeentemuseum (**The Hague**)

See [ER]

4.8.2. For instance, one can define projective spaces of arbitrary dimension n over the skew field $\mathbf{H}$ of quaternions (8.9):

$$P^n(\mathbf{H}) = P(\mathbf{H}^{n+1}).$$

Such spaces are Hausdorff and compact with the natural topology (cf. 4.3). The projective line $P^1(\mathbf{H})$ is homeomorphic to the sphere S^4 (cf. 4.9.7). The algebraic topology of $P^n(\mathbf{H})$ is also completely known; see, for example [SR, 265], or the interesting article [E–K].

4.8.3. It is even possible to define projective spaces over structures more general than fields. An important example is the division ring of *octonions* or *Cayley numbers* $\mathbf{Ca}$, which is an eight-dimensional real vector space with a non-associative multiplicative law. One can construct the projective line $P^1(\mathbf{Ca})$, homeomorphic to the sphere S^8, and the octonion projective plane $P^2(\mathbf{Ca})$; see [PO, chapter XIV], [BO2, 199], [E–K, 12], and especially [BES, chapter 3].

The reason why one cannot define projective spaces of dimension greater than two over $\mathbf{Ca}$ is that such that a projective space (with the standard axioms of intersection of subspaces) must necessarily satisfy Desargues's theorem (see 5.4.3 and 5.4.4), which, in turn, implies that the base structure is associative ([PT, chapter 3] or [HA, 374]; the last chapter of this latter reference, on projective planes, is very interesting.)

4.9. Exercises

4.9.1. Let $P(E)$ be a finite-dimensional real or complex projective space. Show that, for every hyperplane H, the bijection $P(E) \setminus P(H) \to E_H$ defined in 4.3.1 is a homeomorphism.

4.9.2. Prove the last paragraph of 4.3.9.1.

4.9.3. Prove that $P^n(\mathbf{R})$ is homeomorphic to the quotient topological space of the closed ball $B(0,1)$ by the equivalence relation $\sim$ defined by $x \sim y$ if and only if $x = y$ or $x \in S(0,1) = S^{n-1}$ and $y = -x$. In particular, study the cases $n = 1$ and $n = 2$ (cf. 4.3.6 and 4.3.9.1).

* **4.9.4.** For a real projective space of finite dimension n, find the sign of the Jacobian of the transition maps 4.2.4.3. Deduce that the manifold $P^n(\mathbf{R})$ is orientable if n is even and non-orientable if n is odd.

* **4.9.5.** Prove the result in 4.9.4 by studying the connectedness of the projective group $GP\big(P^n(\mathbf{R})\big)$.

4.9.6. Study the images and inverse images of subspaces under morphisms.

4.9.7. Show that $P^1(\mathbf{H})$ is homeomorphic to the sphere S^4, where $\mathbf{H}$ is the skew field of quaternions (cf. 4.8.2).

4.9.8. Determine exactly where the assumption that the base field is commutative intervenes in the proof of 4.5.10. Give a counterexample when the field is non-commutative. A good place to try is $P^n(\mathbf{H})$ (cf. 4.8.2): you can even try to find an infinite set of points with the property that every subset with $n + 1$ points is linearly independent, but which is nonetheless invariant under a homography different from the identity.

* **4.9.9.** Find a model, in the usual three-dimensional space, for the configuration formed by the points, lines and planes of $P^3(\mathbf{Z}_2)$. Draw pictures.

* **4.9.10.** Let H_i be a family of hyperplanes in the projective space $P(E)$ of finite dimension n; find a relation between $\dim \bigcap_i H_i$ in $P(E)$ and $\dim \langle \bigcup_i H_i \rangle$ in $P(E^*)$.

* **4.9.11.** Let K be a field with k elements, and $P(E)$ a projective space of dimension n over K. Show that the cardinality of the set of p-dimensional subspaces of $P(E)$ is equal to

$$\frac{(k^{n+1} - 1)(k^{n+1} - k) \cdots (k^{n+1} - k^p)}{(k^{p+1} - 1)(k^{p+1} - k) \cdots (k^{p+1} - k^p)}.$$

Show that the order of the projective group $GP(E)$ is

$$(k^{n+1} - 1)(k^{n+1} - k) \cdots (k^{n+1} - k^{n-1})k^n.$$

* **4.9.12.** MÖBIUS TETRAHEDRA. Construct, in a three-dimensional projective space, two tetrahedra $\{a, b, c, d\}$ and $\{a', b', c', d'\}$ such that each vertex of the first belongs to a face of the second and vice versa (i.e., $a \in \langle b', c', d' \rangle$, $a' \in \langle b, c, d \rangle$, and so on). For an explanation and a generalization of this phenomenon, see 14.5.5 and 14.8.12. For a study of Möbius tetrahedra formed by mechanical linkages whose vertices describe curves of constant torsion, see [BA].

Figure 4.9.12

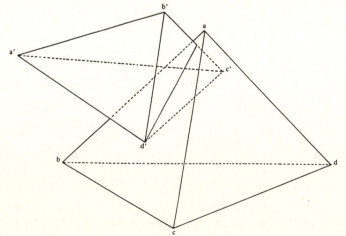

Chapter 5

Affine-projective relationship: applications

In this chapter we add to an affine space X the set $P(\vec{X})$ of directions of its lines, obtaining a projective space in which X is naturally embedded. Conversely, the complement of a hyperplane in a projective space has a natural affine structure. This process of projective completion can be performed in more or less natural ways (section 5.0). In section 5.3 we translate parallelism behavior in X in terms of intersection properties in the completion of X. This has numerous applications, some of which are given in section 5.4.

All fields are assumed commutative.

5.0. Introduction

5.0.1. We have mentioned in 4.0 the shortcomings of affine geometry. In particular, following Desargues, we want to extend an affine space X into a projective space $\tilde{X}$, the union of X and its points at infinity. These are defined as the directions of lines of X, and their set $P(\vec{X})$ is also *written* ∞_X. Thus $\tilde{X} = X \cup \infty_X = X \cup P(\vec{X})$, this being *a priori* just a disjoint union in the set-theoretical sense. The non-trivial part is making $\tilde{X}$ into a projective space. This can be done axiomatically, but this is not the point of view we have been adopting in this book (cf. 4.1.2 and [AN], [HA], [DI], [H–P], [PT]); instead we present three algebraic constructions for the projective completion.

5.0.2. In the coarsest approach, let X have finite dimension n, and let $\{x_i\}_{i=0,1,\ldots,n}$ be an affine frame of X. The desired completion is

$$P^n(K) = P(K^{n+1}),$$

and the embedding $X \to P^n(K)$ is given by

5.0.2.1 $\qquad\qquad x = (\lambda_1, \ldots, \lambda_n) \mapsto p(\lambda_1, \ldots, \lambda_n, 1)$

(cf. 2.2.9 and 4.2). We see that the complement of the image of X in $P^n(K)$ is exactly the hyperplane $P^{n-1}(K) = P(K^n \cong K^n \times \{0\})$, which is indeed identified with $P(\vec{X})$ by 5.0.2.1.

5.0.3. A more sophisticated construction, suggested by 5.0.2, consists in vectorializing X at $a \in X$, and considering the projective space $P(X_a \times K)$, where $X_a \times K$ is the direct product of the two vector spaces X_a and K. The embedding of X into $P(X_a \times K)$ is furnished by composing the identification between X and the section $X_a \times \{1\}$ of $X_a \times K$ with the projection onto the quotient:

5.0.3.1 $\qquad\qquad\qquad\qquad x \mapsto p(x, 1).$

The complement of the image of X in $P(X_a \times K)$ is $P(X_a) \cong P(\vec{X})$.

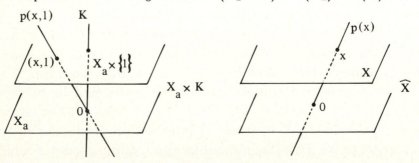

Figure 5.0.3

5.0.4. The most sophisticated approach utilizes the universal vector space $\hat{X}$ associated with X: just take $\tilde{X} = P(\hat{X})$, and embed X in $P(\hat{X})$ by means of

5.0.4.1 $x \mapsto p(x),$

where X is understood as an affine hyperplane of $\hat{X}$ (cf. 3.1.6).

5.0.5. The disadvantage of the first two constructions is that they are not intrinsic. In 5.0.2, for example, we do not know *a priori* that the projective space we obtain is canonically associated with X (that is, "depends only on X" in some sense). The same holds for 5.0.3, though in this case we can already perceive that $P(X_a)$ is always identical with $P(\vec{X})$, since the choice of a only changes things by a translation in X, and translations preserve directions (cf. 2.3.3.4). On the other hand, 5.0.2 and 5.0.3 have two advantages: first, they offer an explicit base in which to calculate; and second, being more elementary, they allow a simpler treatment of the complexification questions which will arise in chapter 7.

5.1. The projective completion of an affine space

5.1.1. Let X be an affine space and $\hat{X}$ the associated universal vector space

$$\hat{X} = \vec{X} \cup (K^* \times X)$$

(cf. 3.1). *Define* $\tilde{X}$ $\tilde{X} = P(\hat{X})$. Recall that X is embedded in $\hat{X}$ via $x \mapsto (x, 1)$, and remark that the canonical projection $p : \hat{X} \to \tilde{X} = P(\hat{X})$ is injective when restricted to X; thus X can be identified with a subset of $\tilde{X}$. This gives rise to a partition of $\tilde{X}$ into two sets, $\tilde{X} = X \cup P(\vec{X})$. We put $\infty_X = P(\hat{X})$, so we can also write $\tilde{X} = X \cup \infty_X$. The conclusion is that the affine space X is embedded in the projective space $\tilde{X}$, the complement of the image of the embedding being a hyperplane $\infty_X = P(\vec{X})$.

5.1.2. Conversely, if $P(E)$ is a projective space and $P(H) \subset P(E)$ is an arbitrary hyperplane, the complement $P(E) \setminus P(H)$ is an affine space in a natural way: $P(E) \setminus P(H) = E_H$ (cf. 2.2.6). By 2.2.7 and 3.1.6, the processes of passing from X to $\tilde{X}$ and from $P(E) \setminus P(H)$ to E_H are inverse to each other.

   We can summarize the aforementioned results from chapters 2 and 3 in the following

5.1.3. THEOREM.

   i) *Let X be an affine space and $\tilde{X} = P(\hat{X})$. Identify X with a subset of $\tilde{X}$, also denoted by X, via $x \mapsto p(x, 1)$. The complement $\tilde{X} \setminus X$ is equal to $P(\vec{X}) = \infty_X$. The space $\tilde{X}$ is called the (projective) completion of X, and $\infty_X = P(\vec{X})$ is called the hyperplane at infinity of X. If X, X' are affine spaces and $\tilde{X}, \tilde{X}'$ are their completions, for each affine morphism $f \in A(X; X')$ there exists a unique projective morphism $\tilde{f} \in M(\tilde{X}; \tilde{X}')$*

such that $\tilde{f}|_X = f$. Moreover, $\tilde{f}(\infty_X) \subset \infty_{X'}$ and $\tilde{f}|_{\infty_X} = \vec{f}$ (cf. 2.3.1 and 4.5.2).

ii) Let (E, H) be a pair formed by a vector space E and a hyperplane $H \subset E$. Then $E_H = P(E) \setminus P(H)$ (cf. 4.1.3.4) has a natural affine structure over $L(E/H; H')$. Let (E', H) be another such pair, and take $g \in M\big(P(E); P(E')\big)$ such that the domain of g contains E_H (cf. 4.5), and that $g(E_H) \subset E'_{H'}$. Then the restriction $g|_{E_H}$ lies in $A(E_H; E'_{H'})$.

iii) The correspondences $X \mapsto (\hat{X}, \vec{X})$ and $(E, H) \mapsto E_H$ are functorial and inverse to one another.

Proof. This essentially follows from 2.2.6, 2.3.7, 3.1.6 and 4.5. For (ii), we must show that $g|_{E_H} \in A(E_H; E'_{H'})$. Take $f \in L(E; E')$ such that $\underline{f} = g$; we must show that, in fact, $f \in L_{H,H'}(E; E')$ (cf. 2.3.6).

By assumption, g is defined on the whole of E_H, which means (cf. 4.5.2) that $f^{-1}(0) \subset H$; and $g(E_H) \subset E'_{H'}$ implies $f(x) \notin H'$ for any $x \notin H$. This shows that $f(H) \subset H'$ and $\underline{f} : E/H \to E'/H'$ is injective.

Functoriality follows from 3.2.1. In order to prove rigorously that the correspondences in (iii) are inverse to each other, we would need to use category theory; in that nomenclature, theorem 5.1.3 says that $(E, H) \to E/H$ is a "completely faithful" functor. $\qquad\square$

For practical applications, (i) and (ii) suffice (in fact, just 5.0.2.1 and 4.2.4.1); (iii) was thrown in for the sake of completeness.

5.2. Examples

5.2.1. CHARACTERIZATION OF DILATATIONS. According to 2.3.3.12 and 4.5.9, one can characterize the dilatations of an affine space as the affine maps f such that $\tilde{f}$ leaves the hyperplane at infinity pointwise invariant:

$$f \in \mathrm{Dil}(X) \Longleftrightarrow \tilde{f}|_{\infty_X} = \mathrm{Id}_{\infty_X} .$$

5.2.2. AFFINE AND PROJECTIVE GROUPS. The maps $g \in \mathrm{GP}(\tilde{X})$ which leave ∞_X globally invariant are exactly those of the form $g = \tilde{f}$, where f is an arbitrary element of $\mathrm{GA}(X)$. In other words, we can identify $\mathrm{GA}(X)$ with a subgroup of $\mathrm{GP}(X)$, namely, $\mathrm{GP}_{\infty_X}(\tilde{X})$. Cayley discovered that all classical geometric groups can be realized as subgroups of the projective group of a suitably chosen projective space: see 9.5.5.2 and 18.10.1.5.

5.2.3. THE CASE $X = K$. In this case $P(\vec{X}) = P(K) = P^0(K)$ consists of a single point, the *point at infinity* of K, and we have $\tilde{K} = K \cup \infty$. Observe that for $K = \mathbf{R}$ the completion $\tilde{\mathbf{R}} = \mathbf{R} \cup \infty$ is distinct from the extended real line used in analysis, $\mathbf{R} \cup \{-\infty, +\infty\}$. In projective geometry there is no distinction between "approaching $+\infty$" and "approaching $-\infty$".

Returning now to K arbitrary, observe that $\hat{K}$ can be naturally identified with K^2, and thus $\tilde{K}$ with $P^1(K)$; under this identification, $x \in K$

corresponds to $p(x,1)$ and ∞ to $p(1,0)$. Putting $\tilde{K} = K \cup \infty$, table 4.5.15 becomes

5.2.4
$$c \neq 0 \begin{cases} t \mapsto \dfrac{at+b}{ct+d}, & t \neq -\dfrac{d}{c} \\ -\dfrac{d}{c} \mapsto \infty \\ \infty \mapsto \dfrac{a}{c} \end{cases} \qquad c = 0 \begin{cases} t \mapsto \dfrac{at+b}{d}, \\ \infty \mapsto \infty \end{cases}$$

where the morphism f under consideration has matrix $M(f) = \left(\begin{smallmatrix} a & c \\ b & d \end{smallmatrix}\right)$.

5.2.5. For $K = \mathbf{R}$ or $\mathbf{C}$ and X finite-dimensional, 4.3.2 shows that the embedding of X into $\hat{X}$ is a homeomorphism onto its image; 4.6.1 shows that X is open and everywhere dense in $\hat{X}$. This, together with 5.2.3, explains the remark at the end of 4.5.14.

By the way, all morphisms in $A(X;X')$ and $M\big(P(E);P(E')\big)$ are continuous with respect to the canonical topologies. In arbitrary dimension, $\tilde{f}$ is the continuous extension of f for $f \in A(X;X')$.

5.2.6. For $K = \mathbf{R}$ or $\mathbf{C}$, 4.3.2 and 4.3.3 show that $\tilde{K}$ is exactly the Alexandroff compactification of K.

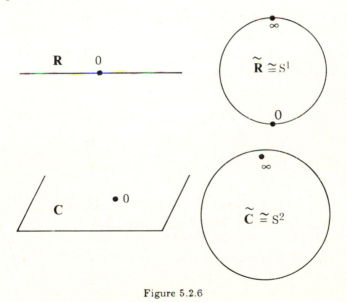

Figure 5.2.6

5.3. Relationship between affine and projective subspaces. Parallelism

We will now show that our construction assigns to parallel affine lines the same point at infinity, as promised.

5.3.1. Let $S \subset X$ be an affine subspace of an affine space X; then S is embedded in $\tilde{X}$. By 2.4.2.3, the projective subspace $\langle S \rangle$ spanned by $S \subset \tilde{X}$ can be identified with the projective completion $\tilde{S}$ of S; thus we have $\langle S \rangle = \tilde{S} = S \cup \infty_S$, where ∞_S is a subset of $\tilde{X}$, namely, $\infty_S = \infty_X \cap \langle S \rangle = \infty_X \cap \tilde{S}$. Note that $\infty_S = P(\vec{S}) \subset P(\vec{X}) = \infty_X$. From definition 2.4.9.1, we get the following result:

5.3.2. PROPOSITION. *The correspondence $S \mapsto \tilde{S}$ is a bijection from the set of affine subspaces of X into the set of projective subspaces of $\tilde{X}$ not entirely contained in ∞_X. We have $\tilde{S} = \infty_S \cup S$, where $\infty_S = \infty_X \cap \tilde{S}$. Moreover, if S, S' are affine subspaces of X, we have*

$$\infty_S = \infty_{S'} \Longleftrightarrow S \parallel S',$$
$$\infty_S \subset \infty_{S'} \Longleftrightarrow S \triangleleft S'.$$

5.3.3. REMARK. For $K = \mathbf{R}$ or $\mathbf{C}$ and X finite-dimensional, 5.2.5 shows that $\tilde{S}$ is just the topological closure of S in $\tilde{X}$.

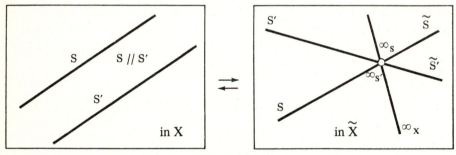

Figure 5.3.3

5.4. Sending objects to infinity; applications

5.4.1. PROPOSITION (PAPPUS' THEOREM—PROJECTIVE VERSION). *Let $P(E)$ be a projective plane, D and D' two distinct lines in $P(E)$, and a, b, c, a', b', c' six distinct points such that $a, b, c \in D \setminus (D \cap D')$, $a', b', c' \in D' \setminus (D \cap D')$. Then the three points $\langle a, b' \rangle \cap \langle a', b \rangle$, $\langle b, c' \rangle \cap \langle b', c \rangle$, $\langle c, b' \rangle \cap \langle c', b \rangle$ are collinear.*

Proof. Remark first that these points exist by 4.6.12. Consider the points $\gamma = \langle a, b' \rangle \cap \langle a', b \rangle$, $\alpha = \langle b, c' \rangle \cap \langle b', c \rangle$, and the projective line $V = \langle \alpha, \gamma \rangle$ joining them. Introduce the affine plane $X = P(E) \setminus V$ (cf. 5.1.3. (ii)); observe that $a, b, c, a', b', c' \in X$. In X we have $\langle a, b' \rangle \parallel \langle a', b \rangle$ and $\langle b, c' \rangle \parallel \langle c', b \rangle$, by 5.3.2 and the construction of V; using the affine version of Pappus (2.5.3), we get $\langle a, c' \rangle \parallel \langle a', c \rangle$, and another application of 5.3.2 gives

$$\beta = \langle a, c' \rangle \cap \langle a', c \rangle \in \langle \alpha, \gamma \rangle. \qquad \square$$

We get the next result by applying 5.4.1 to the completion $\tilde{X}$ of X:

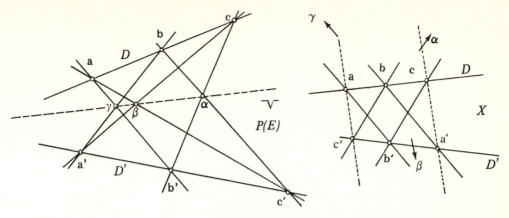

Figure 5.4.1

5.4.2. COROLLARY (PAPPUS' THEOREM—SECOND AFFINE VERSION).
*Let X be an affine plane, D and D' two distinct lines of X, a, b, c, a', b', c' six
distinct points such that $a, b, c \in D \setminus (D \cap D')$, $a', b', c' \in D' \setminus (D \cap D')$. Then
the three points $\langle a, b' \rangle \cap \langle a', b \rangle$, $\langle b, c' \rangle \cap \langle b', c \rangle$, $\langle c, b' \rangle \cap \langle c', b \rangle$ are collinear in
the following sense: if all three exist, they are collinear; if only two of them
exist, the line joining them is parallel to the two lines defining the third (if
none of them exists we are in the case discussed in 2.5.3).*   □

Here we can see how much simpler statements of projective results are than
the statements of corresponding results of affine geometry, which must ac-
count for lots of particular cases. See also 16.8.19.

5.4.3. PROPOSITION (DESARGUES'S THEOREM—PROJECTIVE VERSION)
*Let $P(E)$ be a projective space and s, a, b, c, a', b', c' seven distinct points of
$P(E)$ such that s, a, b, c and s, a', b', c' are projectively independent and that
$a' \in \langle s, a \rangle$, $b' \in \langle s, b \rangle$, $c' \in \langle s, c \rangle$. Then the three points $\langle a, b \rangle \cap \langle a', b' \rangle$,
$\langle b, c \rangle \cap \langle b', c' \rangle$, $\langle c, a \rangle \cap \langle c', a' \rangle$ are collinear. In other words, if the lines joining
the corresponding vertices of two triangles are concurrent, the intersection
points of the corresponding sides are collinear.*

5.4.4. The proof is radically different depending on whether the dimension
of $P(E)$ is equal to or greater than 2. If $\dim(P(E)) \geq 3$, we just use the
intersection properties for subspaces, and the result is valid in any axiomatic
theory of projective spaces (cf. 2.6.7); if $\dim(P(E)) = 2$, one must resort
to theorem 2.5.4, which requires associativity of the base field (cf. 2.5.5 and
4.8.3).

5.4.5. Suppose first that the projective space Z spanned by our seven points
has dimension three; the conclusion follows trivially from 4.6.12, since the
three points being considered belong to the two planes $\langle a, b, c \rangle$ and $\langle a', b', c' \rangle$,
whose intersection is a straight line (else Z would be two-dimensional).
    Now suppose $\dim(P(E)) \geq 3$ but $\dim Z = 2$; we choose an arbitrary
point $m \notin Z$ and two points α, α' on the lines $\langle m, a \rangle$, $\langle m, a' \rangle$ such that

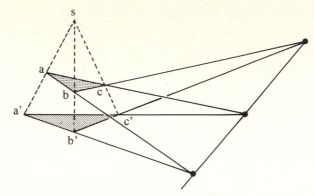

Figure 5.4.5.1

$s \in \langle \alpha, \alpha' \rangle$ and $\alpha \neq a$, $\alpha' \neq a'$ (such points can always be found because a projective line contains at least three points, cf. 4.1.3.7).

Then the seven points $s, \alpha, \alpha', b, b', c, c'$ span a three-dimensional subspace by construction, and thus satisfy the statement of the theorem as shown in the first step. But the projection from $P(E) \setminus \{m\}$ onto Z with center m takes α, α' into a, a' and preserves intersections and collinearities, so the result follows for s, a, a', b, b', c, c'.

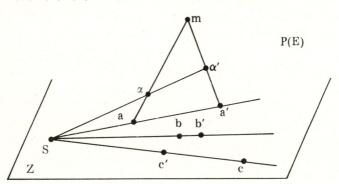

Figure 5.4.5.2

5.4.6. If $\dim\big(P(E)\big) = 2$, we use a proof similar to that of 5.4.1, sending to infinity the points $\langle a, b \rangle \cap \langle a', b' \rangle$ and $\langle b, c \rangle \cap \langle b', c' \rangle$. The conclusion follows from the converse of 2.5.4 (cf. 2.8.6). $\qquad \square$

5.4.7. COROLLARY (DESARGUES'S THEOREM—SECOND AFFINE VERSION). *Let X be an affine space and s, a, b, c, a', b', c' points satisfying the conditions in 5.4.3. Then the three associated intersection points are collinear (allowing for special cases when some of then do not exist, as in 5.4.2).* $\qquad \square$

5.4.8. THEOREM (SECOND FUNDAMENTAL THEOREM OF PROJECTIVE GEOMETRY). *Let $P(E)$ and $P(E')$ be projective spaces of same finite dimension ≥ 2, over fields K and K', respectively. Let $f : P(E) \to P(E')$ be* *a bijection taking triples of collinear points in* $P(E)$ *into collinear points in* $P(E')$. *Then there exists a semilinear bijection* $\hat{f} : E \to E'$ *such that* $f = \hat{f}$.

Proof. As in 2.6.6, one shows that f transforms hyperplanes of $P(E)$ bijectively into hyperplanes of $P(E')$. One then fixes a hyperplane $P(H) \subset P(E)$, whose image is the hyperplane $P(H') \subset P(E')$. The restriction $g = f|_{P(E) \backslash P(H)} : E_H \to E'_{H'}$ is a map from the affine space E_H into the affine space $E'_{H'}$, satisfying the conditions of 2.6.3. Thus g is semiaffine; by an easy extension of 5.1.3 to the semiaffine case, g can be extended to a semimorphism $\tilde{g} : P(E) \to P(E')$, which coincides with f on E_H by construction, and on $P(H)$ because $\tilde{f}|_{P(H)} = \tilde{g}$ (cf. 5.1.3). $\qquad\square$

5.4.9. For various refinements of 5.4.8, see [FL, 83 ff. and 267 ff.]; see also [BI5, 415–416, exercises 16 and 17]. For a generalization of 5.4.8, see [TS, p. VIII].

5.4.10. We shall have ample opportunity to employ this technique of sending points to infinity: see 5.5.3, 6.4.4, 6.4.8, 6.4.10, and a good portion of chapter 17.

5.5. Exercises

5.5.1. State and prove the converses of 5.4.1 and 5.4.3.

* **5.5.2.** Draw the figures for the theorems of Pappus and Desargues (5.4.2 and 5.4.7) when there are points at infinity.

* **5.5.3.** Give a simple proof for the existence of Möbius tetrahedra (cf. 4.9.12 and [B–P–B–S, problem 4.6]) by sending lots of points to infinity.

* **5.5.4.** Suppose you are given a piece of paper with one point marked and segments of two lines which intersect outside the paper. Using only a straightedge (ruler), draw the line that joins the given point with the intersection point of the two lines.

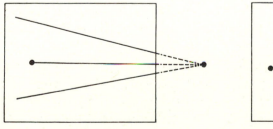
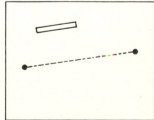

Figure 5.5.4   Figure 5.5.5

* **5.5.5.** Now suppose you are given two points, but your ruler is too short to connect them. Draw the line joining the points.

5.5.6. Study and comment on the proof of the fundamental theorem of projective geometry given in [DX, 28 ff.].

5.5.7. Give necessary and sufficient conditions for the line $\langle \alpha, \beta, \gamma \rangle$ in 5.4.1 to go through $D \cap D'$.

* **5.5.8.** HEXAGONAL WEBS. We shall define a *web* in a real affine plane P as the following set of data: an open set A in P, and for each point a in A three distinct lines $d_i(a)$ $(i = 1, 2, 3)$ in P which go through a and which depend continuously on a. Show that for b on $d_1(a)$ close enough to a, we can define six points $(b_i)_{1,...,6}$ as follows:

$$b_1 = d_3(b) \cap d_2(a), \qquad b_2 = d_1(b_1) \cap d_3(a), \qquad b_3 = d_2(b_2) \cap d_1(a),$$

$$b_4 = d_3(b_3) \cap d_2(a), \qquad b_5 = d_1(b_4) \cap d_3(a), \qquad b_6 = d_2(b_5) \cap d_1(a).$$

A web is said to be *hexagonal* if $b_6 = b$ for all sufficiently close a and b.

Let $(p_i)_{i=1,2,3}$ be three points of P, not on the same line, and let A be the complement of the three lines which connect each pair of points p_i. We define a web by setting $d_i(a) = \langle a, p_i \rangle$ for every $a \in A$. Show that this web is hexagonal.

* **5.5.9.** MORE HEXAGONAL WEBS. Show that the web in figure 5.5.9.1 is hexagonal. More generally, consider a conic section C and a point p not situated on the conic. Assign to any point in the complement of $C \cup p$ the two tangents to C through x and the line xp. Show that the web thus obtained is hexagonal.

For more on webs, see [BL–BO] and [CH–GR]. For figure 5.5.9.2, see [B–P–B–S, remark on p. 166].

Figure 5.5.9.1

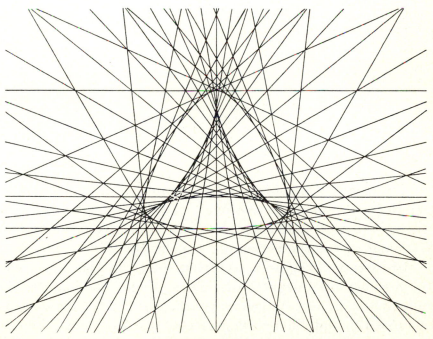

Figure 5.5.9.2

Chapter 6

Projective lines, cross-ratios, homographies

This chapter introduces a homography invariant of quadruples of points on projective lines, called the cross-ratio. This invariant plays a fundamental role in projective geometry; one particular case is the notion of harmonic division.

Cross-ratios will be a recurring theme in this work, coming in sometimes unexpectedly. Examples are the Laguerre formula (8.8.7), hyperbolic geometry (chapter 19), the Ricatti equation (6.8.12), and differential geometry (6.8.20). The ubiquity of cross-ratios illustrates the importance of Cayley's result (5.2.2).

$$\boxed{\text{All fields are assumed commutative.}}$$

6.1. Definition of cross-ratios

6.1.1. Corollary 4.6.9 can be rephrased by saying that triples of distinct points on a projective line are indistiguishable from the point of view of the projective group, or again that the projective group $GP(E)$ is 3-transitive, i.e., transitive on the set of triples of distinct collinear points of $P(E)$ (cf. 1.4.5).

You may have suspected by now that this is no longer true of four points; indeed, we shall use a scalar (in fact, an element of $\tilde{K}$) to classify the orbits under $GP(E)$ of quadruples of collinear points.

6.1.2. DEFINITION. *Let D be a projective line over K and $\{a, b, c, d\}$ a quadruple of points of D such that a, b, c are distinct. The cross-ratio of the four points, denoted by $[a, b, c, d]$, is the element of $\tilde{K} = K \cup \infty$ given by $[a, b, c, d] = f_{a.b.c}(d)$, where $f_{a.b.c}$ denotes the unique element of $\mathrm{Isom}(D; \tilde{K})$ such that $f(a) = \infty$, $f(b) = 0$ and $f(c) = 1$ (cf. 4.6.9). If*

$$\{a, b, c, d\} = \{a_i\}_{i=1.2.3.4},$$

we also write $[a_1, a_2, a_3, a_4] = [a_i]$. If $\{a, b, c, d\}$ are collinear points in a projective space, the first three of which are distinct, their cross-ratio is, by definition, their cross-ratio as points of the line that contains them. If $\{a, b, c, d\}$ are collinear points in an affine space X over K, the first three of which are distinct, their cross-ratio is, by definition, their cross-ratio as points of the projective completion $\tilde{X}$ of X.

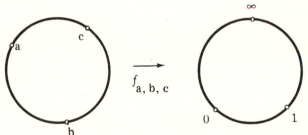

Figure 6.1.2

The following results are a consequence of 4.6.9:

6.1.3. PROPOSITION *If $[a, b, c, d] = \infty$ (resp. 0, 1), then $d = a$ (resp. $d = b$, $d = c$), and conversely. If $[a, b, c, d] \in K \setminus \{0, 1\}$, then $d \neq a, b, c$, and conversely (since $K \setminus \{0, 1\} = \tilde{K} \setminus \{\infty, 0, 1\}$).* □

6.1.4. COROLLARY. *Let D, D' be projective lines and $\{a_i\}_{i=1.2.3.4}$ (resp. $\{a_i'\}_{i=1.2.3.4}$) points in D (resp. D'), the first three elements of each set being distinct. Then there exists $f \in \mathrm{Isom}(D; D')$ such that $f(a_i) = a_i'$ for all i if and only if $[a_i] = [a_i']$.* □

6.1.5. COROLLARY. *Let D be a projective line and a, b, c three distinct points of D. For any $k \in \tilde{K}$ there exists a unique $d \in D$ such that $[a, b, c, d] = k$.* $\square$

6.1.6. PROPOSITION. *Let $D = P(E)$ be a projective line, a, b, c distinct points of D and $x, y \in E$ such that $a = p(x)$, $b = p(y)$ and $c = p(x + y)$ (cf. 4.4.1). Then we have (cf. 5.2.3)*

$$d = p(kx + hy) \iff [a, b, c, d] = p(k, h) \in P(K^2) \cong \tilde{K}.$$

Proof. According to 5.2.3, we have, writing $\{e_1, e_2\}$ for the canonical basis of K^2:

$$p(e_1) = p(1, 0) = \infty, \quad p(e_2) = p(0, 1) = 0, \quad p(e_1 + e_2) = p(1, 1) = 1.$$

Thus $f_{a.b.c} = \underline{f}$, where $f \in L(E; K^2)$ is defined by

$$f(x) = e_1, \quad f(y) = e_2, \quad f(x + y) = f(e_1 + e_2).$$

But then

$$\underline{f}(d) = \underline{f}(p(kx + hy)) = p(f(kx + hy)) = p(k, h). \qquad \square$$

6.1.7. REMARKS. Proposition 6.1.6 is often used as an alternate definition for cross-ratios. It is, in fact, the only possible definition when the base field is not commutative (cf. 6.8.13).skew fields

One can also impose formula 6.2.3 as a definition, proving that it is invariant under the linear group $GL(2; K)$, which is easy. This course is not entirely unjustified if one starts by systematically looking for invariants of homography; this point of view is adopted in [DI–CA, 23]. See also [SF1], [SF2] and 18.10.7.

6.2. Computation of cross-ratios

6.2.1. The question is to compute $[a_i]$, where the a_i are given by their homogeneous coordinates (cf. 4.2.3) in an arbitrary basis $\mathcal{B}$ of E (where $D = P(E)$). Let $a_i = p(x_i) = p(\lambda_i, \mu_i)$, for $i = 1, 2, 3, 4$. Write $x_3 = \alpha x_1 + \beta x_2$ and $x_4 = \gamma x_1 + \delta x_2$; then

$$x_4 = \frac{\gamma}{\alpha}(\alpha x_1) + \frac{\delta}{\beta}(\beta x_2),$$

so that $[a_i] = p(\gamma/\alpha, \delta/\beta)$ (cf. 6.1.6). Using Cramer's rule to compute $\alpha, \beta, \gamma, \delta$, we have

$$\alpha = \frac{\det(x_3, x_2)}{\det(x_1, x_2)}, \quad \beta = \frac{\det(x_1, x_3)}{\det(x_1, x_2)}, \quad \gamma = \frac{\det(x_4, x_2)}{\det(x_1, x_2)}, \quad \delta = \frac{\det(x_1, x_4)}{\det(x_1, x_2)},$$

where the determinants are with respect to the basis $\mathcal{B}$. Hence

6.2.2
$$[a_i] = p\left(\frac{\begin{vmatrix} \lambda_4 & \lambda_2 \\ \mu_4 & \mu_2 \end{vmatrix}}{\begin{vmatrix} \lambda_3 & \lambda_2 \\ \mu_3 & \mu_2 \end{vmatrix}}, \frac{\begin{vmatrix} \lambda_1 & \lambda_4 \\ \mu_1 & \mu_4 \end{vmatrix}}{\begin{vmatrix} \lambda_1 & \lambda_3 \\ \mu_1 & \mu_3 \end{vmatrix}} \right).$$

This expression can be written more compactly as follows (under the convention, compatible with 5.2.4, that division by zero gives $\infty \in \tilde{K}$):

6.2.3
$$[a_i] = \frac{\begin{vmatrix} \lambda_3 & \lambda_1 \\ \mu_3 & \mu_1 \end{vmatrix} \cdot \begin{vmatrix} \lambda_4 & \lambda_2 \\ \mu_4 & \mu_2 \end{vmatrix}}{\begin{vmatrix} \lambda_3 & \lambda_2 \\ \mu_3 & \mu_2 \end{vmatrix} \cdot \begin{vmatrix} \lambda_4 & \lambda_1 \\ \mu_4 & \mu_1 \end{vmatrix}}.$$

Let D be an affine line and $\{a_i\}$ four points, the first three of which are distinct. Take an arbitrary system of affine coordinates, and let the coordinates of x_i be a_i ($i = 1, 2, 3, 4$). Then we can write $a_i = p(x_i, 1)$ in $\tilde{D}$; thus 6.2.3 shows (cf. 2.4.6) that

6.2.4
$$[a_i] = \frac{(x_3 - x_1)(x_4 - x_2)}{(x_3 - x_2)(x_4 - x_1)} = \frac{\overrightarrow{x_3 x_1}}{\overrightarrow{x_3 x_2}} \cdot \frac{\overrightarrow{x_4 x_1}}{\overrightarrow{x_4 x_2}}.$$

In particular, if a, b, c are three points on a line D of an affine space X, we have in $\tilde{X}$ (cf. 5.3.1)

6.2.5
$$[a, b, c, \infty_D] = \frac{\overrightarrow{ca}}{\overrightarrow{cb}}.$$

Formula 6.2.4 furnishes an elementary definition of cross-ratios within the affine framework.

6.2.6. THE MORAL OF THE STORY is that formula 6.2.5 is a geometric confirmation of the fact that the complement $P(E) \setminus P(H) = E_H = X$ of a hyperplane in a projective space has a natural affine structure (cf. 5.1.3 and 2.2.6). For if this is so, one must be able to define geometrically notions such as the sum $b + c$ in a vectorialization X_a of X, and the quotient $\overrightarrow{ca} / \overrightarrow{cb}$ for collinear points. Formula 6.2.5 answers the second problem and the parallelogram rule (cf. 2.6.6.3), interpreted in $P(E)$, answers the first (figure 6.2.6).

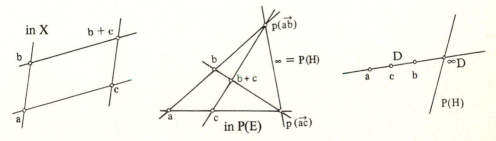

Figure 6.2.6

6.2.7. GEOMETRIC COMPUTATION OF CROSS-RATIOS. See 6.5.6, 6.5.10, and, for the case $K = \mathbf{C}$, 9.6.5.

6.3. Action of permutations

Let $\{a_i\}$ be four distinct points of a projective line D (which forces $K \neq \mathbf{Z}_2!$). Their cross-ratio depends on the order they are considered; the dependence is expressed in the following

6.3.1. PROPOSITION.

$$[a, b, c, d] = [b, a, c, d]^{-1} = [a, b, d, c]^{-1},$$
$$[a, b, c, d] + [a, c, b, d] = 1.$$

Proof. The first two equalities come immediately from 6.2.3. For the last one, put $a = p(x)$, $b = p(y)$, $c = p(x + y)$, $d = p(kx + y)$; then $[a, b, c, d] = k$. Replace the basis $\{x, y\}$ by the basis $\{-x, x + y\}$. Then write $y = -x + (x + y)$ and $kx + y = (1 - k)(-x) + (x + y)$, whence $[a, c, b, d] = 1 - k$ (cf. 6.1.6). $\square$

Since the three formulas of 6.3.1 correspond to three transpositions of $\{a, b, c, d\}$ which generate the group of permutations of this set, it is easy to figure out the effect of any permutation. To make this more precise, we complicate it a bit:

6.3.2. PROPOSITION. *Set $K^{\bullet} = K \setminus \{0, 1\}$ and let S_4 be the symmetric group of $\{1, 2, 3, 4\}$. Let D be a projective line, (a_i) four distinct points of D, $k = [a_i]$ their cross-ratio. Then the cross-ratio $\sigma(k) = [a_{\sigma(i)}] \in K^{\bullet}$ depends only on k and $\sigma \in S_4$, and not on D or the a_i. The map $\phi : S_4 \ni \sigma \mapsto (k \mapsto \sigma^{-1}(k)) \in S_{K^{\bullet}}$ is a S_4-action on $K^{\bullet}$. This action is not faithful; its kernel $\operatorname{Ker} \phi$ is the Klein group $\mathcal{V}_4$ of symmetries of a rectangle (see 0.2), unless K is the field F_4 with four elements, in which case $\operatorname{Ker} \phi = \mathcal{A}_4$, the alternating group. All orbits of K have six elements, with the following exceptions:*

  i) *if K has characteristic three, the orbit of $k = -1$ has a single element;*

  ii) *If K has characteristic different from two or three, the orbit of -1 has three elements $-1, 2, 1/2$, and if K contains the cube roots j, j^2 of 1, the orbit of $-j$ has two elements $-j, -j^2$;*

  iii) *If $K = F_4$, there is a single orbit with two elements.*

Proof. From 6.3.1, we know that $\operatorname{Ker} \phi \supset \mathcal{V}_4$. We investigate the orbit of an arbitrary $k \in K^{\bullet}$; applying 6.3.1, the only possibilities are

$$k, \frac{1}{k}, 1 - k, 1 - \frac{1}{k}, \frac{1}{1 - k}, \frac{k}{k - 1}.$$

All we have to do is check whether these are all different elements of $K^{\bullet}$; this can be reduced to checking the relations $k^2 - 1 = 0$, $k^2 - k + 1 = 0$, $2k - 1 = 0$, $k - 2 = 0$. Bearing in mind that the only root of $X^2 - 1 = 0$ different from 1 is -1, that the only roots of $X^3 - 1 = 0$ different from 1 are j, j^2, and that F_4 is obtaining by adjoining j, j^2 to $\mathbf{Z}_2$, we immediately obtain 6.3.2. $\square$

6.3.3. NOTES. In certain texts, the cross-ratio is *denoted* by $\begin{bmatrix} a & b \\ c & d \end{bmatrix}$. This rectangular notation has the mnemonic advantage of indicating the action of

$\operatorname{Ker} \phi = \mathcal{V}_4$, the group of symmetries of the rectangle, on the matrix itself:

$$\begin{bmatrix} a & b \\ c & d \end{bmatrix} \rightarrow \begin{bmatrix} b & a \\ d & c \end{bmatrix} \rightarrow \begin{bmatrix} c & d \\ a & b \end{bmatrix} \rightarrow \begin{bmatrix} d & c \\ b & a \end{bmatrix}.$$

For the non-commutative case, see 6.8.13.

Figure 6.3.3 shows the graph of the function naturally associated with the cross-ratio function in the case $K = \mathbf{R}$,

$$\lambda \mapsto \frac{4}{27} \frac{(\lambda^2 - \lambda + 1)^3}{\lambda^2 (1 - \lambda)^2}.$$

This function takes the same value on the points of an orbit. [DX, 43–51] and 6.8.11 give some nice facts concerning this function and the cross-ratio.

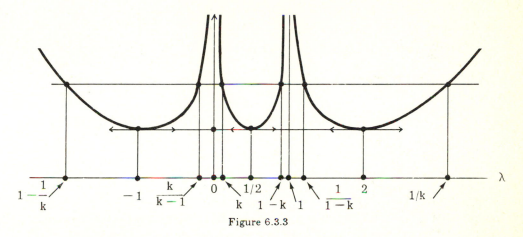

Figure 6.3.3

6.4. Harmonic division

We assume the base field has characteristic $\neq 2$.

The definition of cross-ratios allows two of the four points to be the same. Trying to extend the definition to the case where more than two points are the same leads to hopeless difficulties, except in the case of the orbit $\{-1, 1/2, 2\}$ mentioned in 6.3.2. Whence the following

6.4.1. DEFINITION. *Four points a, b, c, d of a projective space or an affine space are said to be in* harmonic division *if one of the following conditions holds: either three of the points are the same and the fourth is different, or they are all different, collinear and $[a, b, c, d] = -1$.*

Observe that the order matters here: If a, b, c, d are in harmonic division, b, a, c, d are not.

6.4.2. If a, b, c are contained in a line D of an affine space, 6.2.5 shows that a, b, c, ∞_D are in harmonic division if and only if c is the midpoint of $\{a, b\}$.

6.4.3. In an arbitrary set of affine coordinates on an affine line, the points with coordinates a, b, c, d will be in harmonic division if an only if $2(ab + cd) = (a + b)(c + d)$; this is a consequence of 6.2.4. When $a = 0$ (resp. $b = -a$), this reduces to $2/b = 1/c + 1/d$ (resp. $a^2 = b^2 = cd$).

6.4.4. THE COMPLETE QUADRILATERAL. *Let $\{a, b, c, d\}$ be a base for a projective plane P; and put $\alpha = \langle a, b \rangle \cap \langle c, d \rangle$, $\beta = \langle a, d \rangle \cap \langle b, c \rangle$, $\gamma = \langle a, c \rangle \cap \langle b, d \rangle$, $\delta = \langle a, c \rangle \cap \langle \alpha, \beta \rangle$. Then a, c, γ, δ are in harmonic division.*

Proof. Send the line $D = \langle \alpha, \beta \rangle$ to infinity (figure 6.4.4). In the new affine plane $P \setminus D$, we have $ab \parallel cd$ and $ad \parallel bc$ by construction, so a, b, c, d is a parallelogram, and γ is the midpoint of a, c (see figure 2.4.5). Now 6.4.2 shows that $[\alpha, c, \gamma, \delta = \langle a, c \rangle_\infty] = -1$.   □

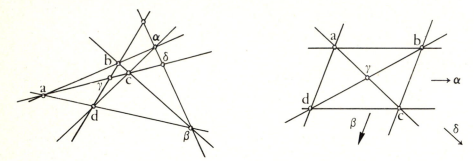

Figure 6.4.4

6.4.5. The preceding result gives a geometric construction for the *harmonic conjugate* n of a point m with respect to two points a, b. (This is defined as the point n such that $[a, b, m, n] = -1$.) One embeds the affine or projective line $\langle a, b \rangle$ in a plane and draws the figure indicated by the construction in 6.4.4.

6.4.6. GEOMETRIC CONSTRUCTION OF MORPHISMS ARISING FROM RE-FLECTIONS. Let E be a vector space and S, T two subspaces such that $E = S \oplus T$ is a direct sum. Let $q : E \to S$ be the projection associated with this direct sum. The *reflection through S and parallel to T* is defined as the map f such that $q(x) = (x + f(x))/2$ for all $x \in E$.

Harmonic division gives a construction for the morphism $g = \underline{f} \in \mathrm{GP}(E)$ as follows: set $V = p(S)$, $W = p(T)$, and take $m \in P(E)$. There exists a unique projective line $D(m)$ going through m and intersecting V and W; $g(m)$ is the point of $D(m)$ such that

$$[m, g(m), D(m) \cap V, D(m) \cap W] = -1.$$

In fact, $D(m)$ arises from a two-dimensional vector space which is well-determined by x and $q(x)$ (unless $x = q(x)$, in which case $g(m) = m$, according to 6.4.1). This plane also contains $f(x)$; taking $\{x, f(x)\}$ as a basis of it, we have

$$p\big(x + f(x)\big) = p\big(2q(x)\big) = V \cap D(m),$$
$$p\big(-x + f(x)\big) = p\big(2q(x)\big) = W \cap D(m).$$

This gives $\big[m, g(m), V \cap D(m), W \cap D(m)\big] = -1$ by 6.1.6.

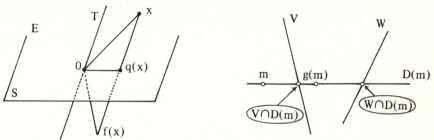

Figure 6.4.6

6.4.7. NOTE. A common case is when S is a hyperplane; then V is a point of $P(E)$ and $D(m)$ is exactly the line containing this point and m. We recall (cf. 4.5.19) that if K is algebraically closed every involution in $GP(E)$ is a hyperplane reflection, and thus very simple from the geometric point of view.

6.4.8. CROSS-RATIOS AND FIELD STRUCTURE. How does one read the field structure on a projective line? More precisely, given, say, five points a, b, c, d, e of a projective line, where is the sixth point f (resp. g) such that $[a, b, c, d] + [a, b, c, e] = [a, b, c, f]$ (resp. $[a, b, c, d][a, b, c, e] = [a, b, c, g]$)? The fundamental theorem of affine geometry (2.6.3) would indicate that f and g can be found by constructions involving only intersections of lines; and in fact the proof of the theorem gives the answer (cf. 2.6.6.4). The construction displayed in figures 2.6.6.4.1 and 2.6.4.4.3 has one of the four original points at infinity; bringing it back to the affine plane we obtain figures 6.4.8. (This of course requires that D be inside an affine plane, but that's necessary anyway, cf. 2.6.1.)

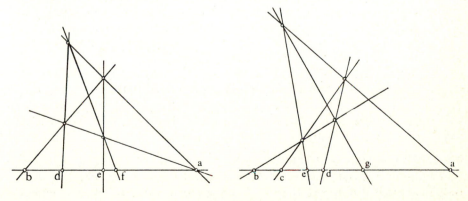

Figure 6.4.8

6.4.9. MAPS THAT PRESERVE HARMONIC DIVISION. Along the same lines, constructions 6.4.5 and 5.4.8 suggest that a bijection $f : D \to D$ from a projective line onto itself preserving the property of four points of being in harmonic division cannot be arbitrary, and possibly has to be a semimorphism. This is indeed the case:

6.4.10. PROPOSITION (VON STAUDT). *Let $D = P(E)$ be a projective line and $f : D \to D$ a bijection. The property that for all distinct $a, b, c, d \in D$ with $[a, b, c, d] = -1$ the images $f(a), f(b), f(c), f(d)$ also satisfy $\left[f(a), f(b), f(c), f(d)\right] = -1$ is equivalent to the existence of a semilinear map $\hat{f} : E \to E$ such that $\hat{\hat{f}} = f$.*

Proof. Let $\sigma : K \to K$ be an arbitrary automorphism of K, and let $\hat{f}$ be semilinear for σ. Then

$$\left[\hat{\underline{f}}(a), \hat{\underline{f}}(b), \hat{\underline{f}}(c), \hat{\underline{f}}(d)\right] = \sigma\big([a, b, c, d]\big);$$

but $\sigma(-1) = -1$ because σ is an automorphism.

To prove the converse, identify $\tilde{K}$ with D and assume, without loss of generality, that f fixes $0, 1$ and ∞ (since the composition $f \circ g$ of f with an appropriate element g of $\mathrm{GP}(\tilde{K})$ fixes those three points and preserves harmonic division if f does). In particular, f can be seen as a map $f : K \to K$. By 6.4.2, we have

$$f\left(\frac{x + y}{2}\right) = \frac{f(x) + f(y)}{2}$$

for all $x, y \in K$, whence $f(x/2) = f(x)/2$ for all $x \in K$. Thus $f(x + y) = f(x) + f(y)$ for all $x, y \in K$. We now use 6.4.3: the points $\{1, x^2, x, -x\}$ are in harmonic division, hence so are $\{f(1) = 1, f(x^2), f(x), f(-x) = -f(x)\}$; again by 6.4.3, this implies $f(x^2) = \big(f(x)\big)^2$. Applying this formula with $x + y$ instead of x gives $f(xy) = f(x)f(y)$, concluding the proof that f is a field automorphism. □

6.4.11. COROLLARY. *A bijection of the real projective line preserving harmonic division is a homography. A continuous bijection of the complex projective line preserving harmonic division is a homography or the complex conjugate of one.*

Proof. This follows immediately from 2.6.4. See also 18.10.2.4. □

6.5. Cross-ratios and duality; applications

What is a line of $P(E^*)$, that is, a "line of hyperplanes"? The answer comes from 2.4.8.1 (recall also 4.1.3.5):

6.5.1. DEFINITION AND PROPOSITION. *A projective line Δ of $P(E^*)$ is called a pencil of hyperplanes of E (a pencil of lines if $\dim\big(P(E)\big) = 2$, a pencil of planes if $\dim\big(P(E)\big) = 3$.) There exists a codimension-two subspace V of $P(E)$ such that $\Delta = \big\{H \in \mathcal{H}(E) \mid H \supset V\big\}$. Conversely, for every* *codimension-two subspace V of $P(E)$ the set $\left\{\, H \in \mathcal{H}(E) \mid H \supset V \,\right\} \subset P(E^*)$ is a line. Moreover, for every $x \in P(E) \setminus V$ there exists a unique $H \in \Delta$ containing x.*   □

The following is a simple result, but has numerous applications:

6.5.2. PROPOSITION. *Let $(H_i)_{i=1.2.3.4}$ belong to a single pencil of hyperplanes of $P(E)$, associated with the codimension-two subspace V. Assume the first three of these hyperplanes are distinct, and let D be a line of $P(E)$ such that $D \cap V = \emptyset$. Then, for any i, the set $D \cap H_i$ consists of a single point h_i, and the two cross-ratios $[h_i]$ and $[H_i]$ are equal, where the first is considered on the line D and the second on the line of $P(E^*)$ which corresponds to the pencil of hyperplanes under consideration.*

Proof. This follows, for example, from the next result.   □

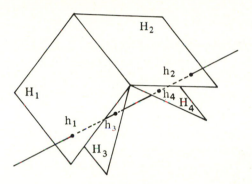

Figure 6.5.2

6.5.3. LEMMA. *Let Δ be a pencil of hyperplanes, V the associated subpace and D a line which does not intersect V. Then the map $\Delta \ni H \mapsto H \cap D \in D$ lies in $\mathrm{Isom}(\Delta; D)$.*

Proof. Let $H, K \in \Delta$ be distinct, set $a = D \cap H$ and $b = D \cap K$, and take $\phi, \psi \in E^*$ such that $H = p(\phi^{-1}(0))$, $K = p(\psi^{-1}(0))$. Normalize ϕ, ψ so that $\phi(y) = -\psi(x) = 1$ for some x, y satisfying $p(x) = a$, $p(y) = b$. By construction, we have

$$D = \left\{\, p(\lambda x + \mu y) \mid (\lambda, \mu) \in K^2 \setminus 0 \,\right\},$$
$$\Delta = \left\{\, ((\xi \phi + \eta \psi)^{-1}(0)) \mid (\xi, \eta) \in K^2 \setminus 0 \,\right\}.$$

The intersection with D of the element of Δ with coordinates (ξ, η) is the point of D of coordinates (λ, μ) given by

$$(\xi \phi + \eta \psi)(\lambda x + \mu y) = 0.$$

This reduces to $\xi \mu = \eta \lambda$, so in $P(K^2)$ we have $p(\lambda, \mu) = p(\xi, \eta)$, and the desired map is in fact the identity in these coordinates.   □

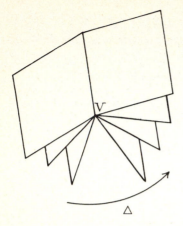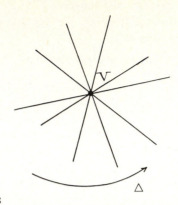

Figure 6.5.3

6.5.4. COROLLARY. *The cross-ratio of the four intersection points of the* H_i *with* D *does not depend on* D. $\qquad\square$

6.5.5. THALES' THEOREM (SECOND PROOF) (cf. 2.5.1). The data are an affine space X and three parallel hyperplanes H, H', H'' in X. In the projective completion $\tilde{X}$ of X, one obtains four hyperplanes $\tilde{H}, \tilde{H}', \tilde{H}'', \infty_X$ which belong to the same pencil. Thales' theorem follows now from 6.5.2 and 6.2.5.

6.5.6. In the same vein, we can calculate the cross-ratio of four collinear points a, b, c, d of an affine space using the construction shown in figure 6.5.6 (after extending the line into a plane, if necessary). Draw a line a', b', c' parallel to $\langle m, d \rangle$; then

$$[a, b, c, d] = \frac{\overrightarrow{c'a'}}{\overrightarrow{c'b'}}.$$

This follows from an application of 6.5.4 to the two lines a, b, c and a', b', c', together with 6.2.5.

One could thus build a purely affine, and, in fact, metric, elementary theory of cross-ratios.

6.5.7. POLAR OF A POINT WITH RESPECT TO TWO LINES. Consider in a projective plane two distinct lines D, D', and take $x \notin D \cup D'$. A line through x intersects D, D' at m, m'; the harmonic conjugate of x with respect to m, m' describes a line through $D \cap D'$, namely, the line F satisfying $\left[D, D', F, \langle x, D \cap D' \rangle\right] = -1$ in $P(E^*)$. This line is called the *polar (line)* of x with respect to $\{D, D'\}$; a reason for the name can be found in 14.5.6. Using this notion, it is easy to reprove 6.4.4, since the polar of a' (figure 6.5.7) with respect to $\{\langle \alpha, d \rangle, \langle \beta, b \rangle\}$ is the same as with respect to $\{\langle \alpha, b \rangle, \langle \beta, d \rangle\}$, and thus goes through a and c.

6.5.8. The following two dual results are proved using the same technique: let $\{a, b, c, d\}$ be distinct and collinear points, and similarly $\{a, b', c', d'\}$.

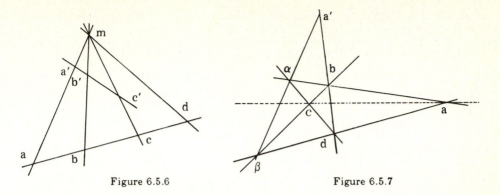

Figure 6.5.6   Figure 6.5.7

Then the lines $\langle b, b' \rangle$, $\langle c, c' \rangle$, $\langle d, d' \rangle$ are concurrent if and only if $[a, b, c, d] = [a, b', c', d']$. Let $\{D, E, F, G\}$ be distinct lines in the same pencil of lines of a projective plane, and similarly for $\{D, E', F', G'\}$. Then the points $E \cap E'$, $F \cap F'$, $G \cap G'$ are collinear if and only if $[D, E, F, G] = [D, E', F', G']$.

6.5.9. Another way to present 6.5.8 is as follows: Let m, m' be points of a projective plane giving to the pencils Δ, Δ', respectively. Let $f : \Delta \to \Delta'$ be a homography, $f \in \mathrm{Isom}(\Delta; \Delta')$, such that $f(\langle m, m' \rangle) = \langle m, m' \rangle$. Then the set $\{ D \cap f(D) \mid D \in \Delta \}$ consists of the line $\langle m, m' \rangle$ union a second line S. (When $f(\langle m, m' \rangle) \neq \langle m, m' \rangle$ the set is a conic going through m and m'; see 16.1.4.) We leave to the reader the statement of the dual result.

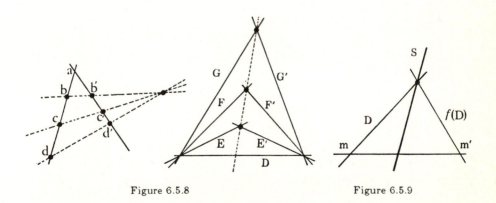

Figure 6.5.8   Figure 6.5.9

6.5.10. CROSS-RATIOS AND PROJECTIVE COORDINATES. Consider a projective base $\{m_i\}_{i=0.1.....n+1}$ and a an arbitrary point m. By 4.4.3, the associated projective coordinates $(x_1, \ldots, x_{n+1})$ of m in this base are only defined up to a scalar, that is, their ratios x_i / x_j determine m. The next proposition allows us to find these ratios geometrically:

6.5.11. PROPOSITION. *Given two indices i, j $(i \neq j)$, and the hyperplanes H_0, H defined by*

$$H_0 = \langle m_1, \ldots, \hat{m}_i, \ldots, \hat{m}_j, \ldots, m_{n+1}, m_0 \rangle,$$
$$H = \langle m_1, \ldots, \hat{m}_i, \ldots, \hat{m}_j, \ldots, m_{n+1}, m \rangle,$$

the ratio x_i/x_j is given by

$$\frac{x_i}{x_j} = \left[m_i, m_j, \langle m_i, m_j \rangle \cap H_0, \langle m_i, m_j \rangle \cap H \right].$$

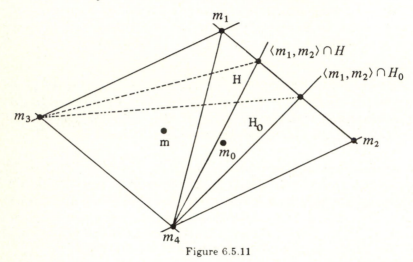

Figure 6.5.11

Proof. Fix a basis $\{e_i\}_{i=1,\ldots,n+1}$ associated with the projective base being used (cf. 4.4.2). Then (λ, μ) are coordinates for $\langle m_i, m_j \rangle$ under the homography

$$p(\lambda, \mu) \longmapsto p(\lambda e_i, \mu e_j).$$

We have $(1,0) \longmapsto p(e_i) = m_i$, $(0,1) \longmapsto p(e_j) = m_j$,

$$(1,1) \longmapsto p(e_i + e_j) = \langle m_i, m_j \rangle \cap H_0,$$
$$(x_i, x_j) \longmapsto p(x_i e_i + x_j e_j) = \langle m_i, m_j \rangle \cap H.$$

This proves 6.5.11, in view of 6.1.6.   □

6.6. Homographies of a projective line

The group $\mathrm{GP}(\tilde{K})$ (or $\mathrm{GP}(D)$, where D is a projective line, the two groups being isomorphic), its subspaces and its elements have been studied in great depth, especially for $K = \mathbf{R}$ and $\mathbf{C}$. They are very simple groups, related to a number of objects in analysis, arithmetic and differential geometry. Here we present only a few elementary results, the barest outline of the theory; on the other hand, we include numerous references, especially to work that in some way uses these groups. For a more systematic study, [GN, chapter IV] is an excellent reference.

6.6.1. PROPOSITION. *Let D be a projective line over an algebraically closed field K, and let $f \in \mathrm{GP}(D)$ be different from the identity. Then f has one or two fixed points. If a, b are distinct fixed points of f, the ratio $\left[a, b, m, f(m)\right]$ is constant for $m \in D \setminus \{a, b\}$, and lies in $K \setminus \{0, 1\}$. Conversely, given $k \in K \setminus \{0, 1\}$ and distinct points $a, b \in D$, the formula $\left[a, b, m, f(m)\right] = k$ defines a homography of D with fixed points a, b. If f has a unique fixed point a, it is a translation of the affine line $D \setminus \{a\}$.*

6.6.2. *Proof.* By 4.5.17, we know that f has at least one fixed (or "double") point. By 4.6.9, there can be at most two. Send one fixed point $a \in D$ to infinity; then the restriction of f to the affine line $D \setminus \{a\}$ is an isomorphism. Since D is one-dimensional, this restriction is a dilatation, and two cases are possible. If it is a homothety, let its center be b; we have $\overrightarrow{bf(m)} = k\overrightarrow{bm}$ for all $m \in D \setminus \{a\}$, and 6.2.5 yields the desired conclusion. Otherwise the restriction of f is a translation and the proposition is proved. $\square$

When there are two distinct fixed points, the constant value of the cross-ratio $\left[a, b, m, f(m)\right]$ is given by the following lemma:

6.6.3. LEMMA. *Put $D = P(E)$, and let $g \in \mathrm{GL}(E)$ be such that $f = \underline{g}$. If $a, b \in D$ are distinct fixed points of D, let λ and μ be the eigenvalues of g associated with the eigenvectors $x, y \in E$ such that $p(x) = a$, $p(y) = b$. Then $\left[a, b, m, f(m)\right] = \lambda / \mu$ for any $m \in D$.*

Proof. Choose a projective frame $\{a, b, m\}$, with $a = p(e_i)$, $b = p(e_2)$, $m = p(e_1 + e_2)$. The matrix of g must be $\left(\begin{smallmatrix} \lambda & 0 \\ 0 & \mu \end{smallmatrix}\right)$, so $f(m) = p(\lambda e_1 + \mu e_2)$, proving the lemma by 6.1.6. $\square$

6.6.3.1. Even if the base field is not algebraically closed, 6.6.1 holds as long as there is one fixed point. It may happen that there are no fixed points, for example $K = \mathbf{R}$, $M(f) = \left(\begin{smallmatrix} 0 & -1 \\ 1 & 0 \end{smallmatrix}\right)$. This can be written, in the notation of 5.2.4:

$$t \mapsto -\frac{1}{t} \quad (t \neq 0), \qquad \infty \mapsto 0, \qquad 0 \mapsto \infty.$$

6.6.4. IMAGE AND OBJECT FOCAL POINTS. Whether or not there are fixed points, one can describe a homography as follows: write $D = X \cup \infty$, where X is an affine line (this is equivalent to choosing a point ∞ on D). The *image* (resp. *object*) *focal point* of $f \in \mathrm{GP}(D)$ is the point $a = f^{-1}(\infty)$ (resp. $b = f(\infty)$). Unless ∞ is fixed, we have $a, b \in X$. By 6.1.4,

$$[\infty, a, m, n] = \left[b, \infty, f(m), f(n)\right]$$

for any $m, n \in D$. Applying 6.2.5 and 6.3.1 we obtain

$$\frac{\overrightarrow{an}}{\overrightarrow{am}} = \frac{\overrightarrow{bf(m)}}{\overrightarrow{bf(n)}},$$

that is,

6.6.5 $$\overrightarrow{am} \cdot \overrightarrow{bf(m)} = \text{constant}.$$

This formula determines f if one knows the focal points and one pair $\big(n, f(n)\big)$. It also works when $a = b$.

The name "focal points" comes from geometrical optics, where one proves that a centered optical system, under the Gauss approximation, associates to each point of the axis of the system (union infinity) another point of the axis, and that this correspondence is a homography. One can start by showing this for a single diopter or mirror, using classical formulas; the extension to arbitrary centered optical systems, which are by definition composed of a finite number of diopters or mirrors having the same axis, follows by taking the composition of the individual homographies (cf. 4.5.9). On the subject of geometrical optics, see, for instance, [A–B, chapter 4].

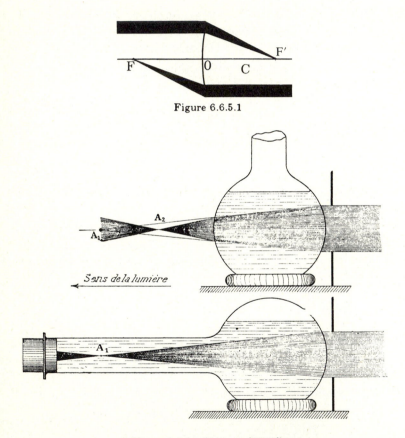

Figure 6.6.5.1

Figure 6.6.5.2 (Source: [BOU])

6.6.6. GEOMETRIC CONSTRUCTIONS. For geometric constructions on the homographies of a line, see 16.3.10.1.

6.6.7. NOTES. Here are some references on homographies in general and their applications in various areas of mathematics: [GN, chapter IV] is a general reference; for differential equations, see 6.8.12; for automorphic functions, see [SA]; hyperbolic geometry is discussed in chapter 19; [VL] has a pleasant miscellany of geometric applications; 18.10.2.2 and [CH2, 183] discuss the homographies of $P(\mathbf{C}^2) = S^2$.

6.7. Involutions

6.7.1. DEFINITION. *An involution of the projective line D is a homography f of D such that $f^2 = \mathrm{Id}_D$ and $f \neq \mathrm{Id}_D$.*

This differs slightly from the standard definition of an involutive map, which allows the case $f = \mathrm{Id}$. From section 6.6, we immediately get the

6.7.2. PROPOSITION. *If an involution f has a fixed point, it has two fixed points a, b, and is determined by the condition $\big[a, b, m, f(m)\big] = -1$ for all m. Whether or not f has a fixed point, its restriction to the affine space $D \setminus \infty$ is characterized by*

$$\overline{am} \cdot \overline{af(m)} = \text{constant},$$

where a is the (image and object) focal point of f. In particular, if K is algebraically closed, f is always of the form $\big[a, b, m, f(m)\big] = -1$; if $K = \mathbf{R}$, either f has fixed points, or it can be written $\overline{am} \cdot \overline{af(m)} = -1$ in some appropriate affine frame.

The following statement collects some simple properties of involutions:

6.7.3. PROPOSITION. *Every homography is the product of at most three involutions. A homography f is an involution if and only if there exists m such that $f^2(m) = m$ and $f(m) \neq m$. A homography with matrix $A = \left(\begin{smallmatrix} a & b \\ c & d \end{smallmatrix}\right)$ is an involution if and only if $\mathrm{Tr}\, A = a + d = 0$.*

Proof. By composing f with an involution if necessary, we can assume that f has a fixed point, which can be taken to be ∞. By 6.6.2, f is either a translation $x \mapsto x + t$ or a homothety $x \mapsto \lambda x$ of an affine line. In the first case, it is the composition of the involutions $x \mapsto -x$ and $x \mapsto t - x$, and, in the second, of the involutions $x \mapsto \lambda/x$ and $x \mapsto 1/x$.

If $f^2(m) = m$ and $f(m) \neq m$, we can take $m = 0$ and $f(m) = \infty$, whence $f(\infty) = 0$. By 5.2.4, we must have $c \neq 0$, $d = 0$ and $a = 0$, so f is of the form $t \mapsto b/ct$, which is indeed an involution.

Writing

$$M(f) = \begin{pmatrix} a & b \\ c & d \end{pmatrix}, \qquad M(f^2) = \begin{pmatrix} a^2 + bc & b(a + d) \\ c(a + d) & bc + d^2 \end{pmatrix}$$

we have $f^2 = \lambda\,\mathrm{Id}$ only if $b(a + d) = c(a + d) = 0$ and $a^2 + bc = d^2 + bc = 1$. If $a + d \neq 0$ we have $b = c = 0$ and $a = d$, hence $f = \mathrm{Id}$. $\qquad\square$

6.7.4. NOTES. The reader can easily check the following facts: an involution is determined by two pairs $\big(a, f(a)\big)$ and $\big(b, f(b)\big)$ (with $a \neq b$); two distinct involutions coincide at exactly one point if K is algebraically closed. These assertions will also follow from the corresponding geometric constructions, given in 16.3.10.1. Involutions will be encountered again when we discuss Desargues's theorem on pencils of conics (16.5.4). Finally, see [GN, chapter IV] or, if unavailable, 14.8.16.

6.8. Exercises

* **6.8.1.** Let x, y, z, u, v be five points on the same projective line. Show that the following always holds:

$$[x, y, u, v][y, z, u, v][z, x, u, v] = 1.$$

6.8.2. Show the theorems of Pappus and Desargues (5.4.1, 5.4.3) by using the statements in 6.5.8.

6.8.3. Consider a homography $x \mapsto \dfrac{ax + b}{cx + d}$ of the real line $\mathbf{R}$. Study whether there is any relation between the existence of fixed points and the sign of the derivative of $x \mapsto \dfrac{ax + b}{cx + d}$ $\left(x \neq -\dfrac{d}{c}\right)$.

6.8.4. Consult a book on geometrical optics (for instance, [A–B, chapter 4]) and study the nature of the homographies associated with: a (plane, convex spherical, concave spherical) mirror, a plane diopter, a thin lens, a thick lens. Study the reduced formulas given in the book.

6.8.5. For $K = \mathbf{R}$ or $\mathbf{C}$, study the behavior of the iterates f^n $(n \in \mathbf{Z})$ of a homography of a projective line, according to the nature of f.

6.8.6. Is every homography the product of two involutions for $K = \mathbf{C}$? For $K = \mathbf{R}$?

* **6.8.7.** Let f be a homography with two distinct fixed points a, b; show that the set $\{k, 1/k\}$, where $k = [a, b, m, f(m)]$ for every m, depends only on f and not on the choice of the order of a, b. If f has matrix $M = \left(\begin{smallmatrix} \alpha & \beta \\ \gamma & \delta \end{smallmatrix}\right)$ show that $\{k, 1/k\}$ are the roots of the equation

$$(\alpha\delta - \beta\gamma)X^2 - (\alpha^2 + 2\beta\gamma + \delta^2)X + (\alpha\delta - \beta\gamma) = 0.$$

* **6.8.8.** The data are those of 6.8.7, and moreover $K = \mathbf{C}$. We say that f is *elliptic* if the complex number k (or $1/k$) has absolute value 1, *hyperbolic* if k is positive real, and *loxodromic* otherwise. Show that after normalizing $M(f)$ by $\alpha\delta - \beta\gamma = 1$, we can characterize these three cases by using the trace t of f, given by $t = \alpha + \delta$:

$$f \text{ is } \begin{cases} \text{elliptic} & \Longleftrightarrow & \alpha + \delta \text{ is real and } |\alpha + \delta| < 2, \\ \text{hyperbolic} & \Longleftrightarrow & \alpha + \delta \text{ is real and } |\alpha + \delta| > 2, \\ \text{loxodromic} & \Longleftrightarrow & \alpha + \delta \text{ is not real.} \end{cases}$$

Study, for the three cases considered, the nature of the iterates $f^n (n \in \mathbf{Z})$. The word "loxodromic" is justified by the fact that the iterates $f^n(z)$ of a point z of the Riemann sphere $\mathbf{C} \cup \infty$ all belong to the same loxodrome of the sphere (cf. 18.1.8.2 and 18.11.3).

6.8.9. Let f be a homography of $\mathbf{C} \cup \infty$, with matrix $\left(\begin{smallmatrix} a & b \\ c & d \end{smallmatrix}\right)$. Show that a necessary and sufficient condition for f to fix the upper half plane $H = \{\, x \in \mathbf{C} \mid \mathrm{Im}(z) > 0 \,\}$ is that a, b, c, d be real and that $ad - bc > 0$. For such an f, study the nature of the fixed points and of the iterates of f, as a particular case of 6.8.8. Show that if commuting homographies f and g both leave H fixed, they have the same fixed points.

6.8.10. In the notation of 6.3.2, show that the stabilizer of $\{4\}$ is isomorphic to S_3 and acts faithfully if $K \neq F_4$.

6.8.11. Let $ax^4 + bx^3 + cx^2 + dx + e = 0$ be a fourth-degree equation over $\mathbf{C}$. Write the condition on a, b, c, d, e for the four roots of this equation to have cross-ratio $-j$ or $-j^2$ $\big($cf. 6.3.2 (ii)$\big)$.

Find the condition for the four roots to be in harmonic division (messy). Find the six possible values of the cross-ratio in terms of a, b, c, d, e (very messy). Cf. [DX, 43–51].

* **6.8.12.** RICATTI DIFFERENTIAL EQUATIONS. For $a, b, c : [\alpha, \beta] \to \mathbf{R}$ continuous functions, consider the differential equation $y'(t) = a(t)y^2 + b(t)y + c(t)$, called a *Ricatti equation*. Show that if y_i $(i = 1, 2, 3, 4)$ are four solutions of this equation, the cross-ratio $\big[y_i(t)\big]$ is independent of t.

6.8.13. Let K be a skew field, E a two-dimensional vector space over K, and a, b, c, d four distinct points of the projective line $P(E)$. The *cross-ratio* of (a, b, c, d), denoted by $\left[\begin{smallmatrix} a & b \\ c & d \end{smallmatrix}\right]$, is the set of $\xi \in K$ such that there exist $u, v \in E$ satisfying $a = p(u)$, $b = p(v)$, $c = p(u + v)$, $d = p(\xi u + v)$. Show that $\left[\begin{smallmatrix} a & b \\ c & d \end{smallmatrix}\right]$ is the set of conjugates of a fixed element of the multiplicative group K^* (under the action of inner automorphisms). State and prove a converse. Find out which results of this chapter involving cross-ratios still hold.

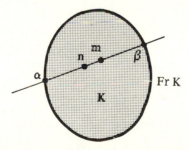

Figure 6.8.14

6.8.14. Let K be a convex compact set of a finite-dimensional real affine space. Assume K has non-empty interior and all its boundary points are extremal. Define the function $d : K \times K \to \mathbf{R}$ as follows: $d(m, m) = 0$ for all $m \in K$ and

$$d(m, n) = \frac{1}{2}\big|\log([m, n, \alpha, \beta])\big|$$

for $m \neq n$, where α, β denote the two intersection points of the line m, n with the frontier of K. Show that d is well-defined and makes K into a metric space. Show that, for any $m, n \in K$, there exists a shortest path from m to n under this metric, and that it coincides with the segment $[m, n]$ (cf. 9.9.5; see also 11.9.4).

6.8.15. Study the homographies of a projective line over a skew field.

6.8.16. Show that every homography of $\mathbf{C} \cup \infty$ which leaves the unit disc $D = \{ z \in \mathbf{C} \mid |z| \leq 1 \}$ globally invariant is of the form

$$z \mapsto e^{i\theta} \frac{z + z_0}{1 + \bar{z}_0 z},$$

where θ is real and $|z_0| < 1$. Show that the group of such homographies is 2-transitive over $\overset{\circ}{D}$ (cf. 1.4.5), and preserves the distance d of 6.8.14. Compare with 19.6.9.

* **6.8.17.** Let $(a_i)_{i=1,\ldots,5}$ be points on a projective plane, so that the first four form a projective base. Denote by d_{ij} the line $\langle a_i, a_j \rangle$. Show that the following holds:

$$[d_{12}, d_{13}, d_{14}, d_{15}][d_{23}, d_{21}, d_{24}, d_{25}][d_{31}, d_{32}, d_{34}, d_{35}] = 1.$$

Show that a necessary and sufficient condition for the existence of a homography taking the $(a_i)_{i=1,\ldots,5}$ into new points $(a'_i)_{i=1,\ldots,5}$ is that the following two equalities be satisfied:

$$[d_{12}, d_{13}, d_{14}, d_{15}] = [d'_{12}, d'_{13}, d'_{14}, d'_{15}]$$

and

$$[d_{23}, d_{21}, d_{24}, d_{25}] = [d'_{23}, d'_{21}, d'_{24}, d'_{25}],$$

where we have put $d'_{ij} = \langle a'_i, a'_j \rangle$.

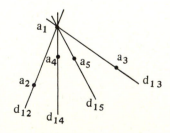

Figure 6.8.17

6.8.18. Study the notion of harmonic division in the non-commutative case. skew field What happens to 6.4.10?

6.8.19. Why is there no point c in figure 6.4.8?

6.8.20. RULED SURFACES. Let S be a *ruled surface* in $\mathbf{R}^3$, that is, the union $\bigcup_{t \in I} D(t)$ of a family $D(t)$ of lines in $\mathbf{R}^3$, parametrized by $t \in I$, where I is an open interval of $\mathbf{R}$. Assume that, at every point m of $D(t_0)$ $(t_0 \in I)$, the surface S is a differentiable submanifold of $\mathbf{R}^3$, and denote by $T(m)$ the tangent plane to S at m. Show that $m \mapsto T(m)$ is either a homography or a constant. Compare with 14.4.4.

* **6.8.21.** Let T be a tetrahedron in a three-dimensional projective space and let D be a line. Show that the cross-ratio of the four intersection points of D with the faces of T is equal to the cross-ratio of the four planes passing through D and the vertices of T.

6.8.22. See exercise 14.8.16.

Chapter 7
Complexifications

The aim of this chapter is to present an intrinsic definition of complexification for real vector, projective and affine spaces. We also study the problem of when morphisms and subspaces can be complexified, and finally (section 7.6) show that completing a real affine space into a projective space and then complexifying gives the same result as first complexifying and then taking the completion. This fact will later be used to introduce, in an intrinsic way, the cyclic points of a Euclidean affine space (9.5.5).

The intrinsic process for complexification is lengthy, though not really difficult if one knows about the universal space $\hat{X}$ associated with an affine space X (chapters 3 and 5). Nonetheless, the exposition is preceded by an introduction giving a much shorter, albeit non-intrinsic, construction.

In this chapter i denotes the point $(0,1)$ of $\mathbf{C}$.

7.0. Introduction

7.0.1. In this book we shall have to complexify several objects, for example, real vector spaces and their endomorphisms (to find eigenspaces, see 7.4.3), Euclidean spaces and the quadratic form defining them (to introduce isotropic vectors and characterize similarities, see 8.8.6.4), and projective spaces (to introduce cyclic points, see 9.5.5). The process of complexifying a real object is natural and fundamental in mathematics; as Hadamard said, the shortest path between two real results often goes through complex terrain.

7.0.2. The simplest process, and one that would be sufficient for us, to complexify a real vector space E of finite dimension n, is the following: Take a basis for E, and embed E in $\mathbf{C}^n$ via the map

$$E \ni x = (x_1, \ldots, x_n) \mapsto (x_1, \ldots, x_n) \in \mathbf{C}^n,$$

where x_i is the i-th coordinate of x with respect to the chosen basis. The same process works for a real affine space of dimension n: work with an affine frame, and consider in $\mathbf{C}^n$ its natural affine structure (cf. 2.2.1). To complexify a Euclidean space and also the quadratic form defining it, we take a basis, write the quadratic form as $\sum_{i,j} a_{ij} x_i x_j$ in this basis, and consider $\mathbf{C}^n$ with the quadratic form $\sum_{i,j} a_{ij} z_i z_j$, for $(z_1, \ldots, z_n) \in \mathbf{C}^n$; E is embedded in $\mathbf{C}^n$ as above. To complexify a real projective space, we complexify the vector space from which is arises, and projectivize the result.

The most complicated situation which we shall discuss is the following: X is an affine vector space, $\tilde{X}$ its projective completion (cf. 5.1), and we desire to complexify both X and $\hat{X}$, in such a way that the complex projective space obtained from $\tilde{X}$ is the projective completion of the affine complex space obtained from X. This is easy to do using an affine frame of X: in the notation of 5.0.2, one starts by introducing the complexification $\mathbf{C}^n$ of X and the embedding $X \to \mathbf{C}^n$ above. The projective completion $\hat{X}$ of X is identified with $P^n(\mathbf{R})$, and the embedding $X \to P^n(\mathbf{R})$ is given by $(x_1, \ldots, x_n) \mapsto p(x_1, \ldots, x_n, 1)$. But $P^n(\mathbf{R})$ can be naturally complexified into $P^n(\mathbf{C})$, with the obvious inclusion $P^n(\mathbf{R}) \subset P^n(\mathbf{C})$; thus, the embedding $\mathbf{C}^n \to P^n(\mathbf{C})$, given by $(z_1, \ldots, z_n) \mapsto p(z_1, \ldots, z_n, 1)$, is indeed the same as the embedding of $\mathbf{C}^n$, considered as a complex affine space, into its projective completion. In other words, we have a commutative diagram

7.0.2.1
$$\begin{array}{ccc} X & \longrightarrow & \tilde{X} \\ \downarrow & & \downarrow \\ \mathbf{C}^n & \longrightarrow & P^n(\mathbf{C}), \end{array}$$

where the vertical arrows are complexifications and the horizontal ones projective completions.

7.0.3. The observations above would be enough for the problems discussed in this book. Nonetheless, the constructions described in 7.0.2 have the fundamental drawback of depending, *a priori,* on the choice of a basis. It might be argued that this flaw is merely esthetic, and that a concern with elegance and naturality is not essential; but such a concern certainly has been around for a long time, and, in the particular case of the constructions of 7.0.2, it gave rise to a long argument around the "Poncelet continuity principle". This principle, as quoted in [GX, 72], goes as follows: "Consider a figure F formed by points, straight lines and curves, and suppose that F moves around, undergoing a continuous deformation. Assume that, for a position F' of the figure, some of its elements have become imaginary; a property of F which does not involve these elements, but which may have been established by using such elements, will nonetheless apply to F' as well." The principle transcends the framework of complexifications and in fact is connected with algebraic geometry, but we have quoted it because of its historical interest and its obvious relation to 7.0.2.

Another quotation, from [DX, 15], is relevant here; it belongs to a period between the inception of the use of imaginary numbers in geometry and the appearance of today's intrinsic complexifications. "Before we continue, it may perhaps be useful to respond to an objection that von Staudt addressed to the analytic theory of imaginary numbers, and which would apply equally well to the theory just presented. This learned geometer maintained that the existence of imaginary points, as we have defined them, somehow depends on the coordinate axes; and this objection would apply to any object studied by means of such points. To rebut this objection, it is sufficient to show that, if an imaginary point belongs to one or more surfaces, it will still belong to them after a change of coordinates. This is certainly the case if one agrees to apply to imaginary points and points at infinity the change of coordinate formulas that have been proved for real points."

7.0.4. A more sophisticated point of view would consist in using 7.0.2 but checking, whenever necessary, that the objects considered in the complexifications depend only on the objects they come from in the real spaces, and not on the bases chosen to perform the complexification. For example, one can check, when complexifying a Euclidean vector space and its quadratic form, that the isotropic cone of the complexification depends only on the Euclidean structure.

7.0.5. This whole discussion shows that an intrinsic formulation for the constructions in 7.0.2 is not at all out of place, which is why we develop it in this chapter. We do it tersely, however, skipping trivial demonstrations and some details in the statements, and, more important, defining only explicit canonical complexifications, and not complexifications in themselves, whose existence and uniqueness up to a canonical isomorphism would have to be proved. For this latter approach, admittedly more elegant but also more costly, the reader is referred to [FL, 144–155].

7.0.6. NOTE. The reader may have suspected that there exist generalizations of the notion of complexification; for an example, see [GN, chapter IV], which deals with quadratic extensions (more general than the extension $\mathbf{C} \supset \mathbf{R}$).

7.1. Complexification of a real vector space

Complexifying a real vector space consists in embedding it in the smallest possible complex vector space, like going from $\mathbf{R}$ to $\mathbf{C}$. This is accomplished as follows:

7.1.1. DEFINITION. *Let E be a real vector space. The complexification of E, denoted by E^C, is the product $E \times E$, endowed with the complex vector structure defined by*

$$(x, y) + (x', y') = (x + x', y + y'),$$
$$(\lambda + i\mu)(x, y) = (\lambda x - \mu y, \lambda y + \mu x).$$

The embedding $E \to E^C$ is given by $x \mapsto (x, 0)$. The involution

$$\sigma : (x, y) \mapsto (x, -y)$$

from E^C into itself is called the conjugation in E^C. We have

$$E = \mathrm{Ker}(\sigma - \mathrm{Id}_{E^C}) = \{\, z \in E^C \mid \sigma(z) = z \,\}.$$

7.1.2. REMARKS.

7.1.2.1. The conjugation map is semilinear: $\sigma(\lambda z) = \bar{\lambda}\sigma(z)$.

7.1.2.2. Once we have 7.1.1, we can observe that $i(y, 0) = (0, y) \cong iy$, so we can write $(x, y) = x + iy$ and $E^C = E \oplus iE$ (a real direct sum). Then $\sigma(x + iy) = x - iy$, which justifies the word conjugation.

7.1.2.3. Other equally good (and equally gratuitous) definitions are:

$$E^C = L_{\mathbf{R}}(\mathbf{C}; E) = \{\, f : \mathbf{C} \to E \mid f \text{ is } \mathbf{R}\text{-linear} \,\},$$

$$E^C = E \otimes_{\mathbf{R}} \mathbf{C}.$$

For a universal definition, see [FL, 146].

7.1.3. BASES. Let $\{e_s\}_{s=1,\dots,n}$ be a basis for E; then $\{e_s\} \subset E^C$ is a basis for E^C (over $\mathbf{C}$, not over $\mathbf{R}$). Thus 7.0.2 is justified *a posteriori*. Conversely, if $\{e'_s\}_{s=1,\dots,n}$ is an arbitrary basis for E^C (over $\mathbf{C}$), it is a fundamental fact that the $2n$ vectors

$$\{e'_s + \sigma(e'_s)\} \cup \left\{ \frac{1}{i}(e'_s - \sigma(e'_s)) \right\}$$

are in E, being invariant under σ. Thus one can choose from among these $2n$ vectors an $\mathbf{R}$-basis for E.

7.2. Functoriality; complexification of morphisms

7.2.1. PROPOSITION. *Let E, E', E'' be real vector spaces, and take $f \in L(E; E')$. There exists a unique $f^C \in L_C(E^C; E'^C)$ such that $f^C|_E = f$. Also $\sigma \circ f^C = f^C \circ \sigma$, where σ denotes conjugation in both E and E'. Finally, complexification is functorial, that is, $(\mathrm{Id}_E)^C = \mathrm{Id}_{E^C}$ and $(g \circ f)^C = g^C \circ f^C$ for $f \in L(E; E'), g \in L(E'; E'')$.*

Proof. It is enough to observe that, for any $x, y \in E$,

$$f^C(x + iy) = f(x) + if(y)$$

(cf. 7.1.2.2), and that the diagram below commutes:

$$
\begin{array}{ccc}
E^C & \xrightarrow{f^C} & E'^C \\
\downarrow{\sigma} & & \downarrow{\sigma} \\
E^C & \xrightarrow{f} & E'^C .
\end{array}
$$

$\square$

7.2.2. If $M(f)$ is the matrix of f with respect to a pair of bases of E, E', then $M(f^C) = M(f)$ with respect to the same bases, considered as bases of E^C, E'^C (cf. 7.1.3).

7.3. Complexification of polynomials

7.3.1. PROPOSITION. *Let $\phi : E \times \cdots \times E \to \mathbf{R}$ be a k-linear map over a real vector space E. There exists a unique k-linear map*

$$\phi^C : E^C \times \cdots \times E^C \to \mathbf{C}$$

(over $\mathbf{C}$) such that $\phi^C|_{E \times \cdots \times E} = \phi$. Moreover, if ϕ is symmetric, so is ϕ^C.

Proof. The value of ϕ^C must necessarily be

7.3.2 $$\phi^C(x_1 + iy_1, \ldots, x_k + iy_k) = \sum_P i^{\#P} \phi(X_1^P, \ldots, X_k^P),$$

where P ranges through the 2^k subsets of $\{1, 2, \ldots, k\}$, $\#P$ is the cardinality of P, and X_s^P is equal to x_s if $s \in P$ and to y_s if $s \notin P$. It is easy to check that ϕ^C satisfies the stated properties. $\square$

7.3.3. EXAMPLE. For $k = 2$,

$$\phi^C(x + iy, x' + iy') = \phi(x, x') - \phi(y, y') + i\big(\phi(x, y') + \phi(x', y)\big).$$

7.3.4. PROPOSITION. *Let $f \in P_k(E)$. There exists a unique $f^C \in P_k(E^C)$ such that $f^C|_E = f$.*

Proof. By 3.3.1, we can assume that f is homogeneous of degree k. Let $f = \phi \circ \Delta$, where $\phi : E^k \to \mathbf{R}$ be a symmetric multilinear map. Put $f^C = \phi^C \circ \Delta^C$, where ϕ^C is the map given in 7.3.1 and $\Delta^C : E^C \to E^C \times \cdots \times E^C$ is the diagonal map. Then f^C satisfies the required properties, and is the only map to do so by the uniqueness parts of 3.3.1 and 7.3.1. $\square$

7.3.5. Fixing a basis for E and working in coordinates, one can easily obtain f^C by taking formulas 3.3.3 and replacing the real numbers λ_s by arbitrary complex numbers.

7.4. Subspaces and complexifications

7.4.1. PROPOSITION. *Let E be a real vector space, F a vector subspace of E. The $\mathbf{C}$-vector subspace of E^C spanned by the subset $F \subset E^C$, denoted by F^C, is equal to $F^C = F + iF = \{\, x + iy \mid x, y \in F \,\}$, and is called the complexification of F (in E^C); it is invariant under σ. Conversely, if $S \subset E^C$ is a (complex) vector subspace of E^C such that $\sigma(S) = S$, the intersection $S \cap E$ is a (real) vector subspace of E, and its real dimension $\dim_{\mathbf{R}}(S \cap E)$ is the same as the complex dimension $\dim_{\mathbf{C}} S$ of S. Moreover, one has $S = (S \cap E)^C$.*

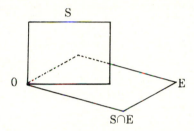

Figure 7.4.1

Proof. The non-trivial part is the second half. It is clear that $S \cap E$ is a vector subspace of E, and that $(S \cap E)^C \subset S$; to prove equality from the fact that $\sigma(S) = S$, we use the same idea as in 7.1.3: for every $x \in S$, the vectors $x + \sigma(x)$ and $(1/i)(x - \sigma(x))$ are in $S \cap E$. Then

$$x = \frac{1}{2}\left((x + \sigma(x)) + i\left(\frac{1}{i}(x - \sigma(x))\right)\right) \in (S \cap E) + i(S \cap E) = (S \cap E)^C. \ \square$$

7.4.2. NOTES. It is difficult to illustrate this proposition with a drawing, since the first non-trivial case requires $\dim_{\mathbf{C}} S = 1$, $\dim_{\mathbf{C}} E = 2$, hence $\dim_{\mathbf{R}} E = 4$. Notice that 7.4.1 is false if $\sigma(S) \neq S$; take, for example, $E = \mathbf{R}^2$, $E^C = \mathbf{C}^2$, $S = \{\, (z, iz) \mid z \in \mathbf{C} \,\}$. Then $S \cap \mathbf{R}^2 = 0$, so $(S \cap E)^C = 0$.

Observe how we have been systematically using σ (cf. 7.1.1, 7.2.1, 7.4.1); this shows we have to carry around the conjugation σ whenever we discuss the space E^C. It can, in fact, be proved that giving E^C and σ is equivalent to giving E and E^C; cf. [FL, 146], where this is the basis for the definition of a complexification.

7.4.3. COROLLARY. *Let $f \in L(E; E)$ be an endomorphism of a finite-dimensional real vector space. Then f admits a one- or two-dimensional eigenspace P.*

Proof. Let V be a one-dimensional eigenspace of E^C for the endomorphism f^C, with eigenvalue λ. Then $\sigma(V)$ is a one-dimensional eigenspace, with eigenvalue $\overline{\lambda}$ (cf. 7.1.2.1). Thus the complex vector subspace $S = V + \sigma(V)$ is invariant under σ, and has dimension one or two. The desired subspace is now $E \cap S$. $\square$

7.4.4. This result, the real version of the theorem of existence of eigenvectors for complex endomorphisms, is very useful (cf. 8.2.15, 13.5).

7.5. Complexification of a projective space

7.5.1. DEFINITION. *Let $P = P(E)$ be a real projective space. The complexification of P, denoted by P^C, is the complex projective space $P^C = P(E^C)$. We identify P with a subset of P^C by taking the quotient of the inclusion $E \setminus 0 \to E^C \setminus 0$. The map $P^C \to P^C$ obtained from $\sigma : E^C \setminus 0 \to E^C \setminus 0$ by passing to the quotient is also denoted by σ, and called the conjugation in P^C.*

7.5.2. Observe that P is not a projective subspace of P^C any more than E is a (complex) vector subspace of E^C. Observe also that $\sigma : P^C \to P^C$ takes collinear points into collinear points, and is, in fact, a semimorphism (cf. 5.4.8). Finally, P has the same real dimension as the complex dimension of P^C, just like E and E^C.

7.5.3. PROPOSITION. *Let E be a vector space. Given $f \in \mathrm{GP}(E)$, there exists a unique $f^C \in \mathrm{GP}(E^C)$ such that $f^C|_{P(E)} = f$.* $\square$

7.5.4. PROPOSITION. *Let P be a real projective space, $S \subset P$ a projective subspace. The (complex) projective subspace of P^C spanned by the subset S of P^C, denoted by S^C, satisfies $\sigma(S^C) = S^C$, $S = S^C \cap P$, and is called the complexification of S (in P^C). Conversely, let T be a projective subspace of P^C such that $\sigma(T) = T$. Then $T \cap P$ is a (real) projective subspace of P whose real dimension is the same as the complex dimension of T. Furthermore, $(T \cap P)^C = T$.* $\square$

7.5.5. Calculations are performed using homogeneous coordinates associated with a basis of E or a projective base of $P(E)$, which remain, respectively, a basis of E^C and a (complex) projective base of $P(E)$. It is enough to use complex coordinates instead of real ones. Again, this furnishes an *a posteriori* justification for 7.0.2.

7.6. Complexification of an affine space

7.6.1. Let $(X, \vec{X})$ be a real affine space. Embed X in its universal vector space $\hat{X}$ (cf. 3.1.7), and consider the inclusions $X \subset \hat{X} \subset \hat{X}^C$, where $\hat{X}^C$ denotes the complexification of the real vector space $\hat{X}$. Recall (cf. 3.1.7) that $X = M^{-1}(1)$, where M is a linear form over $\hat{X}$. By 7.2.1, there exists a well-defined linear form $\hat{X}^C \to \mathbf{C}$, the complexification of $M : \hat{X} \to \mathbf{R}$, which we *denote* by M^C. Finally, put

$$X^C = (M^C)^{-1}(1) \subset \hat{X}^C.$$

Then X^C is a (complex) affine hyperplane of the complex affine space $\hat{X}^C$ considered with its natural affine structure (cf. 2.2.1). In particular, X^C is a complex affine space.

7.6.2. DEFINITION AND PROPOSITION. *The complexification of the real affine space X is, by definition, the complex affine space $X^C = (M^C)^{-1}(1)$.*

The vector space underlying X^C is $\overrightarrow{X^C} = \vec{X}^C$, and we have a natural inclusion $X \subset X^C$. The conjugation σ of $\hat{X}^C$ leaves X^C invariant, and we have $X = \left\{ z \in X^C \mid \sigma(z) = z \right\}$. The restriction of σ to X^C, still denoted by σ, is called the conjugation of X^C. Vectorializing at an arbitrary $a \in X$, we have $(X^C)_a = (X_a)^C$, and a natural isomorphism $\hat{X}^C \cong \widehat{X^C}$.

Proof. The vector subspace underlying X^C is $(M^C)^{-1}(0)$, which is indeed the same as $\vec{X}^C$ because $M^{-1}(0) = \vec{X}$. Moreover,

$$\sigma(X^C) = \sigma\big((M^C)^{-1}(1)\big) = (M^C)^{-1}\big(\sigma(1)\big) = (M^C)^{-1}(1) = X^C$$

by 7.2.1. On the other hand, for $x \in X^C$, $\sigma(z) = z$ is equivalent to $z \in X^C \cap \hat{X} = X$. The equality $(X^C)_a = (X_a)^C$ follows from the last part of 3.1.7. As for the isomorphism $\hat{X}^C \cong \widehat{X^C}$, use the definition of universal spaces. $\square$

7.6.3. PROPOSITION. *Let X, X' be real affine spaces and $f \in A(X; X')$ a morphism. There exists a unique $f^C \in A_{\mathbf{C}}(X^C; X'^C)$ such that $f^C|_X = f$. This f^C is called the complexification of f. We have $\sigma \circ f^C = f^C \circ \sigma$.*

Proof. Apply 7.2.1 to the vectorializations X_a and $X'_{f(a)}$, and use the last result in 7.6.2. $\square$

7.6.4. For X a real affine space and S a subspace of X, one has a result analogous to 7.4.1, namely, $\vec{S}^C = \overrightarrow{S^C}$.

7.6.5. Similarly, one can extend 7.3 to polynomials on X; it suffices to apply the criterion given in 3.3.14, together with 7.6.2 and 7.3.4.

7.6.6. We can now discuss 7.0.2 as a whole, especially diagram 7.0.2.1, which is the central point of this chapter, and will be used in 9.5.5 and chapter 17. Let X be a real affine space and $\hat{X}$ its projective completion (cf. chapter 5):

$$\tilde{X} = P(\hat{X}) = X \cup P(\vec{X}) = X \cup \infty_X.$$

Introduce the affine spaces X^C, $\hat{X}^C$, and the complex projective space $(\tilde{X})^C = P(\hat{X}^C)$, the complexification of the real projective space $P(\hat{X}) = \tilde{X}$. Inside $(\tilde{X})^C$ is embedded the complexification ∞_X^C f the real subspace ∞_X (cf. 7.5.4). The inclusion $X^C \subset P(\hat{X}^C) = \tilde{X}^C$ gives the following result:

7.6.7. LEMMA. $X^C = \tilde{X}^C \setminus \infty_X^C$.

Proof. We have $\infty_X = P(\vec{X})$, thus $\infty_X^C = P(\vec{X}^C)$, and $\overrightarrow{X^C} = \vec{X}^C$ (cf. 7.6.4).
$\square$

7.6.8. PROPOSITION. *The inclusion* $X^C \subset P(\hat{X}^C) = \tilde{X}^C$ *is exactly the natural inclusion of* X^C *into its (complex) projective completion. In particular,*
$\tilde{X}^C = \widetilde{X^C}$.
$\square$

7.6.9. An application of 5.2.2 shows that, for any $f \in \mathrm{GA}(X)$, there exists a unique $\tilde{f}^C \in \mathrm{GP}_{\tilde{X}^C}(X^C)$ such that $\tilde{f}^C|_{X^C} = f^C$.

7.6.10. One can state similar results about subspaces.

7.6.11. Proposition 7.6.8 justifies *a posteriori* the constructions and the diagram in 7.0.2.

7.7. Exercises

7.7.1. State and prove a converse for the formula $\sigma \circ f^C = f^C \circ \sigma$ (7.2.1).

7.7.2. Given $f \in L(E; E')$, give conditions for the existence of an extension $\overline{f} : E^C \rightarrow E'^C$ of f such that $\overline{f}$ is semilinear with respect to conjugation in **C** (cf. 2.6.2).

7.7.3. How can 7.5.3 be extended to arbitrary morphisms from $P(E)$ into $P(E')$?

7.7.4. Fill in the details of 7.6.4, 7.6.5 and 7.6.10.

* **7.7.5.** Show that the tensor product $E \otimes_\mathbf{R} \mathbf{C}$ is a natural complexification of E, and so is $\mathrm{Hom}_\mathbf{R}(\mathbf{C}; E)$.

Chapter 8

Euclidean vector spaces

Our universe is very well approximated by a Euclidean affine space, or, more precisely, by a Euclidean space up to a multiplicative scalar, since there is no natural unit of length. Euclidean affine spaces are the main framework in this book, and we start studying them in chapter 9. Here we introduce this study by discussing Euclidean vector spaces, which underlie Euclidean affine spaces.

The treatment is largely algebraic, both in appearance and in content. Also, one must make a choice of what topics to cover, since the amount of material accumulated since the time of the Greeks is immense; in the interest of coherence, we have chosen a unifying thread, the orthogonal group $O(\cdot)$, and built around this (see a plan of study in 8.2.14).

After introducing the necessary definitions and discussing some generalities, in sections 8.1 and 8.2, we study the structure of individual elements of $O(E)$ and their eigenspaces (section 8.4) and the structure of $O(E)$ as a whole (sections 8.5, 8.3, 8.9). A detailed analysis is made possible in dimension two by the use of complex numbers (section 8.3), and in dimensions three and four by the use of quaternions (section 8.9). Section 8.10 is informational only, and is meant to bridge the gap between the basic material and current research. Section 8.7 is quite technical, dealing with oriented angles, a pretty fine invariant in the two-dimensional case which allows one to obtain nice results about circles (for instance, 10.9.7). The coarser notion of (non-oriented) angles between half-lines and between lines is more fundamental, and is developed in 8.6. Section 8.8, on similarities, is motivated, among other things, by the already mentioned fact that our physical universe is only Euclidean up to a scalar. Finally, section 8.11 introduces the classical cross product (useful in many areas) and the canonical volume form of a Euclidean vector space, which will be used to define the canonical measure of a Euclidean affine space.

If you are familiar with the way mathematicians have of generalizing everything, you will have suspected that one can generalize the notion of Euclidean vector spaces, by leaving aside some of their properties. We only discuss such generalizations very superficially (chapter 13); the interested reader can consult [DE1], [PO], [S–T], [BI2], [AN].

> Unless we mention otherwise, E is a Euclidean vector space, whose dimension we denote by n.

8.1. Definition and basic properties

8.1.1. DEFINITION. *A Euclidean vector space E is a finite-dimensional vector space of $\mathbf{R}$, together with a positive definite symmetric bilinear form ϕ (i.e., $\phi : E \times E \to \mathbf{R}$ is symmetric and bilinear, and $\phi(x, x) > 0$ for all $x \neq 0$). We write $\phi(x, y) = (x \mid y)$, and call this number the scalar product of x and y. The norm of x is, by definition, $\|x\| = \sqrt{\phi(x, x)} = \sqrt{(x \mid x)}$. If $(x \mid y) = 0$, we say that x and y are orthogonal. A set $\{e_i\}_{i=1,\ldots,k} \subset E$ is called orthogonal if $(e_i \mid e_j) = 0$ for all pairs $i \neq j$, and orthonormal if it is orthogonal and $\|e_i\| = 1$ for all i.*

8.1.2. NOTES.

8.1.2.1. A vector subspace F of a Euclidean vector space E has a natural Euclidean structure given by the symmetric bilinear form $\phi|_F$.

8.1.2.2. The standard example of a Euclidean vector space is $E = \mathbf{R}^n$, with

$$\phi\big((x_1, \ldots, x_n), (y_1, \ldots, y_n)\big) = \sum_{i=1}^{n} x_i y_i.$$

8.1.2.3. The map ϕ is a quadratic form on E (cf. 3.3.2), and one has, for every $x, y \in E$,

8.1.2.4
$$\begin{cases} \|x + y\|^2 = \|x\|^2 + 2(x \mid y) + \|y\|^2 \\[2mm] (x \mid y) = \tfrac{1}{2}(\|x + y\|^2 - \|x\|^2 - \|y\|^2). \end{cases}$$

8.1.2.5. An orthogonal set all of whose vectors are non-zero is linearly independent.

8.1.2.6. If $\{e_i\}$ is an orthonormal basis for E, the coefficients of the decomposition $x = \sum_i x_i e_i$ are given by $x_i = (x \mid e_i)$. Moreover,

$$\left(\sum_i x_i e_i \,\Big|\, \sum_j y_j e_j\right) = \sum_i x_i y_i.$$

8.1.3. PROPOSITION. *For every $x, y \in E$, we have $|(x \mid y)| \leq \|x\|\|y\|$. Equality holds if and only if x and y are linearly dependent. For every $x, y \in E$, we have $\|x + y\| \leq \|x\| + \|y\|$; in particular, the map $d : E \times E \to \mathbf{R}$ defined by $d(x, y) = \|x - y\|$ is a metric, and generates the canonical topology on E (cf. 2.7.1).* □

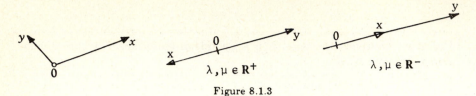

Figure 8.1.3

8.1.4. PROPOSITION (GRAM-SCHMIDT ORTHONORMALIZATION). *Let*
$\{b_i\}_{i=1,\ldots,k}$ *be a linearly independent set in* E. *There exists an orthonormal set* $\{a_i\}_{i=1,\ldots,k}$ *with the following properties:* $\{a_i\}_{i=1,\ldots,k}$ *is homotopic to* $\{b_i\}_{i=1,\ldots,k}$ *(cf. 2.7.2.7) and, for every* $h = 1,\ldots,k$, *the span of* $\{a_i\}_{i=1,\ldots,h}$ *is equal to the span of* $\{b_i\}_{i=1,\ldots,h}$. *In particular, given a (possibly empty) orthonormal set in* E, *one can always complete it into an orthonormal basis, and every basis is homotopic to an orthonormal one.*

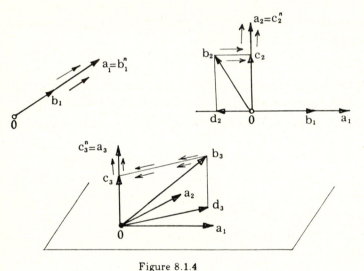

Figure 8.1.4

Proof. Use recurrence over k. For $k = 1$, take $a_1 = b_1^n = \|b_1\|^{-1}b_1$; here we have introduced the *notation* $x^n = \|x\|^{-1}x$, that is, x^n is the vector obtained by normalizing x. It is easy to see that x^n is homotopic to x: the homotopy can be taken as the segment from x to x^n, namely, $x(t) = tx^n + (1 - t)x$ for $t \in [0, 1]$.

Now assume we have found $\{a_i\}_{i=1,\ldots,h-1}$ with the desired properties. Put

$$d_h = \sum_{i=1}^{h-1} (a_i \mid b_h)a_i, \qquad c_h = b_h - d_h, \qquad a_h = c_h^n.$$

One verifies immediately that $\{a_i\}_{i=1,\ldots,h}$ satisfies our conditions, the homotopy being taken as a composition of which the two last steps lead from b_h to c_h via $b_h(t) = tc_h + (1 - h)b_h$, and from c_h to a_h as above. $\square$

8.1.5. PROPOSITION. *Let E, E' be Euclidean spaces of same dimension, and $f : E \to E'$ a map. The following conditions are equivalent:*

i) *$f \in L(E; E')$ and $\|f(x)\| = \|x\|$ for all $x \in E$;*

ii) *$\big(f(x) \mid f(y)\big) = (x \mid y)$ for all $x, y \in E$.*

Such a map is necessarily bijective and is called an isometry. The set of all such is denoted by $O(E; E')$.

Proof. (i) $\Rightarrow$ (ii) follows from linearity and 8.1.2.4. (ii) $\Rightarrow$ (i): take an orthonormal basis $\{e_i\}$ for E. By (ii) and the fact that E, E' have same dimension, $\big\{f(e_i)\big\}$ is an orthonormal basis for E'; now 8.1.2.6 and (ii) show that $f\big(\sum_i x_i e_i\big) = \sum_i x_i f(e_i)$, so that f is linear. $\square$

8.1.6. COROLLARY. *Every n-dimensional Euclidean vector space is isometric to $\mathbf{R}^n$ (cf. 8.1.2.2).* $\square$

8.1.7. Some take advantage of 8.1.6 to talk about *the* Euclidean space of dimension n. This is not a crime, as long as one knows what one's doing.

8.1.8. DUALITY.

8.1.8.1. Lemma. *The map $\flat : E \ni x \mapsto x^\flat = \big\{y \mapsto (x \mid y)\big\} \in E^*$, where E is a Euclidean vector space and E^* its dual as a real vector space, is a vector space isomorphism. The inverse map is denoted by $\sharp = \flat^{-1} : E^* \to E$.*

Proof. We know that $\dim E = \dim E^*$ (by considering dual bases, for example); since $\flat$ is linear, we only have to check that its kernel is zero. But if $(x \mid y) = 0$ for all y, in particular $(x \mid x) = \|x\|^2 = 0$, so $x = 0$. $\square$

8.1.8.2. Lemma 8.1.8.1 gives a natural Euclidean structure to E^*, and we will consider E^* with this structure when necessary.

The next result follows from 2.4.8.1 and 8.1.8.1:

8.1.8.3. Definition and proposition. *If $A \subset E$ is a non-empty subset, the set*

$$A^\perp = \big\{\, x \in E \mid (x \mid y) = 0 \text{ for all } y \in A \,\big\}.$$

is a vector subspace of E, called the orthogonal subspace to A. If $\langle A \rangle$ is the span of A, we have $A^\perp = \langle A \rangle^\perp$. If A is a subspace, we have a direct sum $A \oplus A^\perp = E$, and $\dim A + \dim A^\perp = \dim E$, $A^{\perp\perp} = A$. If A, B are two subspaces, we have $(A + B)^\perp = A^\perp \cap B^\perp$, $(A \cap B)^\perp = A^\perp + B^\perp$. Two subsets A, B of E are called orthogonal if $A \subset B^\perp$, or, equivalently, $B \subset A^\perp$; in this case we write $A \perp B$.

8.1.8.4. Notation. If $E = V \oplus W$ is a direct sum such that $W = V^\perp$, we talk about an *orthogonal direct sum*, and write $E = V \oplus^\perp W$. More generally, one can have an orthogonal direct sum of a family of subspaces:

$$E = V_1 \overset{\perp}{\oplus} \cdots \overset{\perp}{\oplus} V_k = \overset{\perp}{\underset{i}{\bigoplus}} V_i.$$

8.1.8.5. In the particular case dim $E = 3$, 8.1.8.3 recovers results from Euclidean geometry which are otherwise taken as axioms, for instance: the set of lines perpendicular to a given line at a point forms a plane. As you may have noticed, this result is quite important in practice: it is thanks to it that doors can open and close without theoretical difficulty. Another corollary: if two lines D, D' are respectively perpendicular to planes P, P' at the same point, the plane spanned by D, D' is perpendicular to the line $P \cap P'$.

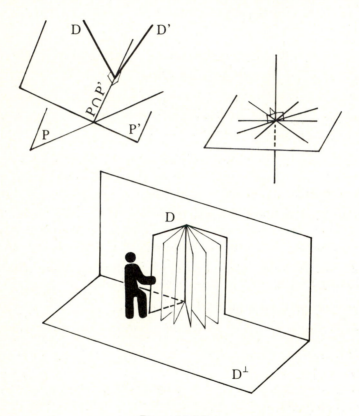

Figure 8.1.8.5

8.1.8.6. Lemma 8.1.8.1 allows one to define the *adjoint* endomorphism to $f : E \to E$, denoted by $^t f$. To do this, consider the *transpose* $f^* : E'^* \to E^*$, and set

$$^t f = \sharp \circ f^* \circ \flat : E \to E.$$

The adjoint satisfies

$$\left(^t f(x) \mid y \right) = \left(x \mid f(y) \right)$$

for all $x, y \in E$.

8.2. The orthogonal group: first properties and plan of attack

8.2.1. PROPOSITION. *The group $O(E) = O(E; E)$ is called the orthogonal group of E; we write $O(n) = O(\mathbf{R}^n)$. The condition $f \in O(E)$ is equivalent to $^tAA = I$, where A is the matrix of f in some (or any) orthonormal basis, ^tA is the transpose of A and I is the identity matrix. In particular, $\det f = \pm 1$; we set $O^+(E) = \{ f \in O(E) \mid \det f = 1 \}$ and $O^-(E) = \{ f \in O(E) \mid \det f = -1 \}$. The elements of $O^+(E)$ are called rotations. Finally, $f \in O(E)$ if and only if $^tff = \mathrm{Id}_E$ (cf. 8.1.8.6).*

This follows from the lemma below, which is a consequence of 8.1.2.6 and the definition of the product of two matrices:

8.2.2. LEMMA. *Let $\{e_i\}$ be an orthonormal basis for E, and A, B two square matrices with rank equal to the dimension of E. Then, letting x_k (resp. y_k) be the k-th column vector of A (resp. B), we have*

$$^tAB = ((x_i \mid y_j)). \qquad \square$$

8.2.3. REMARKS

8.2.3.1. If n is the dimension of E, we have an isomorphism $O(E) \cong O(n)$ by 8.1.6. Thus the study of orthogonal groups (at least as far as the linear group structure is concerned) can be restricted to the $O(n)$.

8.2.3.2. The subgroup $O^+(E)$ is normal in $O(E)$. We have the obvious multiplication rules:

$$O^-(E)O^-(E) \subset O^+(E), \qquad O^-(E)O^+(E) \subset O^-(E).$$

8.2.3.3. In the topology induced from $GL(E)$ (cf. 2.7.1), $O(E)$ is compact. In fact, take an arbitrary basis for the finite-dimensional vector space E. If we denote by $A(f)$ the matrix of f in this basis, the map $f \mapsto \left(\mathrm{Tr}(^tA(f)A(f)) \right)^{\frac{1}{2}}$ is the norm of a Euclidean structure over $GL(E)$. If, in addition, E is Euclidean and the basis taken is orthonormal, all elements of $O(E)$ have norm equal to $n = \dim E$ when expressed in this basis, by 8.2.1, and $O(E)$ is bounded in $GL(E)$. Finally $O(E)$ is closed since it is defined by the condition $^tAA = I$.

Another way of seeing that $O(E)$ is bounded is to assign elements of $GL(E)$ their norm as linear maps from the Euclidean space E into itself (watch out, the norm thus obtained in not Euclidean!). We have, for $f \in O(E)$,

$$\|f\| = \sup \{ \|f(x)\| \mid x \in E \text{ and } \|x\| = 1 \} = 1,$$

by the definition of $O(E)$.

8.2.4. The fact that $O(E) \subset GL(E)$ is compact has the following very useful converse:

8.2.5. THEOREM. *Let G be a compact subgroup of* $GL(E)$, *where E is a finite-dimensional real vector space. There exists on E at least one Euclidean structure ϕ such that $G \subset O(E)$, where $O(E)$ is the orthogonal group for this Euclidean structure.*

8.2.5.1. First proof. The shortest proof uses the existence of the Haar measure on a compact group (see, for example, [DE4, volume II, p. 225 ff.]). Let dg be the Haar measure on G, and let ϕ be an arbitrary Euclidean structure over E. For $g \in G$, define a new Euclidean structure $g^*\phi$ by $(g^*\phi)(x, y) = \phi\big(g(x), g(y)\big)$ for every $x, y \in E$, and then take the average of the $g^*\phi$ for the measure dg:

$$\overline{\phi} = \int_{g \in G} g^*\phi \, dg,$$

that is,

$$\overline{\phi}(x, y) = \int_{g \in G} \phi\big(g(x), g(y)\big) \, dg$$

for all x, y. It is obvious that $\overline{\phi}$ is still a Euclidean structure. Further, $\overline{\phi}$ is invariant under G because dg is invariant under translations of G. But saying that $g^*\overline{\phi} = \overline{\phi}$ for all $g \in G$ is the same as saying that G is contained in the orthogonal group of $(E, \overline{\phi})$.   □

8.2.5.2. Second proof. The second demonstration uses only Lebesgue measures on finite-dimensional real vector spaces, via 2.7.5.7. Consider the real vector space $P_2^\bullet(E)$ of quadratic forms over the n-dimensional real vector space E; this has dimension $n(n+1)/2$. In $P_2^\bullet(E)$, the subset $Q(E)$ of positive definite quadratic forms is an open convex cone, invariant under the action of G defined in 8.2.5.1. Starting from an arbitrary ϕ in $Q(E)$, the orbit $K = \{\, g^*\phi \mid g \in G \,\}$ of ϕ under G is compact in $P_2^\bullet(E)$, and is contained in $Q(E)$. In particular, its center $\mathrm{cent}(K)$ is in $Q(E)$ and is fixed under G by 2.7.5.7.   □

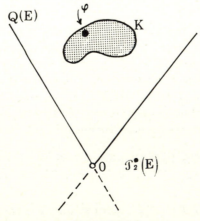

Figure 8.2.5.2

8.2.5.3. Third proof. We shall see in 11.8.10.8 a geometric proof for 8.2.5, based on convexity and not using measure theory at all. □

8.2.6. REMARKS. If G is finite, $G = \{ g_i \mid i \in I \}$, just take $\overline{\phi} = \sum_i g_i^* \phi$.

It is generally not true that there is only one Euclidean structure invariant under G; suffice it to take G containing only the identity! But the following nice criterion holds: G has a unique invariant Euclidean structure (up to a scalar) if and only if G is *irreducible* in $\mathrm{GL}(E)$, that is, there is no subspace of E (other than $\{0\}$ and E itself) left invariant by every $g \in G$. See 8.12.1.

For recent applications of 8.2.5, see, for example, [PA], [BO3, chapter VIII], and 8.12.1.

8.2.7. PROPOSITION. *Assume* $\dim E \geq 2$. *The group* $O^+(E)$ *acts transitively on the unit sphere* $S(E)$ *of* E, *given by* $S(E) = \{ x \in E \mid \|x\| = 1 \}$. *It also acts transitively on the set of* p-*dimensional grassmannians* $G_{E,p}$ *of* E ($0 \leq p \leq \dim E$). *The group* $O(E)$ *acts simply transitively on orthonormal bases of* E. *The group* $O^+(E)$ *acts simply transitively on homotopic orthonormal bases (cf. 2.7.2.7).* □

8.2.8. Proposition 8.2.7 allows us to write the standard n-dimensional unit sphere $S^n = S(\mathbf{R}^{n+1})$ as the homogeneous space

$$S^n = O(n+1)/O(n)$$

(cf. 1.5.5), where $O(n)$ is naturally embedded in $O(n+1)$ as the subgroup of $O(n+1)$ formed by maps f leaving the vector $(0,\ldots,0,1) \in \mathbf{R}^{n+1}$ invariant. Similarly, the grassmannians can be expressed as a homogeneous space

$$G_{n,p} = O(n)/\big(O(p) \times O(n-p)\big),$$

where $O(p)$ is the subgroup of $O(n)$ consisting of maps f leaving invariant the last $n - p$ vectors $e_{p+1}, \ldots, e_n$ of the canonical basis of $\mathbf{R}^n$, and $O(n-p)$ leaves invariant $e_1, \ldots, e_p$.

8.2.9. PROPOSITION. *Let* $E = S \oplus T$ *be a direct sum decomposition of* E, *and let* σ *be the reflection through* S *and parallel to* T *(cf. 6.4.6). Then* $\sigma \in O(E)$ *if and only if* $E = S \oplus^\perp T$ *(or, equivalently,* $T = S^\perp$*); in this case* σ *is denoted by* σ_S, *and is called an (orthogonal) reflection (or symmetry) of the Euclidean space* E. *Every reflection is an involution, and, conversely, every* $f \in O(E)$ *such that* $f^2 = \mathrm{Id}_E$ *is a reflection. We have* $\sigma_S \in O^+(E)$ *if* $\dim S^\perp$ *is even, and* $\sigma_S \in O^-(E)$ *if* $\dim S^\perp$ *is odd.* □

8.2.10. It is useful to write down the explicit formula for σ_H in the case when H is a hyperplane. Fix a non-zero vector x in $H^\perp$ (which has dimension 1); then

$$\sigma_H(y) = y - 2\frac{(y \mid x)}{\|x\|^2}.$$

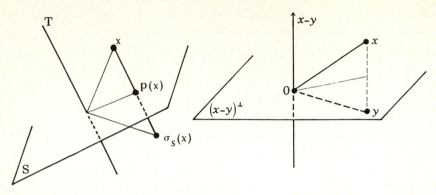

Figure 8.2.9

8.2.11. For $x, y \in E$ with $\|x\| = \|y\|$, there exists a hyperplane H such that $\sigma_H(x) = y$, and H is unique if $x \neq y$, $x \neq 0$. If $x = y$, every hyperplane containing x works, and if not $(x - y)^\perp$ is the only hyperplane that works. Observe that H is just the hyperplane equidistant from x and y, defined (cf. 9.7.5) as $H = \{\, z \in E \mid d(x, z) = d(y, z) \,\}$.

8.2.12. THEOREM. *Every $f \in O(E)$ is the product of at most $n = \dim E$ hyperplane reflections.*

Proof. It is enough to use recurrence: by 8.2.11, one can reduce f to a map fixing $x \neq 0$, and this map leaves invariant $S = x^\perp$, which has dimension strictly less than E. □

There is good reason to call 8.2.12 a theorem, in spite of the simplicity of its proof, because it is a fundamental result with many applications. If, instead of a Euclidean structure, we consider an arbitrary non-degenerate quadratic form on a finite-dimensional vector space over a commutative field, the result is still true, but its proof is much more difficult, and is, in fact, one of the central objectives of chapter 13.

8.2.13. COROLLARY. *If $\dim E = 1$, we have $O(E) = \mathrm{Id}_E \cup -\mathrm{Id}_E$. If $\dim E = 2$, every $f \in O^-(E)$ is a reflection through a line, and every $f \in O^+(E)$ is the product of two such reflections.* □

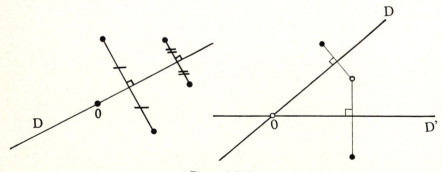

Figure 8.2.13

8.2.14. PLAN OF STUDY FOR $O(E)$.

Structure of an element of $O(E)$. In 8.4.5 and 8.4.6 we study the minimal number of reflections through hyperplanes (resp. codimension-two subspaces) which make up an arbitrary $f \in O(E)$ (resp. $O^+(E)$). In 8.2.15, we decompose $f \in O(E)$ into a product of isometries of one- or two-dimensional subspaces, and this leads to a study of $O(2)$ (since $O(1)$ is trivial, cf. 8.2.13). The essential fact here is that $O^+(2)$ is abelian.

Global structure of $O(E)$. Besides showing that $O^+(2)$ is abelian and studying the center of $O(E)$ and $O^+(E)$ (easy), we show in 8.5 that $O^+(n)$ is simple for $n = 3$ and $n \geq 5$. The exceptional $O(4)$, which is indeed non-simple, is studied in 8.9.10 with the help of quaternions. Quaternions are a fine instrument in the study of $O(3)$ and $O(4)$.

Topology and algebraic topology of $O(E)$. We shall see in 8.4.3 that $O^+(E)$ is path-connected, and thus that $O^+(E)$ and $O^-(E)$ are the two connected components of $O(E)$. In 8.10.3 we find that the fundamental group $\pi_1(O(n))$ is $\mathbf{Z}$ for $n = 2$ and $\mathbf{Z}_2$ for $n \geq 3$. These are fundamental facts in mathematics: $\pi_1(O(2)) = \mathbf{Z}$ is a cornerstone of complex analysis and geometry of plane curves, and $\pi_1(O(n)) = \mathbf{Z}_2$ has been the source of unexpected developments (cf. 8.10.3).

8.2.15. LEMMA. *Let $f \in O(E)$. There exists an orthogonal direct sum $E = \bigoplus_i^\perp P_i$ such that $f(P_i) = P_i$ and $\dim P_i = 1$ or 2 for all i.*

Proof. By 7.4.3 there exists a one- or two-dimensional subspace $P_1 \subset E$ such that $f(P_1) = P_1$. $\qquad\qquad\square$

8.2.16. THE CENTER OF $O(E)$ AND OF $O^+(E)$. Assume that $\dim E \geq 3$ (the two-dimensional case is studied in the next section) and that $f \in O(E)$ commutes with every $g \in O^+(E)$. Taking $g = \sigma_V$ to be a codimension-two reflection, we have $f(V) = V$ because g fixed every point of V. Now $f(V) = V$ for every codimension-two subspace V implies $f(D) = D$ for every vector line D, so f is a homothety, $f = \lambda \operatorname{Id}_E$ (cf. 4.5.3). Since $f \in O(E)$, we have $\lambda = \pm 1$ and $f = \pm \operatorname{Id}_E$. Thus:

8.2.17. PROPOSITION. *The center of $O(E)$ is $\{\operatorname{Id}_E, -\operatorname{Id}_E\}$. The center of $O^+(E)$ is $\{\operatorname{Id}_E\}$ if $\dim E$ is even, and $\{\operatorname{Id}_E, -\operatorname{Id}_E\}$ if $\dim E$ is odd.* $\qquad\square$

8.3. The two-dimensional case

In this section E is a Euclidean plane, non-oriented unless we say otherwise.

8.3.1. LEMMA. *Fixing an orthonormal basis for E and letting $M(f)$ be the matrix of f with respect to this basis, we have*

i) $f \in O^+(E) \iff M(f) = \begin{pmatrix} a & -b \\ b & a \end{pmatrix}$ and $a^2 + b^2 = 1,$

ii) $f \in O^+(E) \iff M(f) = \begin{pmatrix} a & b \\ b & -a \end{pmatrix}$ and $a^2 + b^2 = 1.$

Furthermore, for a fixed $f \in O^+(E)$, the real numbers a and $|b|$ do not depend on the basis.

Proof. Let $\mathcal{B} = \{e_1, e_2\}$ be the basis, and put $M(f) = \begin{pmatrix} a & c \\ b & d \end{pmatrix}$, that is, $f(e_1) = (a, b)$ and $f(e_2) = (c, d)$. By assumption, $f(e_2) \in \big(f(e_1)\big)^\perp$; but we have $\dim\big(f(e_1)\big)^\perp = 1$ and $(-b, a) \in \big(f(e_1)\big)^\perp$. Thus $f(e_2) = (-kb, ka)$ for some $k \in \mathbf{R}$; but $\|f(e_2)\| = \|f(e_1)\| = 1$ implies $a^2 + b^2 = 1$, so $k = \pm 1$. Since $\det f = k(a^2 + b^2)$, the case $k = -1$ corresponds to $f \in O^-(E)$, and the case $k = 1$ corresponds to $f \in O^+(E)$. Either way the scalar a is independent of the basis because $\operatorname{Tr} f = 2a$, and $|b|$ because $b^2 = 1 - a^2$. $\square$

8.3.2. REMARK. This proof shows that, given unit vectors e_1 and u, there exists a unique $f \in O^\pm(E)$ such that $f(e_1) = u$. For then $f(e_2)$ is determined up to sign, as it is contained in $\big(f(e_1)\big)^\perp$; one choice gives $f \in O^+(E)$, and the other $f \in O^-(E)$. This remark about simple transitivity also follows from 1.4.4.1.

8.3.3. THEOREM. *The set $O^-(E)$ consists of reflections through lines of E. The group $O^+(E)$ is abelian and acts simply transitively on the circle $S(E) = \{x \in E \mid \|E\| = 1\}$.*

Proof. The first assertion ensues from 8.2.13, simple transitivity in the second from 8.3.2. Commutativity follows from the calculation below:

8.3.4 $\begin{pmatrix} a & -b \\ b & a \end{pmatrix}\begin{pmatrix} a' & -b' \\ b' & a' \end{pmatrix} = \begin{pmatrix} aa' - bb' & -(ab' + a'b) \\ ab' + a'b & aa' - bb' \end{pmatrix}.$ $\square$

8.3.5. COROLLARY. *If E is oriented and we choose a positively oriented basis as in 8.3.1, b does not depend on the basis. For $f \in O^+(E)$ and $g \in O^-(E)$ we have $fg = gf^{-1}$. Every $f \in O^+(E)$ is the product of two reflections through lines; one of the two lines can be chosen arbitrarily.*

Proof. Let $\mathcal{B}, \mathcal{B}'$ be positive orthonormal bases and consider $f \in O^+(E)$. The matrices $M_{\mathcal{B}}(f)$ and $M_{\mathcal{B}'}(f)$ are similar, since $M_{\mathcal{B}'}(f) = \big(M_{\mathcal{B}}(t)\big)^{-1} M_{\mathcal{B}}(f) \cdot M_{\mathcal{B}}(t)$, where t is defined by $\mathcal{B}' = t(\mathcal{B})$. For $f \in O^+(E)$ and $g \in O^-(E)$, we have $fg \in O^-(E)$ by 8.2.3.2, hence fg is a reflection by 8.3.3. Reflections are involutive, so $fgfg = \mathrm{Id}_E$; since $g \in O^-(E)$ is also involutive, $fg = gf^{-1}$ follows. Finally, if $f \in O^+(E)$ and g is an arbitrary reflection in $O^-(E)$, one just writes $f = (fg)g$. $\square$

8.3.6. DEFINITION AND PROPOSITION. *The cosine function $\cos : O^+(E) \to \mathbf{R}$ is defined by $f \mapsto a$, where a is as in 8.3.1. If E is oriented, the sine function $\sin : O^+(E) \to \mathbf{R}$ is defined by $f \mapsto b$, where b is as in 8.3.5. In this case the map*

$$\Theta : O^+(E) \ni f \mapsto \cos f + i \sin f \in \mathbf{C}$$

is an isomorphism from $O^+(E)$ onto the group $\mathbf{U}$ of complex numbers with absolute value 1. The map Θ is also a homeomorphism for the induced topologies on $O^+(E) \subset \mathrm{GL}(E)$ and $\mathbf{U} \subset \mathbf{C}$.

Proof. Θ is a homomorphism by 8.3.4, and it is surjective since $a^2 + b^2 = 1$ and clearly injective. It is obviously continuous and $O^+(E)$ is compact, so it is also a homeomorphism. □

8.3.7. RECAP (see [FL, 181–183] or [CH2, 30], for example). *The map $\Lambda : \mathbf{R} \ni t \mapsto e^{it} \in \mathbf{C}$, arising from the complex exponential map $z \mapsto e^z = \sum_{n=0}^{\infty} \dfrac{z^n}{n!}$, is a homomorphism from the additive group $\mathbf{R}$ onto the multiplicative group $\mathbf{U}$. The kernel $\Lambda^{-1}(1)$ is of the form $\xi\mathbf{Z}$, where $\xi > 0$; the number $\xi/2$ is called π, so $\Lambda^{-1}(1) = 2\pi\mathbf{Z}$. The circle $\mathbf{U}$ is homeomorphic to the quotient topological space $\mathbf{R}/2\pi\mathbf{Z}$. We also call cosine and sine the maps $\mathbf{R} \to \mathbf{R}$ defined by $\cos t = \mathrm{Re}(e^{it})$ and $\sin t = \mathrm{Im}(e^{it})$, respectively.*

$$O^+(E) \xrightarrow{\ \Theta\ } \mathbf{U}$$

with $\Lambda : \mathbf{R} \to \mathbf{U}$ □

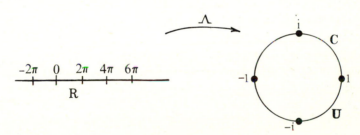

Figure 8.3.7

8.3.8. This implies, among other things, that $\mathbf{U}$, and thus the circle $S(E)$ for every Euclidean plane E, is path-connected, because $\mathbf{R}$ is. See also 8.12.4.

8.3.9. DEFINITION. *Take $f \in O^+(E)$, where E is oriented. A measure of f is an arbitrary element $\theta \in \Lambda^{-1}(\Theta(f))$ (cf. 8.3.6). If θ is a measure of f, every measure of f is of the form $\theta + 2k\pi$ for $k \in \mathbf{Z}$.*

8.3.10. If θ is a measure of f, the matrix of f in any positively oriented orthonormal basis is $\begin{pmatrix} \cos\theta & -\sin\theta \\ \sin\theta & \cos\theta \end{pmatrix}$, in the notation of 8.3.7.

8.3.11. NOTES. The restriction of the function cos to $[0, \pi]$ is a bijection onto $[-1, 1]$; its inverse is *called* Arccos. The restriction of sin to $[0, \pi]$ is not injective, but the restriction to $[0, \pi/2]$ is a bijection onto $[0, 1]$, and its inverse is *called* Arcsin.

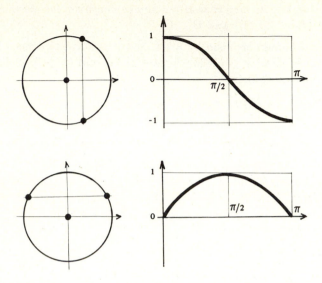

Figure 8.3.11

8.3.12. QUESTION AND ANSWER. One often identifies $\mathbf{R}^2$ with $\mathbf{C}$, but it would be nice, as suggested by 8.3.6, to identify an arbitrary oriented Euclidean plane E with $\mathbf{C}$ in an intrinsic way. This cannot be done, but it is possible to make E into a complex line using only the Euclidean structure of E and its orientation. In fact, let $\partial \in O^+(E)$ be a rotation of E having measure $\pi/2$, and define multiplication by complex scalars on E by

$$(\lambda + i\mu)x = \lambda x + \mu\partial(x).$$

Since $\partial^2 = -\operatorname{Id}_E$ and $\partial \in \operatorname{GL}(E)$, it follows that E is indeed a complex vector space with this multiplication and the preexisting vector addition.

We will encounter ∂ again in 8.7.3.5, this time without reference to measures of rotations. Observe that the elements of $O^+(E)$ correspond to the linear maps $z \mapsto az$ for $|a| = 1$, and (real) homotheties correspond to maps $z \mapsto bz$ for $b \in \mathbf{R}^*$.

8.3.13. NOTES. It may seem a waste to have to introduce complex exponentials in order to "measure" angles, but this measurement does in fact embed a fundamental difficulty, and it simply cannot be done without some cost. The reader can convince himself of this by consulting [FL, 178–186], [DE2, appendix I], [BI1, V, §2 and VIII, §2].

A much easier task is to show that every surjective continuous homomorphism $\mathbf{R} \to \mathbf{U}$ is of the form $t \mapsto \Lambda(kt)$, with $k \in \mathbf{R}^*$ (see [FL, 184]).

8.4. Structure of elements of $O(E)$. Generators for $O(E)$ and $O^+(E)$

8.4.1. PROPOSITION. *Take $f \in O(E)$. There exists an orthonormal basis of E in which the matrix of f has the form*

$$\begin{pmatrix} I_p & & & & \\ & -I_q & & 0 & \\ & & A_1 & & \\ & 0 & & \ddots & \\ & & & & A_r \end{pmatrix},$$

where I_p and I_q are the identity matrices of order p and q, and

$$A_i = \begin{pmatrix} \cos\theta_i & -\sin\theta_i \\ \sin\theta_i & \cos\theta_i \end{pmatrix},$$

with $\theta_i \in \mathbf{R} \setminus \pi\mathbf{Z}$ for all $i = 1, \ldots, r$.

Proof. This result follows from 8.2.15 and 8.3.10. $\qquad\square$

8.4.2. This decomposition is not unique, not only because the angles θ_i are only determined mod $2\pi\mathbf{Z}$, but, more importantly, because if some of the angles of the A_i are equal, the basis is not unique (see example in 18.8). On the other hand, the structure of the pieces I_p and $-I_q$ is well determined, both the numbers p and q and the elements of the basis that correspond to them being determined by the proper subspaces $\mathrm{Ker}(\mathrm{Id}_E - f)$ and $\mathrm{Ker}(\mathrm{Id}_E + f)$ corresponding to the eigenvalues 1 and -1. In particular, p, q and r depend only on f.

8.4.3. COROLLARY. *The group $O(E)$ has two connected components, $O^+(E)$ and $O^-(E)$, which are path-connected. If E is a finite-dimensional real vector space, the group $\mathrm{GL}(E)$ has two connected components, $\mathrm{GL}^+(E)$ and $\mathrm{GL}^-(E)$, which are path-connected.*

Proof. To show the latter assertion, endow E with an arbitrary Euclidean structure; by 8.1.4, it will be enough to show that $O^+(E)$ is path-connected. Take $f \in O^+(E)$, and apply 8.4.1, using the same notations. The integer q is even if and only if $\det f = 1$; a continuous path from f to Id_E in $O^+(E)$ can be defined by the matrices

$$\begin{pmatrix} I_p & & & & & & \\ & B_1(t) & & & 0 & & \\ & & \ddots & & & & \\ & & & B_{q'}(t) & & & \\ & & & & A_1(t) & & \\ & 0 & & & & \ddots & \\ & & & & & & A_r(t) \end{pmatrix},$$

where $t \in [0, 1]$ and

$$B_1(t) = \cdots = B_q(t) = \begin{pmatrix} \cos \pi t & -\sin \pi t \\ \sin \pi t & \cos \pi t \end{pmatrix}$$

$$A_i(t) = \begin{pmatrix} \cos \theta_i t & -\sin \theta_i t \\ \sin \theta_i t & \cos \theta_i t \end{pmatrix} \quad (i = 1, \ldots, r). \qquad \square$$

One can prove that $O^+(E)$ is path-connected using more elementary means, involving neither 8.4.1 or the measure of angles (whose definition, as we recall, resorted to the complex exponential, cf. 8.3.7). The problem is the same as finding a continuous path between two orthonormal bases $\mathcal{B}$, $\mathcal{B}'$ of E with same orientation. Take $\mathcal{B} = \{e_i\}$, $\mathcal{B}' = \{e_i'\}$; we will take e_1 into e_1' by a continuous deformation, and this will prove the result by recurrence, if we consider the orthogonal complement $e_1^\perp$.

Thus, assume that $e_1 \neq e_1'$, and let P be the plane containing e_1 and e_1'. The circle $S(P)$ is path-connected (cf. 8.3.8 and especially 8.12.4), and it is easy to extend a continuous path from e_1 to e_1' into a continuous path from $\mathcal{B}$ and $\mathcal{B}''$, where the first vector of $\mathcal{B}''$ is e_1'.

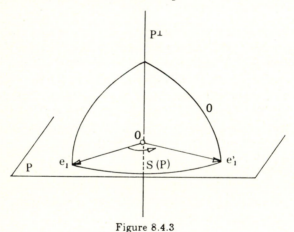

Figure 8.4.3

8.4.4. We will now find exactly the smallest number of hyperplane reflections into which one can decompose an arbitrary $f \in O(E)$ (cf. 8.2.12). We will also prove that codimension-two reflections generate $O^+(E)$ (just like hyperplane reflections generate $O(E)$), and will find the minimum number of such reflections in a decomposition of $f \in O^+(E)$.

8.4.5. PROPOSITION. *Let $f \in O(E)$, and put*

$$s = \dim E - \dim\big(\mathrm{Ker}(f - \mathrm{Id}_E)\big).$$

Then f can be written as a product of s, but no less than s, hyperplane reflections. If f is a product of s reflections through hyperplanes, the intersection of these hyperplanes is exactly $\mathrm{Ker}(f - \mathrm{Id}_E)$.

Proof. If $f = \sigma_{H_1} \cdots \sigma_{H_k}$, we have $H_1 \cap \cdots \cap H_k \subset \mathrm{Ker}(f - \mathrm{Id}_E)$, so $k \geq s$, and if $k = s$ this inclusion is actually an equality. Conversely, take $f \in O(E)$ and apply 8.4.1. The s hyperplane reflections that make up f are obtained as follows: for each basis vector $e_j \in \mathrm{Ker}(f + \mathrm{Id}_E)$, take the reflection through the hyperplane $e_j^{\perp}$; for each matrix A_i associated to a plane P_i, decompose the rotation of P_i into two reflections through lines of P_i (cf. 8.2.13), and extend these reflections into hyperplane reflections of E making them the identity on $P_i^{\perp}$. $\qquad\qquad\square$

8.4.6. PROPOSITION. *Let* $f \in O^+(E)$*, with* $\dim E \geq 3$*, and set*

$$s = \dim E - \dim\big(\mathrm{Ker}(f - \mathrm{Id}_E)\big).$$

Then f *is the product of* s *codimension-two reflections.*

Proof. This result is not used in this book, except in 8.5.3.1, and its proof is left as an exercise (cf. [FL, 193]), except in the case $\dim E = 3$, when it follows immediately from 8.4.7.1 and 8.2.13. $\qquad\qquad\square$

The proposition is false in dimension two, since then the only codimension-two reflection of E is $-\mathrm{Id}_E$, which is of course very far from generating $O^+(E)$.

8.4.7. EXAMPLES.

8.4.7.1. $O^+(E)$ *in dimension 3.* By 8.4.1, every $f \in O^+(E) \setminus \mathrm{Id}_E$ has a unique associated line D, called the axis of D; we say that f is a *rotation with axis D*. In fact, $f|_D \in O^+(D^{\perp})$, and $D^{\perp}$ is a plane. Such an f has also an associated angle (cf. 8.6) $\theta \in [0, \pi]$; if $\theta = \pi$, f is the reflection through D. The practical calculation of θ is accomplished by remarking that, by 8.4.1, the trace of f is $1 + 2\cos\theta$, and that the trace does not depend on the basis. Thus, in an arbitrary orthonormal basis, the matrix

$$\begin{pmatrix} 0 & 0 & 1 \\ 1 & 0 & 0 \\ 0 & 1 & 0 \end{pmatrix}$$

is a rotation by an angle $2\pi/3$ around the axis spanned by $(1, 1, 1)$; observe that this rotation has order three. The uniqueness of the axis of $f \in O^+(E)$ is used in an essential way in 1.8.3.1.

8.4.7.2. $O^-(E)$ *in dimension 3.* By 8.4.1, we see that every map $f \in O^-(E)$ is the composition of a hyperplane reflection σ_H (here $\dim H = 2$) and a rotation whose axis is the line $D = H^{\perp}$.

8.4.7.3. The first example, other than $O^+(2)$, where $f \in O(E)$ may not have any eigenvector occurs in dimension 4. In the notation of 8.4.1,

$$M(f) = \begin{pmatrix} \cos\theta_1 & -\sin\theta_1 & 0 & 0 \\ \sin\theta_1 & \cos\theta_1 & 0 & 0 \\ 0 & 0 & \cos\theta_2 & -\sin\theta_2 \\ 0 & 0 & \sin\theta_2 & \cos\theta_2 \end{pmatrix}$$

and $\theta_1, \theta_2 \notin \pi\mathbf{Z}$.

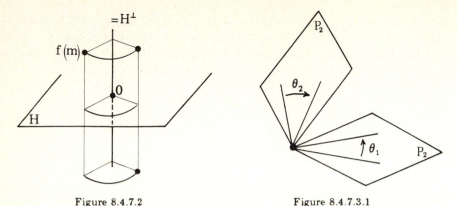

Figure 8.4.7.2   Figure 8.4.7.3.1

Two interesting cases, in some sense extreme ones, are the following: first, $\theta_1 = \theta_2$, which is exactly the example already met in 1.2.9 and 4.3.6.2, and which will occupy a whole section (18.8). Second, θ_1 and θ_2 are "incommensurable", in the classical Euclidean terminology; this means $\theta_1/\theta_2 \notin \mathbf{Q}$. Then the orbit of $m \in E \setminus 0$ under the action of the group $G = \{\, f^n \mid n \in \mathbf{Z} \,\}$ generated by f is an interesting object to study. Its closure is a differentiable submanifold of E homeomorphic to a torus (cf. 18.9). Figure 8.4.7.3.2 represents something closely related to the orbit of f, the orbit of the group of rotations

$$M(f) = \begin{pmatrix} \cos t\theta_1 & -\sin t\theta_1 & 0 & 0 \\ \sin t\theta_1 & \cos t\theta_1 & 0 & 0 \\ 0 & 0 & \cos t\theta_2 & -\sin t\theta_2 \\ 0 & 0 & \sin t\theta_2 & \cos t\theta_2 \end{pmatrix},$$

where t runs over $\mathbf{R}$ (and θ_1, θ_2 are fixed, with $\theta_1/\theta_2 \notin \mathbf{Q}$). Observe that the figure represents only the topological structure of the orbit and its closure, not the orbit itself in the four-dimensional space E (no wonder!).

Figure 8.4.7.3.2

8.5. Simplicity of $O(E)$

This is a typical example of "geometric algebra", the achievement of a purely algebraic result through geometric methods. We have met an example of this in 1.6.7.2 and the whole of chapter 13 is in this spirit; but these are but momentary reversals of the general policy of this book, which is using algebra when necessary as a tool to solve geometric problems.

We recall that a group G is called *simple* if it has no non-trivial normal subgroup (non-trivial means different from the identity and the whole of G).

8.5.1. THEOREM. *The group $O^+(3)$ is simple.*

Proof. It suffices to show that if $G \subset O^+(3)$ is a normal subgroup different from the identity, G contains at least one codimension-two reflection. In fact, if g is the reflection through D and $f \in O^+(3)$ is arbitrary, fgf^{-1} is the reflection through $f(D)$ and still belongs to G. But $O^+(3)$ acts transitively on lines of $\mathbf{R}^3$ (cf. 8.2.7), showing that G contains every codimension-two reflection and is thus equal to $O^+(3)$ by 8.4.6.

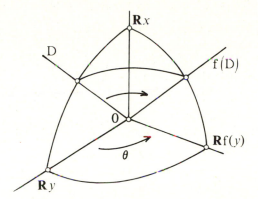

Figure 8.5.1

Now let $f \in G$ be arbitrary, $f \neq \mathrm{Id}_E$. By 8.4.7.1, f is a rotation with axis $\mathbf{R}x$, for $x \in \mathbf{R}^3 \setminus 0$, and angle $\theta \in {]0, \pi[}$ (if $\theta = \pi$, f is already a codimension-two reflection). Since $f^n \in G$, $n \in \mathbf{N}$, there exists n such that $n\theta \in [\pi/2, \pi[$, so we may as well assume $\theta \in [\pi/2, \pi]$. Now there exists a line D such that the two lines D, $f(D)$ are orthogonal; to see this, take coordinate axes $\{x, y, z\}$ in $\mathbf{R}^3$, where $\mathbf{R}x$ is the axis of f, y is an arbitrary vector orthogonal to x and $z = f(y)$. When D runs through all lines in the plane $\mathbf{R}x + \mathbf{R}y$ between $\mathbf{R}y$ and $\mathbf{R}x$, the angle between D and $f(D)$ varies between $\theta > \pi/2$ to 0, and thus must be equal to $\pi/2$ somewhere. Choose $g = \sigma_D$ as the reflection through D, and $h = gfg^{-1}f^{-1} \in G$; the action of h on $f(D)$ is given by

$$f(D) \ni m \overset{f^{-1}}{\mapsto} f^{-1}(m) \overset{g^{-1}}{\mapsto} f^{-1}(m) \overset{f}{\mapsto} m \overset{g}{\mapsto} -m \in f(D),$$

so that $h|_{f(D)} = -\mathrm{Id}_{f(D)}$, and by 8.4.7.1 h must be a reflection (through a line perpendicular to $f(D)$). □

8.5.2. NOTE. It may seem disappointing that we used properties of the real numbers, in particular the axiom for archimedean fields (to find n such that $n\theta \geq \pi/2$) to prove this result which is, after all, purely algebraic; but the truth is that we could not have avoided doing so. There is an example of an orthogonal group of a three-dimensional space with a positive definite quadratic form over a non-archimedean field which is not simple ([AN, 179 ff.]).

8.5.3. THEOREM. *The group $O^+(n)$, modded out by its center* (cf. 8.2.17), *is simple for every $n \geq 5$.*

8.5.3.1. *Proof.* Let G be a simple subgroup of $O^+(n)$, and take $f \in O^+(n) \setminus$ center. Our plan is to use f to construct $g \in G$ of the form $g' \oplus^\perp \mathrm{Id}_V$, with $\dim V = n-3$ and $g' \in O^+(V^\perp)$, that is, g acts as the identity on V and as an isometry on $V^\perp$. We have an orthogonal direct sum decomposition $V \oplus^\perp V' = \mathbf{R}^n$. By the proof of 8.5.1, we can find a codimension-two reflection of $V^\perp$. This reflection can be extended by the identity on V into a codimension-two reflection of $\mathbf{R}^n$; since this extension lies in G, we have $G = O^+(n)$ by 8.4.6.

8.5.3.2. There exists a plane P such that $f(P) \neq P$, otherwise every line of $\mathbf{R}^n$ would be invariant under f, and f would be in the center of $O^+(E)$ (cf. 8.2.16). Let $S = P + f(P)$ be the subspace spanned by P and $f(P)$; since $n \geq 5$, we have $S^\perp \neq \{0\}$. Consider the reflection $h = \sigma_{P^\perp}$ through $P^\perp$, and write $k = hfh^{-1}f^{-1}$. Then $k = \sigma_{P^\perp}\sigma_{f(P^\perp)}$, and in particular $k|_{S^\perp} = \mathrm{Id}_{S^\perp}$, so $k \in G \setminus$ center. Fix $x \in S^\perp \setminus 0$ and y such that $z = k(y) \neq y$. For every $u \in \mathbf{R}^n \setminus 0$, denote by $t_u = \sigma_{u^\perp}$ the hyperplane reflection through $u^\perp$. Then we have $l = t_y t_x \in O^+(n)$,

$$g = klk^{-1}l^{-1} \in G \setminus \text{center},$$

and

$$g = kt_y t_x k^{-1} t_x t_y = kt_y k^{-1} t_x t_x t_y = t_{k(y)} t_y = t_z t_y$$

(because $kt_x k^{-1} = t_{k(x)} = t_x$, so $kt_x = t_x k$). This $g = t_z t_y$ is of the desired form, $g = g' \oplus \mathrm{Id}_V$, with $\dim V^\perp = 3$ (in fact here $V^\perp \subset S$). $\qquad\square$

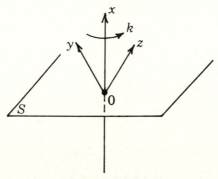

Figure 8.5.3.2

8.5.3.3. What happens for $O(4)$? The answer will be given in 8.9.10. The basic reference on the simplicity of the so-called classical groups is [DE1]; see also 13.6.8 and 13.7.14.

8.6. Angles between lines and half-lines

$$\boxed{\text{In this section dim } E \geq 2.}$$

This section discusses non-oriented angles, as opposed as oriented angles between lines or half-lines of a Euclidean plane, which are the topic of the next section. Non-oriented angles are defined in a Euclidean space of arbitrary dimension ≥ 2, possibly a plane. Lines here are vector lines; the affine case will be treated in the next chapter (cf. 9.2.1).

8.6.1. HALF-LINES AND ORIENTED LINES. Let E be an arbitrary real vector space. A *half-line* of E is a subset of E of the form $\mathbf{R}_+^* x$ for $x \in E \setminus 0$. The set of half-lines is *denoted by* $\tilde{D}(E)$. The set of vector lines of E, alias the grassmannian $G_{E,1}$, will be *denoted* here by $D(E)$. There is an obvious map $p : \tilde{D}(E) \rightarrow D(E)$, under which every point has two inverse images. Given $\Delta \in \tilde{D}(E)$, its *opposite* half-line $-\Delta$ is the other inverse image of $p(\Delta)$, that is $p(-\Delta) = p(\Delta) = \mathbf{R}x$ for any $x \in \Delta$.

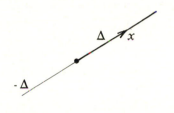 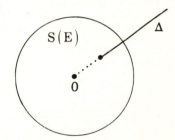

Figure 8.6.1

Recall that $D(E)$ is simply the projective space $P(E)$ (cf. 4.1.3.4). If E is a Euclidean space, $\tilde{D}(E)$ also has a nice interpretation as the unit sphere $S(E)$ of E, given by the bijection $\tilde{D}(E) \ni x \mapsto \dfrac{x}{\|x\|} \in S(E)$. Finally, $\tilde{D}(E)$ is equivalent to the space of oriented vector lines of E (cf. 2.7.2).

8.6.2. The definition below arises partly from the measurement of angles (cf. 8.3.7 and 8.3.11) and partly from 8.6.6:

8.6.3. DEFINITION. *Let E be a Euclidean vector space and*

$$\Delta, \Delta' \in \tilde{D}(E), \quad D, D' \in D(E).$$

i) *The scalar* $\dfrac{|(x \mid x')|}{\|x\|\|x'\|}$ *depends only on D and D', and not on the points $x \in D \setminus 0$ and $x' \in D' \setminus 0$. The (non-oriented) angle between D and D', denoted by $\overline{DD'}$ or $\overline{D, D'}$, is the real number $\in [0, \pi/2]$ defined as*

$$\overline{DD'} = \text{Arccos}\left(\frac{|(x \mid x')|}{\|x\|\|x'\|}\right)$$

for $x \in D \setminus 0$ and $x' \in D' \setminus 0$ (cf. 8.3.11).

ii) *The scalar* $\dfrac{(x \mid x')}{\|x\|\|x'\|}$ *depends only on* Δ, Δ', *and not on the points* $x \in \Delta \setminus 0$, $x' \in \Delta' \setminus 0$. *The (non-oriented) angle between* Δ *and* Δ', *denoted by* $\overline{\Delta\Delta'}$ *or* $\overline{\Delta, \Delta'}$, *is the real number* $\in [0, \pi]$ *defined by*

$$\overline{\Delta\Delta'} = \text{Arccos}\left(\frac{(x \mid x')}{\|x\|\|x'\|}\right).$$

8.6.4. In other terms,

$$\cos(\overline{DD'}) = \frac{|(x \mid x')|}{\|x\|\|x'\|}, \quad \cos(\overline{\Delta\Delta'}) = \frac{(x \mid x')}{\|x\|\|x'\|}.$$

By 8.3.11, we just have to verify that $\dfrac{|(x \mid x')|}{\|x\|\|x'\|} \in [-1, 1]$, and this follows from 8.1.3.

8.6.5. NOTES. From 8.1.3 we see that $\overline{DD'} = 0$ if and only if $D = D'$, and $\overline{DD'} = \pi/2$ if and only if D, D' are orthogonal. Analogously, $\overline{\Delta\Delta'} = 0$ if and only if $\Delta = \Delta'$, $\overline{\Delta\Delta'} = \pi$ if and only if $\Delta = -\Delta$, and $\overline{\Delta\Delta'} = \pi/2$ if and only if Δ and Δ' are orthogonal.

The value of angles is a *hereditary property*: If $F \subset E$ is a subspace containing Δ, Δ', the angle $\overline{\Delta\Delta'}$ is the same, whether measured in F or in E, and ditto for $\mathcal{D}(E)$.

Angles in $\tilde{\mathcal{D}}(E)$ or $\mathcal{D}(E)$ obviously satisfy two of the axioms for a metric; as we shall see in 9.9.8, chapter 18 and section 19.1, they satisfy the third as well. The metrics they define are, respectively, the intrinsic metric of the sphere $S(E)$ and the elliptic metric on the projective space $P(E)$.

8.6.6. PROPOSITION. *Consider the obvious action of $O(E)$ on $\tilde{\mathcal{D}}^2(E)$ and $\mathcal{D}^2(E)$ given by $g(\cdot, \cdot) = (g(\cdot), g(\cdot))$. Then*
  i) *under the action of $O(E)$, two pairs (D, D') and (D_1, D_1') belong to the same orbit if and only if $\overline{DD'} = \overline{D_1 D_1'}$, and two pairs (Δ, Δ') and (Δ_1, Δ_1') belong to the same orbit if and only if $\overline{\Delta\Delta'} = \overline{\Delta_1\Delta_1'}$;*
  ii) *under the action of $O^+(E)$, both statements in (i) hold if $\dim E \geq 3$, but for $\dim E = 2$ they are both false.*

In other words, the orbits of $O(E)$ in $\tilde{\mathcal{D}}^2(E)$ (resp. $\mathcal{D}^2(E)$) are parametrized by a scalar $\in [0, \pi]$ (resp. $\in [0, \pi/2]$), and for $\dim E \geq 3$ the orbits of $O^+(E)$ are the same as those of $O(E)$.

Proof. We choose to demonstrate the proposition for $\tilde{\mathcal{D}}(E)$; the case $\mathcal{D}(E)$ is no more difficult. By 8.2.7, we can assume that $\Delta, \Delta', \Delta_1$ and Δ_1' are all in the same plane of E, and also that $\Delta = \Delta_1$. Pick an orthonormal basis $\{x, y\}$ of this plane such that $x \in \Delta = \Delta_1$, and vectors $x' \in \Delta'$ and $x_1' \in \Delta_1'$ such that $\|x'\| = \|x_1'\| = 1$. By definition and the assumption that $\overline{\Delta\Delta'} = \overline{\Delta_1\Delta_1'}$, the two vectors x', x_1' have same x-coordinate; since their norm is 1, their

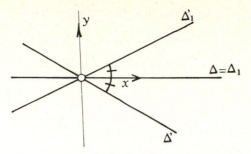

Figure 8.6.6

y-coordinates have same absolute value. If they are equal we have $\Delta' = \Delta'_1$ and we are done; if they are symmetric, the reflection σ_Δ satisfies

$$\sigma_\Delta(\Delta') = \Delta'_1.$$

This reflection of the plane can be extended into a codimension-two reflection of E if $E \geq 3$, proving part (i). As for part (ii), figure 8.6.6 gives the desired counterexample.   □

8.6.7. QUESTION. From the above, a real number in $[0, \pi/2]$ classifies equivalence classes of pairs of lines under isometries; a natural question is to classify pairs of subspaces (V, W) of E, where V and W have fixed dimensions.

When $\dim E = 3$, the classification by one parameter in $[0, \pi/2]$ works nicely. For $\dim V = \dim W = 2$, the parameter is the angle between the orthogonal complements $V^\perp$, $W^\perp$; for $\dim V = 1$ and $\dim W = 2$, one can take the infimum of $\overline{VD}$ for $D \subset W$, which is achieved when D is the orthogonal projection of V onto W.

In higher dimensions, starting with $\dim E = 4$ and $\dim V = \dim W = 2$, a single real parameter is no longer enough to classify the orbits of pairs (V, W) under the action of $O(E)$. For a complete discussion, see [FL, 310–316].

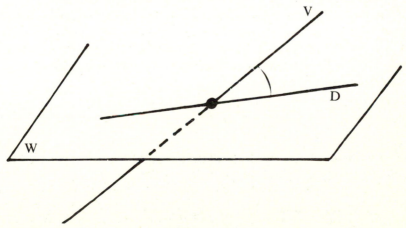

Figure 8.6.7

8.7. Oriented angles in the plane

$$\boxed{\text{In this section } \dim E = 2.}$$

8.7.1. It follows from 8.6.6 (ii) that angles (in $[0, \pi]$ or $[0, \pi/2]$) do not suffice to classify the orbits of $\tilde{D}^2(E)$ or $D^2(E)$ under $O^+(E)$ in dimension 2; a finer invariant is required. This finer invariant, the oriented angle between lines, is well worth the trouble of defining it, since it will allow us to derive quick and elegant results in later chapters (see for example 10.9).

It may appear that the machinery used to introduce something as simple as angles between lines is unnecessarily heavy, but there is simply no lighter or less cumbersome method, as the reader can verify by consulting [FL, 160–186] and comparing it with older texts like [I–R] and [D–C1], where all relations between oriented angles are formulated "mod $k\pi$" or "mod $2k\pi$". The gain in clarity in our treatment is obvious.

8.7.2. ORIENTED ANGLES BETWEEN ORIENTED LINES

8.7.2.1. Since $O^+(E)$ acts simply transitively on $\tilde{D}^2(E)$ (cf. 8.3.3), we can consider the map $\Phi : \tilde{D}^2(E) \to O^+(E)$. This map gives an equivalence relation on $\tilde{D}^2(E)$ (equivalent points have the same image). Another equivalence relation is obtained by considering the orbits (cf. 1.6) of $D^2(E)$ under $O^+(E)$. We make the following agreeable remark:

8.7.2.2. Lemma. *The two equivalence relations above coincide.*

Proof. The first identifies two pairs (Δ, Δ') and (Δ_1, Δ'_1) if there exists $g \in O^+(E)$ such that $\Delta' = g(\Delta)$ and $\Delta'_1 = g(\Delta_1)$; the second, if there exists $f \in O^+(E)$ such that $\Delta_1 = f(\Delta)$ and $\Delta'_1 = f(\Delta')$. The lemma follows from the commutativity of $O^+(E)$. □

Thus, calling $\sim$ either of these relations, we can consider the quotient $\tilde{A}(E) = \tilde{D}^2(E)/ \sim$, the canonical projection $\Phi : \tilde{D}^2(E) \to \tilde{A}(E)$ and the map $\underline{\Phi} : \tilde{A}(E) \to O^+(E)$ obtained from Φ by passing to the quotient. Since Φ is surjective, $\underline{\Phi}$ is by construction a bijection:

$$\begin{array}{ccc} \tilde{D}^2(E) & \overset{p}{\longrightarrow} & \tilde{A}(E) \\ {\scriptstyle \Phi}\searrow & & \swarrow{\scriptstyle \underline{\Phi}} \\ & O^+(E) & \end{array}$$

8.7.2.3. Definition. *We call $\tilde{A}(E)$ the set of oriented angles between oriented lines of E. This set has a group structure deriving from that of $O^+(E)$; we denote the group operation additively. The oriented angle between $\Delta, \Delta' \in \tilde{D}(E)$ is defined as $\Phi\big((\Delta, \Delta')\big)$, and denoted by $\widehat{\Delta\Delta'}$ or $\widehat{\Delta, \Delta'}$.*

We list a number of mostly trivial but useful properties of oriented angles:

8.7.2.4. Proposition. *If Δ and Δ' are arbitrary oriented lines,*

i) *$\Delta' = f(\Delta)$ with $f \in O^+(E)$ if and only if $\Phi(\widehat{\Delta\Delta'}) = f$; in particular,*

$$\widehat{\Delta\Delta'} = 0 \Longleftrightarrow \Delta = \Delta';$$

ii) *$\widehat{\Delta\Delta'} = \widehat{\Delta_1\Delta'_1}$ if and only if there exists $f \in O^+(E)$ such that $f(\Delta) = \Delta_1$ and $f(\Delta') = \Delta'_1$;*

iii) *$\widehat{\Delta\Delta'} = \widehat{\Delta_1\Delta'_1}$ if and only if $\widehat{\Delta\Delta_1} = \widehat{\Delta'\Delta'_1}$;*

iv) *$\widehat{\Delta\Delta'} + \widehat{\Delta'\Delta''} = \widehat{\Delta\Delta''}$ (Chasles's relation); in particular, $\widehat{\Delta\Delta'} = -\widehat{\Delta'\Delta}$;*

v) *$\widehat{f(\Delta)f(\Delta')} = \widehat{\Delta\Delta'}$ for any $f \in O^+(E)$, and $\widehat{f(\Delta)f(\Delta')} = -\widehat{\Delta\Delta'}$ for any $f \in O^-(E)$.*

Proof. The only property that does not follow from the definitions and 8.7.2.2 is the second statement in (v), and this follows from 8.3.5. $\qquad\square$

8.7.2.5. Remarks. We see that it is not necessary to orient E in order to define oriented angles; but in order to measure them, one does need to fix an orientation.

Observe the relation between some of the properties above and those for affine spaces, for example, Chasles' relation and the parallelogram rule $(\overrightarrow{ab} = \overrightarrow{cd} \Longleftrightarrow \overrightarrow{ac} = \overrightarrow{bd})$. These relations hold for every simply transitive abelian group.

8.7.3. BISECTORS. We look for solutions of the equation $2x = a$ in the group $\tilde{A}(E)$. We first solve the equation $2x = 0$, or, equivalently, $f^2 = \mathrm{Id}_E$ in $O^+(E)$. If $M(f) = \begin{pmatrix} a & -b \\ b & a \end{pmatrix}$ we know from 8.3.4 that $a^2 = 1$, so $a = \pm 1$ and $b = 0$. This implies $f = \pm \mathrm{Id}_E$. (This can also be seen using our knowledge about involutions, cf. 8.2.9.)

8.7.3.1. Proposition. *The equation $2x = 0$ has exactly two solutions, 0 and $\Phi^{-1}(-\mathrm{Id}_E)$. We denote this solution by ϖ, and call it the flat angle. The condition $\widehat{\Delta\Delta'} = \varpi$ is equivalent to $\Delta' = -\Delta$, where $-\Delta$ is the half-line opposite to Δ (cf. 8.6.1).* $\qquad\square$

This result 8.7.3.1 is also a consequence of what follows now.

8.7.3.2. Definition. *An oriented line Σ bisects the two oriented lines Δ and Δ' if $\widehat{\Delta\Sigma} = \widehat{\Sigma\Delta'}$. We say that Σ is a bisector of Δ and Δ'.*

By 8.7.2.4 (iv) and (v), an oriented line Σ bisects Δ and Δ' if the reflection σ_Σ through Σ (cf. 8.2.9) satisfies $\sigma_\Sigma(\Delta) = \Delta'$. But we know from 8.2.11 that there is a unique line supporting such a Σ, because an oriented line determines a unique positive unit vector. Thus Δ and Δ' have exactly two bisectors $\pm\Sigma$, opposite to each other. Further, $\widehat{\Delta\Sigma} = \widehat{\Sigma\Delta'}$ implies $\widehat{\Delta\Delta'} = 2\widehat{\Delta\Sigma}$, whence the following result:

8.7.3.3. Proposition. *Two oriented lines Δ, Δ' have exactly two bisectors, supported by the same line and opposite to each other. For every $a \in \tilde{A}(E)$, the equation $2x = a$ has exactly two solutions, of the form $\{b, b + \varpi\}$. If $\widehat{\Delta\Delta'} = a$, the solutions x of $2x = a$ are related to the bisectors Σ of Δ, Δ' by $x = \widehat{\Delta\Sigma}$.*

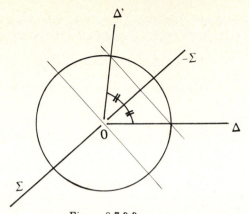

Figure 8.7.3.3

8.7.3.4. The equation $2x = \varpi$ is particularly interesting, because if $\widehat{\Delta\Delta'} = x$, where x satisfies $2x = \varpi$, the lines supporting Δ and Δ' are orthogonal, and conversely. One can see this algebraically or geometrically: the condition on the matrix $\left(\begin{smallmatrix} a & -b \\ b & a \end{smallmatrix}\right)$ is that $a = 0$, whence $b = \pm 1$; and, geometrically, the bisectors of two lines Δ, Δ' are orthogonal to Δ because the line D which supports Δ must be invariant under σ_Σ and distinct from D. Moreover, the solutions of $2x = \varpi$ are connected with orientations as follows:

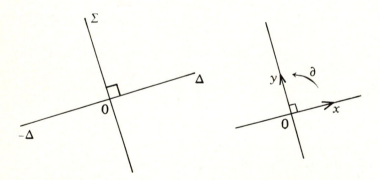

Figure 8.7.3.4   Figure 8.7.3.5

8.7.3.5. Proposition. *The solutions of $2x = \varpi$ are called the right angles of $\tilde{A}(E)$. Choosing one of these angles is equivalent to orienting E, as follows: if ∂ is the chosen angle, the orthonormal base $\{x, y\}$ is positively oriented if and only if $y = \underline{\Phi}(\partial)(x)$.*   □

Observe that ∂ here is indeed the same as in 8.3.12.

8.7.4. MEASURING ORIENTED ANGLES. We assume in this paragraph that the Euclidean plane E is oriented, and apply 8.3.6 and 8.3.9:

8.7.4.1. Definition. *The measure (or a measure) of an oriented angle $\alpha \in \tilde{A}(E)$ is a measure of $\underline{\Phi}(\alpha)$, i.e., an element of $\Delta^{-1}(\Theta(\underline{\Phi}(\alpha)))$:*

$$\tilde{A}(E) \xrightarrow{\Phi} O^+(E) \xrightarrow{\Theta} \mathbf{U} \xleftarrow[\Delta]{} \mathbf{R}$$

8.7.4.2. Examples. The real number π is a measure of ϖ, and $\pi/2$ is a measure of ∂. If t is a measure of α, every measure of α is of the form $t + 2k\pi$, where $k \in \mathbf{Z}$.

Using measures, one can show that the equation $nx = a$ in $\tilde{A}(E)$, where n is an integer ≥ 1 and $a \in \tilde{A}(E)$, has exactly n distinct solutions (cf. 8.12.7).

8.7.5. RELATIONSHIP BETWEEN $\overline{\Delta\Delta'}$ AND $\widehat{\Delta\Delta'}$. We now assume that E is oriented. Let Δ, Δ' be two oriented lines, seen here as half-lines (cf. 8.6.1). The angle $\widehat{\Delta\Delta'}$ always has a measure in the interval $[0, 2\pi[$. On the other hand, we can consider the basis $\{x, x'\}$, where $x \in \Delta$ and $x' \in \Delta'$ (excluding the case $\Delta' = \pm\Delta$); the orientation of this basis does not depend on the choice of x and x'.

8.7.5.1. Proposition. *Let t be the measure of $\widehat{\Delta\Delta'}$ that lies in $[0, 2\pi[$. Bases defined by (Δ, Δ') are positively (resp. negatively) oriented if and only if $t \in [0, \pi[$ (resp. $t \in [\pi, 2\pi[$), and then $\overline{\Delta\Delta'} = t$ (resp. $\overline{\Delta\Delta'} = 2\pi - t$), except for the trivial cases $\Delta = \Delta'$, $t = 0$ and $\Delta = -\Delta'$, $t = 2\pi - t = \pi$.*

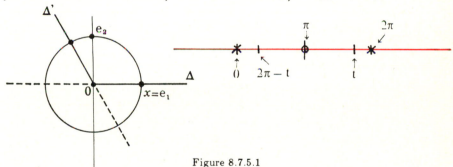

Figure 8.7.5.1

Proof. Leaving aside the trivial cases, we can assume $\Delta' \neq \pm\Delta$. Changing the orientation switches the sign of Δ', so t becomes $2\pi - t$; this reduces the proof to the case of positively oriented bases. More precisely, let $x \in \Delta$, $x' \in \Delta'$ be unit vectors, and $\{e_1 = x, e_2\}$ be the positively oriented orthonormal basis with first vector x. The key point is that the coordinate of x' along the e_2 direction is positive; but that coordinate is $\sin t$, so $t \in [0, \pi[$, and 8.3.11 completes the proof. $\square$

8.7.5.2. Definition. *Let $\Delta, \Delta', \Delta''$ be three half-lines of a (not necessarily oriented) Euclidean plane. We say that Δ' lies between Δ and Δ'' if one of the following conditions obtains: either $\Delta = \Delta' = \Delta''$, or $\Delta = -\Delta''$, or $\Delta' \neq \pm\Delta$ and Δ' belongs to the intersection of the half-plane defined by Δ and containing Δ'' with the half-plane defined by Δ'' and containing Δ.*

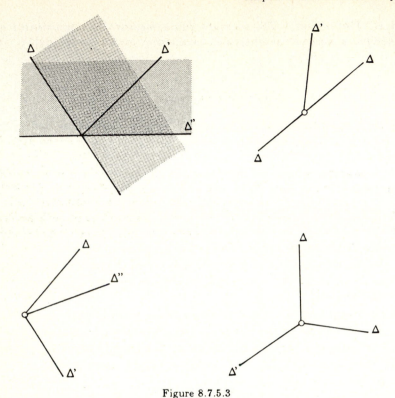

Figure 8.7.5.3

8.7.5.3. Corollary. *If Δ' is between Δ and Δ'', we have $\overline{\Delta\Delta'} + \overline{\Delta'\Delta''} = \overline{\Delta\Delta''}$.*

Proof. We only consider the cases when the three half-lines are distinct, the other being trivial. Choose the orientation of E in such a way that bases defined by (Δ, Δ') are positively oriented; the hypotheses show that the same is true of bases defined by (Δ', Δ'') or (Δ, Δ'') (excluding the case $\Delta'' = \Delta$). Let s, t be the measures of $\widehat{\Delta\Delta'}$ and $\widehat{\Delta'\Delta''}$ that lie in $[0, 2\pi[$; by 8.7.5.1, $s, t \in [0, \pi[$ and $s = \overline{\Delta\Delta'}$, $t = \overline{\Delta'\Delta''}$. But $s + t$ is a measure of $\Delta\Delta''$ and $s + t \in [0, 2\pi[$; if $\Delta'' = -\Delta$ we have $s + t = \pi$, and if $\Delta'' \neq -\Delta$ bases defined by (Δ, Δ'') are positively oriented, as remarked above, which proves the corollary by 8.7.5.1, since $s + t = \Delta\Delta''$. □

8.7.5.4. Corollary 8.7.5.3 is very important. It is a very intuitive result when interpreted in the following physical way: define the *sector* determined by Δ and Δ' as the set of half-lines lying between Δ and Δ' if $\Delta \neq \Delta'$, or either of the half-planes determined by the line supporting Δ, if $\Delta = \Delta'$. Then the non-oriented angle $\overline{\Delta\Delta'}$ is the measure of the sector determined by Δ and Δ', and if Δ' lies between Δ and Δ'' the sector determined by Δ and Δ'' is the union of the sectors determined by Δ, Δ' and Δ', Δ''.

Another approach starts from the *arc of circle* determined by two oriented lines, that is, the part of the unit circle $S(E)$ of E that intersects the corresponding sector. Then 8.7.5.3 says that the length of an arc of circle is the sum of the lengths of the two arcs into which it is divided. We shall come back (9.9.8) to this viewpoint, which originates in 8.6.5. More about sectors in [DE2, 65–67] and [FL, 183].

8.7.6. NOTES. The preceding discussion illustrates the difficulty in measuring angles, stemming from the fact that the union of two sectors is not always a sector (figure 8.7.5.3), and, consequently, one cannot add sectors arbitrarily (see also [FL, 185–186]).

There is an interesting construction for an order relation on $\tilde{A}(E) \setminus \varpi$: look at $\tilde{A}(E)$ as a circle (not intrinsically, cf. 8.3.6), and consider the stereographic projection of this circle onto $\mathbf{R}$ with pole ϖ (cf. 18.1.4). This gives a bijection between $\tilde{A}(E) \setminus \varpi$ and $\mathbf{R}$, and we can transfer to $\tilde{A}(E) \setminus \varpi$ the order relation on $\mathbf{R}$. For more details, see [FL, 176] or [DE2, 97–98]. One could also use the restriction of Δ to $]-\pi, \pi[$.

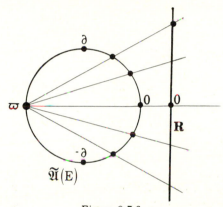

Figure 8.7.6

8.7.7. ORIENTED ANGLES BETWEEN LINES

8.7.7.1. Our object here is to extend the preceding discussion to lines and their set $\mathcal{D}^2(E)$. This presents no difficulties, so we restrict ourselves to a quick presentation of the results, without any demonstrations.

The starting point is either of the following two equivalent remarks: a line D determines exactly two opposite oriented lines (neither of them having precedence over the other, of course); and the group $O^+(E)$ does not act simply transitively on $\mathcal{D}^2(E)$, but the quotient group

$$\mathrm{PO}^+(E) = O^+(E)/\{\mathrm{Id}_E, -\mathrm{Id}_E\}$$

does. The name $\mathrm{PO}^+(E)$ stands for *projective orthogonal group*, because $\mathcal{D}(E) \cong P(E)$. We will discuss projective orthogonal groups in more generality in 14.7.2 and 19.1.

8.7.7.2. Following 8.7.2.1, we define $\Psi : \mathcal{D}^2(E) \to \mathrm{PO}^+(E)$ as the map canonically associated to the simply transitive action of the group $\mathrm{PO}^+(E) =$

$O^+(E)/\{\pm \operatorname{Id}_E\}$ on $\mathcal{D}^2(E)$. The equivalence relation $\sim$ is the same as the one given by the action of $\mathrm{PO}^+(E)$. The quotient $\mathcal{D}^2(E)/\sim$ is *denoted* by $\mathcal{A}(E)$, the canonical projection by $p : \mathcal{D}^2(E) \to \mathcal{A}(E)$, and the induced map $\mathcal{A}(E) \to \mathrm{PO}^+(E)$ by $\underline{\Psi}$. This map is a bijection:

$$
\begin{array}{ccc}
\mathcal{D}^2(E) & \xrightarrow{\;p\;} & \mathcal{A}(E) \\
{\scriptstyle\Psi}\searrow & & \swarrow{\scriptstyle\underline{\Psi}} \\
& O^+(E) &
\end{array}
\qquad
\begin{array}{ccccc}
\tilde{\mathcal{D}}^2(E) & \xrightarrow{\;p\;} & \tilde{\mathcal{A}}(E) & \xrightarrow{\;\Phi\;} & O^+(E) \\
{\scriptstyle q}\downarrow & & {\scriptstyle q}\downarrow & & {\scriptstyle p}\downarrow \\
\mathcal{D}^2(E) & \xrightarrow{\;p\;} & \mathcal{A}(E) & \xrightarrow{\;\underline{\Psi}\;} & \mathrm{PO}^+(E)
\end{array}
$$

Here $q : \tilde{\mathcal{D}}^2(E) \to \mathcal{D}^2(E)$ comes from the homonymous map $q : \tilde{\mathcal{D}}(E) \to \mathcal{D}(E)$ which takes an oriented line into the line supporting it.

   We call $\mathcal{A}(E)$ the group of *oriented angles* between lines of E, its abelian group structure being lifted from $\mathrm{PO}^+(E)$ via $\underline{\Psi}$. The *oriented angle* between the lines $D, D' \in \mathcal{D}(E)$ is the element $p\big((D, D')\big) \in \mathcal{A}(E)$, and is *denoted* by $\widehat{DD'}$ or $\widehat{D, D'}$.

8.7.7.3. We have the following relations:

$$\widehat{DD'} = \widehat{D_1 D_1'} \iff \widehat{DD_1} = \widehat{D'D_1'}; \qquad \widehat{DD'} = 0 \iff D = D';$$

$$\widehat{DD'} + \widehat{D'D''} = \widehat{DD''}; \qquad \widehat{D'D} = -\widehat{DD'};$$

$$f(\widehat{D})f(D') = \pm\widehat{DD'} \qquad \text{for } f \in O^\pm(E).$$

8.7.7.4. The equation $2x = 0$ in $\mathcal{A}(E)$ has two solutions, one being 0. The second, *denoted* by δ and called the *right angle*, is the image under q of the two right angles in $\tilde{\mathcal{A}}(E)$. The equation $2x = a$ in $\mathcal{A}(E)$ has always two solutions, of the form $\{x, x + a\}$; if $a = \widehat{DD'}$, the two lines S such that $x = \widehat{DS}$, for $2x = \widehat{DD'}$, are orthogonal, and called the *bisectors* of D, D'. They are characterized by $\widehat{DS} = \widehat{SD'}$, or $\sigma_S(D) = D'$.

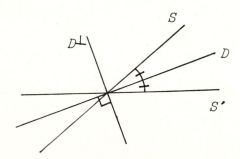

Figure 8.7.7.4

8.7.7.5. Let D, D', S, S' be lines such that the cross ratio $[D, D', S, S']$ in the projective line $P(E^*)$ is equal to -1. Saying that $D \perp D'$ is equivalent to saying that D and D' are the bisectors of S and S'.

8.7.7.6. In order to measure oriented angles between lines, we have to give an orientation to E. Define maps $\underline{\Theta}$ and $\underline{\Lambda}$ so the diagram below (cf. 8.3.6,

8.3.7, 8.7.4) commutes. By definition, a *measure* of $a \in \mathcal{A}(E)$ is an arbitrary element of $\underline{\Lambda}^{-1}(\Theta(\underline{\Psi}(a)))$. All measures of an element of $\mathcal{A}(E)$ are of the form $t + \pi k$, for $k \in \mathbf{Z}$; 0 is a measure of 0, and $\pi/2$ is a measure of δ.

$$\begin{array}{ccccccc}
\tilde{\mathcal{A}}(E) & \overset{\Phi}{\longrightarrow} & O^+(E) & \overset{\Theta}{\longrightarrow} & \mathbf{U} & \overset{\Lambda}{\longleftarrow} & \mathbf{R} \\
\downarrow{\scriptstyle q} & & \downarrow{\scriptstyle p} & & \downarrow{\scriptstyle p} & & \downarrow{\scriptstyle \mathrm{Id}_{\mathbf{R}}} \\
\mathcal{A}(E) & \overset{\Psi}{\longrightarrow} & PO^+(E) & \overset{\Theta}{\longrightarrow} & \mathbf{U}/\{\pm 1\} & \overset{\Lambda}{\longleftarrow} & \mathbf{R}
\end{array}$$

If E is oriented and $D, D' \in \mathcal{D}(E)$ are distinct lines, $\widehat{DD'}$ has a unique measure t in the interval $]0, \pi[$, and we either have $\overline{DD'} = t$ or $\overline{DD'} = \pi - t$ (cf. 8.6.6). If D and D' are not orthogonal, we can choose a unique orientation for E so that $t \in]0, \pi/2[$. Observe that there is no analogue for corollary 8.7.5.3 here; for example, if D, D', D'' are as in figure 8.7.7.6, we have $\overline{DD'} = \overline{D'D''} = \overline{DD''} = \pi/3$, and no matter what in order we consider the lines, we can never have $\overline{DD'} + \overline{D'D''} = \overline{DD''}$!

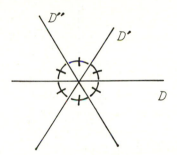

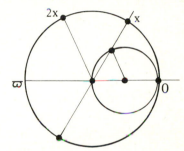

Figure 8.7.7.6   Figure 8.7.7.7

8.7.7.7. The proof of 4.3.6 hints that there is a natural bijection between $\tilde{\mathcal{A}}(E)$ and $\mathcal{A}(E)$. In fact, consider the map $\tilde{\mathcal{A}}(E) \ni x \mapsto 2x \in \tilde{\mathcal{A}}(E)$, with kernel $\{0, \varpi\}$. By 8.7.3.3, the map $\mathcal{A}(E) \to \tilde{\mathcal{A}}(E)$ induced on the quotient is a bijection which, by abuse of notation, is still written $\mathcal{A}(E) \ni x \mapsto 2x \in \tilde{\mathcal{A}}(E)$. Its inverse is written $\tilde{\mathcal{A}}(E) \ni x \mapsto \frac{1}{2}x \in \mathcal{A}(E)$.

Figure 8.7.7.7 gives a geometric interpretation for these maps, explaining why in trigonometry cosines, sines and tangents can be expressed as functions of the tangent of the half-arc.

8.7.7.8. Proposition. *Let D, D' be lines in E. Then*

$$\sigma_{D'} \circ \sigma_D = \underline{\Phi}(2\widehat{DD'}),$$

that is, $\sigma_{D'} \circ \sigma_D$ is a rotation by the angle $2\widehat{DD'}$.

Proof. One knows (8.2.3.2) that $f = \sigma_{D'} \circ \sigma_D \in O^+(E)$; we have $f(D) = \sigma_{D'}(D) = D''$ and $\widehat{DD''} = 2\widehat{DD'}$ by 8.7.7.4. Since $O^+(E)$ is simply transitive, the proposition is proved. $\qquad\square$

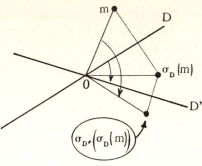

Figure 8.7.7.8

8.7.7.9. The maps between $O^+(E)$ and $\mathrm{PO}^+(E)$ corresponding to the doubling and halving of angles will be *denoted* by $\cdot^2$ and $\sqrt{\cdot}$, respectively.

8.7.8. TRIGONOMETRY. The relation $e^{z+z'} = e^z e^{z'}$ implies the following formulas:

8.7.8.1
$$\cos(a+b) = \cos a \cos b - \sin a \sin b,$$
$$\sin(a+b) = \sin a \cos b + \sin b \cos a,$$

which hold for all $a, b \in \mathbf{R}$. From 8.3.6 it follows that for E oriented the same formulas hold in $O^+(E)$, with the product fg of maps $f, g \in O^+(E)$ instead of the sum $a+b$.

8.7.8.2. From the definitions of the group structures of $\tilde{A}(E)$ and $A(E)$, it follows that, always assuming E to be oriented, formulas 8.7.8.1 hold without change in $\tilde{A}(E)$ and $A(E)$, where the cosine and sine of an oriented angle are defined as the cosine and sine of any of its measures.

8.7.8.3. One deduces from 8.7.8.1, through various algebraic concoctions and decoctions, a number of other formulas, some of which are very useful and nice; for a sizable list, see 8.12.8. For applications of such formulas, see 10.3.10 and 18.6.13.

8.7.8.4. Conversely, oriented angles can be used in the geometric study of complex numbers. Consider $\mathbf{C}$ as an oriented Euclidean plane, that is, $\mathbf{R}^2$ with its canonical Euclidean structure and orientation. For $z \in \mathbf{C} \setminus \{0\}$, we can write $z = |z|(z/|z|)$, where $z/|z| \in \mathbf{U}$. The oriented angle $(\Theta \circ \underline{\Psi})^{-1}(z/|z|) \in \tilde{A}(\mathbf{C}) = \tilde{A}(\mathbf{R}^2)$ is called the *argument* of the complex number z, and is *denoted* by $\arg(z)$. If $z = 0$ we put $\arg(z) =$ any element of $\tilde{A}(\mathbf{C})$. The number $z \in \mathbf{C}$ is determined by $|z|$ and $\arg(z)$, and the pair $\bigl(|z|, \arg(z)\bigr)$ is called the *polar form* of z. One has

8.7.8.5 $$|zz'| = |z||z'|, \quad \arg(zz') = \arg(z) + \arg(z').$$

In calculations it is ofter easier to use measures, that is, to write $z = |z|e^{it}$, where t is a measure of $\arg(z)$. This expression is also called a *polar form* of z, and is sometimes written in the form $z = |z|(\cos t + i \sin t)$. In 9.6.5.1 we will find some classical applications of this formula; see also 12.4.2.

8.8. Similarities. Isotropic cone and lines

8.8.1. We return to arbitrary-dimensional Euclidean spaces E. A similarity of E is a map that preserves not lengths, but ratios between lengths. The interest of this notion lies not only in the nice applications it gives rise to (9.6.7, 9.6.8, 9.6.9), but also in its connection with our physical universe which, not having a privileged unit of length, is Euclidean only up to an arbitrary scalar constant.

8.8.2. DEFINITION. *A similarity of E, of ratio μ, is a map $f \in \mathrm{GL}(E)$ satisfying $\|f(x)\| = \mu\|x\|$ for all $x \in E$. The set of similarities is denoted by $\mathrm{GO}(E)$; orientation-preserving (resp. reversing) similarities are the elements of $\mathrm{GO}^+(E) = \mathrm{GO}(E) \cap \mathrm{GL}^+(E)$ (resp. $\mathrm{GO}^-(E) = \mathrm{GO}(E) \cap \mathrm{GL}^-(E)$).*

8.8.3. A homothety $\lambda\,\mathrm{Id}_E$ is a similarity, of ratio $\mu = |\lambda|$. If $\lambda < 0$, $\lambda\,\mathrm{Id}_E$ preserves orientation if and only if $\dim E$ is even. If $f \in \mathrm{GO}(E)$ has ratio μ, we have $\big(f(x) \mid f(y)\big) = \mu^2(x \mid y)$ for all $x, y \in E$; to see this, write f as $f = (\mu^{-1}f) \circ (\mu\,\mathrm{Id}_E)$, and apply 8.1.5 to $\mu^{-1}f \in \mathrm{GL}(E)$. This also shows that $\mathrm{GO}(E)$ is isomorphic to the direct product of groups $\mathrm{GO}(E) \cong O(E) \times (\mathbf{R}_+^* \,\mathrm{Id}_E)$; similarly, $\mathrm{GO}^+(E) \cong O^+(E) \times (\mathbf{R}_+^* \,\mathrm{Id}_E)$. For $n = 1$, we evidently have $\mathrm{GO}(E) = \mathrm{GL}(E)$; this trivial case will often be excluded in what follows.

8.8.4. PLANE SIMILARITIES AND COMPLEX NUMBERS. Let E be an oriented Euclidean plane, which we identify with a complex vector line as in 8.3.12. The elements of the complex linear group $\mathrm{GL}_{\mathbf{C}}(E)$ of E satisfy definition 8.8.2, so $\mathrm{GL}_{\mathbf{C}}(E) \subset \mathrm{GO}(E)$. In fact, one has the equality $\mathrm{GO}^+(E) = \mathrm{GL}_{\mathbf{C}}(E)$, as shown by 8.8.3 and 8.3.12. To recover $\mathrm{GO}^-(E)$, fix an arbitrary $s \in O^-(E)$ and write $\mathrm{GO}^-(E) = s\,\mathrm{GO}^+(E)$. Since s is reflection through some line, we can take a basis of E lying on this line (a complex basis of E has a single element!) and deduce the following

8.8.4.1. Proposition. *In an arbitrary complex basis of E, an orientation-preserving (resp. reversing) similarity of E is a map $z \mapsto az$ (resp. $z \mapsto a\bar{z}$), where a is any element of $\mathbf{C}^*$.*

8.8.5. SIMILARITIES AND ANGLES

8.8.5.1. Proposition. *Similarities preserve angles. More precisely, for $\Delta, \Delta' \in \tilde{\mathcal{D}}(E)$ and $D, D' \in \mathcal{D}(E)$, we have:*

i) *if $f \in \mathrm{GO}(E)$, then $\overline{f(\Delta)f(\Delta')} = \overline{\Delta\Delta'}$ and $\overline{f(D)f(D')} = \overline{DD'}$;*

ii) *if $\dim E = 2$ and $f \in \mathrm{GO}^+(E)$ (resp. $\mathrm{GO}^-(E)$), then $\widehat{f(\Delta)f(\Delta')} = \widehat{\Delta\Delta'}$ (resp. $-\widehat{\Delta\Delta'}$) and $\widehat{f(D)f(D')} = \widehat{DD'}$ (resp. $-\widehat{DD'}$);*

iii) *if $f \in \mathrm{GO}(E)$, the conditions $D \perp D'$ and $f(D) \perp f(D')$ are equivalent.*

Conversely, if $\dim E \geq 2$ and $f \in \mathrm{GL}(E)$ is such that $f(D)$ and $f(D')$ are orthogonal whenever $D, D' \in \mathcal{D}(E)$ are, f is a similarity.

Proof. The only non-trivial part is the converse, since parts (i), (ii) and (iii) follow from 8.6.6, 8.7.2.4 and 8.7.7.3. If f preserves orthogonal lines, we have $\big(f(x)\mid f(x')\big) = 0$ for all x, x' such that $(x\mid x') = 0$. Start by fixing $x \in E \setminus 0$ and consider the linear form $\phi : y \mapsto \big(f(x)\mid f(y)\big)$; this form vanishes on the hyperplane $x^\perp$ by assumption, so there exists by 2.4.8.6 a scalar $k(x) \in \mathbf{R}^*$ such that

$$\big(f(x)\mid f(y)\big) = k(x)(x\mid y)$$

for all $y \in E$. To show that $k(x)$ does not depend on x, consider $k(x + x')$, where x, x' are linearly independent; we have

$$k(x+x')(x+x'\mid y) = \big(f(x+x')\mid f(y)\big) = \big(f(x)\mid f(y)\big) + \big(f(x')\mid f(y)\big)$$
$$= k(x)(x\mid y) + k(x')(x'\mid y)$$

for every $y \in E$, whence $\big(k(x+x') - k(x)\big)x + \big(k(x+x') - k(x')\big)x' = 0$. By linear independence, this implies $k(x) = k(x') = k(x+x')$, and it follows that k is a constant and f is a similarity. $\qquad\square$

8.8.6. SIMILARITIES AND COMPLEXIFICATIONS. Let E^C be the complexification of E, $f^C \in \mathrm{GL}(E^C)$ the complexification of $f \in \mathrm{GL}(E)$, and N^C, a quadratic form on E^C, the complexification of the quadratic form $N : x \to \|x\|^2$ on E (cf. 7.1.1, 7.2.1, 7.3.4).

8.8.6.1. Definition. *The subset $(N^C)^{-1}(0)$ of E^C is called the isotropic cone of E if $\dim E \geq 3$. If $\dim E = 2$, this set consists of two distinct lines of E^C, called the isotropic lines of E; this pair of lines is denoted by $\{I, J\}$. The conjugation σ on E^C (cf. 7.1.1) satisfies $\sigma\big((N^C)^{-1}(0)\big) = (N^C)^{-1}(0)$. For $\dim E = 2$, conjugation permutes I and J.*

The word cone is justified because, from the homogeneity of N, we have $N^C(\lambda x) = \lambda^2 N^C(X)$. The assertion about $(N^C)^{-1}(0)$ consisting of two distinct lines follows from the non-degeneracy of N, because then N^C is also non-degenerate (cf. 7.3.5, 13.2.1, 13.2.3.1). We give a direct proof, that does not depend on 13.2: let $\{e_1, e_2\}$ be an orthonormal basis for E; then

$$N^C(z_1, z_2) = z_1^2 + z_2^2$$

(cf. 7.3.5), and $z_1^2 + z_2^2 = 0$ is equivalent to $z_1 = \pm i z_2$. The other assertions are immediate:

$$\sigma\big((N^C)^{-1}(0)\big) = (N^C)^{-1}\big(\sigma(0)\big) = (N^C)^{-1}(0)$$

by 7.2.1, and $\sigma(I) = J$, $\sigma(J) = I$ by taking, $I = \big\{(z, -iz)\big\}$ and $J = \big\{(z, iz)\big\}$, for example. We say "for example" because there is no distinction between the two isotropic lines of a Euclidean plane; more precisely:

8.8.6.2. Proposition. *Picking out one of the isotropic lines of the Euclidean plane E is equivalent to orienting E. The correspondence is given by choosing the orientation of E with respect to which the chosen line I has slope $-i$ in an orthonormal basis.*

Proof. We must show that $\mathrm{GO}^+(E)$ leaves I fixed. Intuitively, this comes from the fact that $\mathrm{GO}^+(E)$ is connected, so I cannot suddenly turn into J. Calculating in an arbitrary positively oriented orthonormal basis and putting $M(f) = \left(\begin{smallmatrix} a & -b \\ b & a \end{smallmatrix}\right)$, we have, for $\epsilon = \pm 1$:

$$\begin{pmatrix} a & -\epsilon b \\ b & \epsilon a \end{pmatrix} \begin{pmatrix} 1 \\ -1 \end{pmatrix} = \begin{pmatrix} a + \epsilon ib \\ b - \epsilon ia \end{pmatrix} = (a + \epsilon ib)\begin{pmatrix} 1 \\ -\epsilon i \end{pmatrix}. \qquad \square$$

8.8.6.3. Note. This calculation shows also that the eigenvalue of I (resp. J) under $f \in O^+(E)$ is equal to $a + bi = \Theta(f)$ (resp. $a - bi = \overline{\Theta(f)}$), where Θ is as in 8.3.6. Also, by 8.3.1 (ii), we have $f^C(I) = J$ for $f \in O^-(E)$.

8.8.6.4. Proposition. *Let* $f \in \mathrm{GL}(E)$. *Then*
  i) $f \in \mathrm{GO}(E)$ *if and only if* $f^C\big((N^C)^{-1}(0)\big) = (N^C)^{-1}(0)$, *that is, if and only if* f^C *leaves the isotropic cone of* E *globally invariant;*
  ii) *if* $\dim E = 2$,

$$f \in \mathrm{GO}^+(E) \Longleftrightarrow f^C(I) = I, \qquad f \in \mathrm{GO}^-(E) \Longleftrightarrow f^C(I) = J.$$

Proof. The implications $\Rightarrow$ are trivial. Assume that f^C leaves invariant the nullspace of N^C; then the inverse image $(f^C)^*(N^C)$ of N^C under f^C (cf. 13.1.3.9) is a quadratic form with same nullspace as N^C. By 14.1.6.2, this implies that there exists $k \in \mathbf{C}^*$ such that $(f^C)^*(N^C) = kN^C$; in particular, restricting to $E \subset E^C$, we get $\|f(x)\|^2 = k\|x\|^2$ for all $x \in E$. This shows that k is real and f is a similarity of ratio $\sqrt{k}$, proving (i). For (ii), observe first that $f(\{I, J\}) = \{I, J\}$ by 8.8.6.1; thus we can apply (i) and observe that if $f \in \mathrm{GO}^+(E)$ (resp. $f \in \mathrm{GO}^-(E)$) we have $f(I) = I$ (resp. $f(I) = J$) by 8.8.6.2 and 8.8.6.3. $\qquad \square$

8.8.7. LAGUERRE'S FORMULA. This result is the fruit of the excogitations of Laguerre, who, not satisfied with his college teacher's geometry course, took to himself to clarify it.

8.8.7.1. Assume first that E is an oriented Euclidean plane and (I, J) are its isotropic lines in the order specified in 8.8.6.2. Let D and D' be lines in E, and D^C and D'^C their complexifications in E^C (cf. 7.4).

In the complex projective line $P(E^C) \cong \mathcal{D}(E^C)$, it is natural to consider the cross-ratio of the four points D^C, D'^C, I, J. Let $f \in O^+(E)$ be such that $f(D) = D'$; by 6.6.3 and 8.8.6.3, we have

$$[D^C, D'^C, I, J] = \Theta(f)/\overline{\Theta(f)}.$$

But since $\Theta(f) \in \mathbf{U}$, we have $\overline{\Theta(f)} = \big(\Theta(f)\big)^{-1}$; thus an application of 8.7.5.1 and 8.7.7.7 gives an expression of $[D^C, D'^C, I, J]$ in terms of oriented angles only:

8.8.7.2. Theorem (Laguerre's formula). *We have*

$$[D^C, D'^C, I, J] = \big[\Theta(\Psi(\widehat{DD'}))\big]^2,$$

and, if t *is an arbitrary measure of* $\widehat{DD'}$, *we have* $[D^C, D'^C, I, J] = e^{2it}$. $\quad \square$

For conscience's sake, we verify that the formula behaves as it should when the orientation of E is switched. In fact, I and J are permuted by 8.8.6.4, $\Theta(\cdot)$ becomes $\left(\Theta(\cdot)\right)^{-1}$ by 8.3.6, and correct behavior is assured by 6.3.1. As for switching D and D', apply 8.7.7.3 and 6.3.1. For applications, see 17.4.2.2.

8.8.7.3. Assume now that E is not oriented; the lines I and J are indistinguishable. For two lines D, D' in E, the cross-ratio $[D^C, D'^C, I, J]$ can still be defined up to inversion in $\mathbf{C}$. Define the bijection $\log : \mathbf{U} \setminus \{-1\} \to \,]-\pi, \pi[$ as the inverse of the restriction of Λ to $\,]-\pi, \pi[$. Let t be a measure of $\widehat{DD'}$ for an arbitrary orientation of E; by 8.7.7.6, $\widehat{DD'}$ is equal to t or $\pi - t$, and in either case we have

8.8.7.4 $$\overline{DD'} = \frac{1}{2}\left|\log([D^C, D'^C, I, J])\right|,$$

where the absolute value serves exactly to neutralize the ambiguity between $\log(e^{2it})$ and $\log(e^{2i(\pi-t)}) = \log(e^{-2it}) = -\log(e^{2it})$.

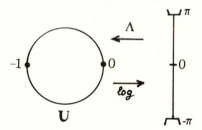

Figure 8.8.7.3

8.8.7.5. Formula 8.8.7.4, due to Cayley, is a good illustration of his principle (cf. 5.2.2): it displays the geometry of angles between lines of E, or, equivalently, the geometry of the metric of the projective space $P(E)$, as a subgeometry of an appropriately chosen projective space (cf. 19.2).

8.9. Quaternions and orthogonal groups

Complex numbers played a fundamental role in the study of $O^+(2)$; similarly, the skew field $\mathbf{H}$ of quaternions will allow a fine study of $O^+(3)$, and especially $O^+(4)$.

8.9.1. REVIEW OF QUATERNIONS. The *quaternions* form a skew field, *denoted* by $\mathbf{H}$, and constructed by giving $\mathbf{R}^4$ a multiplication defined in the following fashion:

$$1u = u \quad (u = 1, i, j, k); \qquad i^2 = ii = j^2 = jj = k^2 = kk = -1;$$
$$ij = -ji = k; \qquad jk = -kj = i; \qquad ki = -ik = j.$$

Here $\{1, i, j, k\}$ denotes the canonical basis of $\mathbf{R}^4$, and the multiplication is extended to the whole of $\mathbf{R}^4$ by linearity. This operation is associative, but not commutative. The subspace $\mathbf{R} \cdot 1$ is identified with $\mathbf{R}$, and the subspace

$$\mathbf{R}i + \mathbf{R}j + \mathbf{R}k \subset \mathbf{H}$$

is identified with $\mathbf{R}^3$ and called the set of *pure quaternions*. Every quaternion $q \in \mathbf{H}$ can be written $q = \mathcal{R}(q) + \mathcal{P}(q)$, where $\mathcal{R}(q) \in \mathbf{R}$ and $\mathcal{P}(q) \in \mathbf{R}^3$. By definition, the *conjugate* $\bar{q}$ of q is the quaternion $\mathcal{R}(q) - \mathcal{P}(q)$. One has

$$\bar{\bar{q}} = q, \qquad q = \bar{q} \Longleftrightarrow q \in \mathbf{R}, \qquad \bar{q} = -q \Longleftrightarrow q \in \mathbf{R}^3,$$
$$\overline{q + r} = \bar{q} + \bar{r}, \qquad \overline{qr} = \bar{r}\bar{q} \quad \text{(watch out!)},$$
$$q \in \mathbf{R}^3 \Longleftrightarrow q^2 \in \mathbf{R}_-, \qquad q \in \mathbf{R} \Longleftrightarrow q^2 \in \mathbf{R}_+.$$

The canonical scalar product in $\mathbf{R}^4$ is recovered through the formula

$$(q \mid r) = \frac{1}{2}(\bar{q}r + \bar{r}q);$$

in particular, $q\bar{q} \in \mathbf{R}_+$, and $\|q\| = \sqrt{q\bar{q}}$ is called the *norm* of q. One has $\|qr\| = \|q\|\|r\|$; this shows that $\mathbf{H}$ is indeed a field, for we can take $q^{-1} = \bar{q}/\|q\|^{-2}$. We define

$$S^3 = \{ q \mid \|q\| = 1 \},$$

the unit sphere of $\mathbf{R}^4 = \mathbf{H}$, and embed S^2 in the space of pure quaternions as

$$S^2 = \{ q \in \mathbf{R}^3 \mid \|q\| = 1 \}$$

We see that S^3 has a multiplicative group structure, just like $S^1 = \mathbf{U} = \{ z \mid |z| = 1 \}$ is also a group (but S^3 is not abelian). In this context it is natural to ask whether higher-dimensional spheres can have a group structure, good in a specified sense; the answer, obtained with the aid of algebraic topology, is no: only S^1 and S^3 have good group structures (see references in [PO, 284] or [HU, chapter 15]). The sphere S^7 has an interesting structure, it's almost a group; this is obtained using Cayley numbers (cf. [PO, 278]). Nor should we expect to find reasonable field structures on $\mathbf{R}^n$ for n other than 2 or 4; the only field structures are really $\mathbf{C}$ on $\mathbf{R}^2$ and $\mathbf{H}$ on $\mathbf{R}^4$ (cf. [PO, 284]), although $\mathbf{R}^8$ again has a reasonable structure, the Cayley numbers. For $\mathbf{R}^n$, see 8.10.3 and the references contained there.

Observe that if q, q' are pure quaternions, the pure part $\mathcal{P}(qq')$ of qq' is exactly the vector product $q \times q'$ in $\mathbf{R}^3$ (see 8.11.13).

Contrary to the case of $\mathbf{C}$, the automorphisms of $\mathbf{H}$ are simple and well-known; see 8.12.10 and compare 2.6.4.

8.9.2. THEOREM.

i) *Let $s \in \mathbf{R}^3 \backslash 0$. Then $q \mapsto -sqs^{-1}$ leaves $\mathbf{R}^3$ invariant, and its restriction to $\mathbf{R}^3$ is the reflection $\sigma_{s^\perp}$ through the plane $s^\perp$ of $\mathbf{R}^3$.*

ii) *Let $s \in \mathbf{H}^*$ and $\rho'_s : q \mapsto sqs^{-1}$. Then ρ'_s leaves $\mathbf{R}^3$ invariant, and its restriction $\rho_s = \rho'_s|_{\mathbf{R}^3}$ belongs to $O^+(3)$. Moreover, $\rho_s = \rho_{s'}$ if and only if there exists $\lambda \in \mathbf{R}^*$ such that $s' = \lambda s$.*

iii) *Conversely, for every $f \in O^+(3)$ there exists $s \in \mathbf{H}^*$ such that $f = \rho_s$.*

Proof. To see that $\mathbf{R}^3$ is invariant, use the criterion $q \in \mathbf{R}^3 \iff q^2 \in \mathbf{R}_-$:

$$(-sqs^{-1})^2 = sq^2 s^{-1} \in \mathbf{R}_-.$$

The map $q \mapsto -sqs^{-1}$ is certainly linear and length-preserving:

$$\| -sqs^{-1}\| = \|q\|$$

for all q. Finally, $s \mapsto -sss^{-1} = -s$ and if $q \in \mathbf{R}^3$ is orthogonal to s we have

$$0 = (q \mid s) = \frac{1}{2}(\bar{q}s + \bar{s}q) = \frac{1}{2}(-qs - sq),$$

so that $sq = -qs$ and

$$q \mapsto -sqs^{-1} = ss^{-1}q = q.$$

To study ρ_s' and ρ_s, one proceeds as above and concludes that $\rho_s \in O(3)$; there remains to see that $\rho_s \in O^+(3)$. This is clear topologically, since $\mathbf{H}^*$ is connected and the map taking $s \in \mathbf{H}^*$ into the map $q \mapsto sqs^{-1}$ in $O(3)$ is continuous, so ρ_s is in the same connected component of $O(3)$ as $\mathrm{Id}_{\mathbf{R}^3}$, which is the image of $s = 1$ under this map. The "moreover" clause follows because the only elements $s \in \mathbf{H}$ that commute with all elements of $\mathbf{R}^3$ are obviously the real numbers. Finally, (iii) follows from (i) and 8.2.12. $\square$

8.9.3. COROLLARY. *The map $\rho : S^3 \ni s \mapsto \rho_s \in O^+(3)$ is a continuous surjective group homomorphism. Its kernel is $\{\pm 1\}$. In particular, $O^+(3)$ is isomorphic to $P^3(\mathbf{R})$.*

Proof. Recall that $S^3 = \{\, q \in \mathbf{H} \mid \|q\| = q \,\}$. The kernel is the set of $s \in \mathbf{R}$ such that $\|s\| = 1$, by 8.9.2 (ii). The quotient of S^3 by the subgroup $\{\pm 1\}$, or, equivalently, by the equivalence relation $x \sim y \iff y = \pm x$, is exactly $P^3(\mathbf{R})$ (cf. 4.3.3.2). $\square$

8.9.4. PROPOSITION. *Let $s = \alpha + t$ with $t \in \mathbf{R}^3 \setminus 0$ and $\alpha \in \mathbf{R}$. The axis of the rotation $\rho_s \in O^+(3)$ is the line $\mathbf{R}t$, and the angle θ of ρ_s, $\theta \in [0, \pi]$, is given by $\tan \theta/2 = \|t\|/|\alpha|$ if $\alpha \neq 0$, and $\theta = \pi$ if $\alpha = 0$, in which case ρ_s is the reflection through $\mathbf{R}t$.*

Proof. It is clear that $\mathbf{R}t$ is invariant. The calculation of θ is more delicate; observe first that $\rho_{zsz^{-1}} = \rho_z \rho_s (\rho_z)^{-1}$ and ρ_s have the same angle, and since $O^+(3)$ acts transitively on S^2, there exists z with $zsz^{-1} = \beta i$, for $\beta \in \mathbf{R}$. This z can be calculated for an an arbitrary $s = \alpha + \beta i$.

The angle of ρ_s will be read in $i^{\perp} \cap \mathbf{R}^3$, that is, $\theta = \widehat{j, \rho_s j}$. But we have

$$\rho_j = (\alpha + \beta i)j(\alpha + \beta i)^{-1} = (\alpha^2 + \beta^2)^{-1}(\alpha + \beta i)j(\alpha - \beta i)$$
$$= (\alpha^2 + \beta^2)^{-1}((\alpha^2 - \beta^2)j + 2\alpha\beta k).$$

Thus

$$\tan \theta = \frac{2\alpha\beta}{\alpha^2 - \beta^2} = \frac{2(\beta/\alpha)}{1 - (\beta/\alpha)^2},$$

and $\tan \theta/2 = \beta/\alpha = \|t\|/|\alpha|$.

Finally, $\alpha = 0$ implies $\rho_s(j) = -j$. $\square$

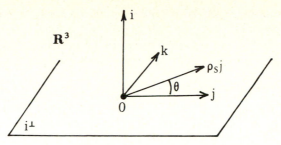

Figure 8.9.4

8.9.5. The importance of 8.9.3 and 8.9.4 is that it gives a parametric representation for $O^+(3)$, the group of rotations of our physical space; such parametrizations are fundamental in mechanics and physics, but their are not easy to find or deal with. Consider, for example, Euler angles ([SY, 286–288]), and try to write the formula for the composition of two rotations! The results in 8.9.3 and 8.9.4, on the other hand, allow us to deal easily and neatly with composition. There is just one drawback one would like to get rid of: the ambiguity caused by the non-trivial kernel $\{\pm 1\}$. This, however, cannot be avoided, in view of the following result:

8.9.6. PROPOSITION. *No right inverse $f : O^+(3) \to S^3$ for ρ is either continuous or a group homomorphism.*

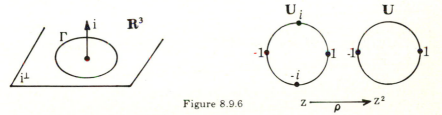

Figure 8.9.6

Proof. Let Γ be the circle consisting of rotations around the axis $\mathbf{R}i$. By 8.9.4, we have $\rho^{-1}(\Gamma) = \Gamma' = S(\mathbf{R} + \mathbf{R}i)$, the unit circle in the plane $\mathbf{R} + \mathbf{R}i \subset \mathbf{H}$. The proof of 8.9.4 also shows that the restriction $\rho|_{\Gamma'} : \Gamma' \to \Gamma$ is exactly the map $z \mapsto z^2$, under the identification $\mathbf{R} + \mathbf{R}i = \mathbf{C}$. Suppose there exists $f : O^+(3) \to S^3$ with $\rho \circ f = \mathrm{Id}_{O^+(3)}$, and look at the restriction $f|_\Gamma$. If f is a homomorphism, we must have $f(\mathrm{Id}) = 1$, then $f(-1) = \pm i$ because $\rho(f(1)) = (f(-1))^2 = -1$ by the above. Choosing $f(-1) = i$, for example, we get

$$(f(-1))^2 = f((-1)^2) = f(1) = 1 = i^2 = -1,$$

a contradiction.

  Now suppose that f is continuous and $f(\mathrm{Id}) = 1$ (the case -1 is treated analogously). By continuity, $f(z) = \sqrt{z}$, the usual determination of the square root, on $\mathbf{C} \setminus \mathbf{R}_-$; but then, when z approaches -1 within the upper (resp. lower) half-plane the square root $f(z)$ approaches i (resp $-i$), contradicting continuity.

8.9.7. As in the proof of 8.9.2, we see that, for $s, r \in \mathbf{H}^*$ satisfying $\|s\|\|r\| = 1$ (and, in particular, if $\|s\| = \|r\| = 1$), the map $\mathbf{H} \ni q \mapsto sqr \in \mathbf{H}$ is an isometry of $\mathbf{H}$, and, in fact, belongs to $O^+(4)$, as in 8.9.2 (ii). We actually have more:

8.9.8. THEOREM. *The map*

$$\tau : S^3 \times S^3 \ni (s, r) \mapsto \{q \mapsto sq\bar{r}\} \in O^+(4)$$

is a continuous, surjective group homomorphism. Its kernel consists of two points $(1, 1)$ *and* $(-1, -1)$.

8.9.9. *Proof.* The set $S^3 \times S^3$ is obviously being considered with its product group structure. We have $\tau(ss', rr')(q) = ss'q\overline{rr'} = s(s'q\bar{r'})\bar{r} = \big(\tau(s, r) \circ \tau(s', r')\big)(q)$. The kernel is formed of pairs (s, r) such that $sq\bar{r} = q$ for every q; we must have $\bar{r} = s^{-1}$, so $sqs^{-1} = q$ for every q, and $s \in \mathbf{R}$, so $s = \pm 1$. To show surjectivity, take $f \in O^+(4)$ and $q_0 = f(1)$. Then $\tau(q^{-1}, 1) \in O^+(4)$ and $\tau(q_0^{-1}, 1)(q_0) = 1$, so $\tau(q_0^{-1}, 1) \circ f \in O^+(3)$. But by 8.9.2 (iii) we have $\tau(q_0^{-1}, 1) \circ f = \rho'_s$ for $s \in \mathbf{H}^*$, so $f = \tau(q_0, 1) \circ \rho'_s = \tau(q_0 s, s^{-1})$.

8.9.10. COROLLARY. The group $O^+(4)$ has normal subgroups other than itself and its center $\{\pm \mathrm{Id}_{\mathbf{R}^4}\}$—for example, $\tau(\{1\} \times S^3)$ and $\tau(S^3 \times \{1\})$. In particular, $O^+(4)/\{\pm \mathrm{Id}_{\mathbf{R}^4}\}$ is not simple.   □

8.9.11. REMARKS. This concludes our investigation of the simplicity of $O(n)$, started in 8.5.

We see that $O^+(4)$ is almost a direct product. The real direct product is the orthogonal projective group $\mathrm{PO}^+(4)$:

8.9.12 $\mathrm{PO}^+(4) = p\big(\tau(S^3 \times \{1\})\big) \times p\big(\tau(\{1\} \times S^3)\big).$

Equation 8.9.12 has nice geometric consequences, cf. 18.8.8. For other applications of quaternions to algebra, see [VL], a very interesting book.

8.9.13. QUESTION. The complex numbers $\mathbf{C}$ have afforded a detailed study of $O^+(2)$, and the quaternions a study of $O^+(3)$ and $O^+(4)$. Thus we have worked our way up to dimension four. How does one parametrize $O^+(n)$? The answer is given by Clifford algebras, cf. 8.10.3.

8.10. Algebraic topology and orthogonal groups

8.10.1. This section closes the plan announced in 8.2.14. It assumes some familiarity, at least, with algebraic topology. We mention *en passant* that each $O^+(n)$ is a smooth manifold and a Lie group; in fact, it is exactly using Lie group techniques that one answers the problems in 8.10.4, among others.

8.10.2. PARTICULAR CASES. We have seen that $O^+(2)$ is the circle S^1, $O^+(3)$ is the real projective space $P^3(\mathbf{R})$, and $O^+(4)$ can be interpreted as $(S^3 \times S^3)/\mathbf{Z}_2$ by 8.9.8. We mention that $O^+(5)$ is related to $\mathbf{H}^2$ ([PO, corollary 13.60]), and $O^+(6)$ is related to $\mathbf{C}^4$ ([PO, prop. 13.61]). See also [DE1, 106–116]. Finally, $O^+(8)$ gives rise to a "triality" phenomenon, due to the fact that the quotient between outer automorphisms $\mathrm{Aut}\big(O^+(8)\big)$ and inner automorphisms $\mathrm{Int}\big(O^+(8)\big)$ is isomorphic to the symmetric group S_3 (cf. [CE2, 119]):

$$\mathrm{Aut}\big(O^+(8)\big)/\mathrm{Int}\big(O^+(8)\big) \cong S_3.$$

8.10.3. FUNDAMENTAL GROUP. Since $O^+(2)$ is homeomorphic to the circle S^1, the fundamental group $\pi_1\big(O^+(2)\big)$ is isomorphic to $\mathbf{Z}$. On the other hand, $\pi_1\big(O^+(n)\big) = \mathbf{Z}_2$ for every $n \geq 3$. For $n = 3$ this follows from 8.9.3, since ρ is a twofold cover and S^3 is simply connected; the case $n > 3$ follows from $n = 3$ and the exact homotopy sequence of the fiber bundle obtained by writing the sphere S^n as the homogeneous space $S^n = O^+(n+1)/O^+(n)$ (cf. 1.5.9 or 8.2.8). It is interesting from the geometric point of view to find the non-zero element of $\pi_1\big(O^+(3)\big)$ (and hence of $\pi_1\big(O^+(n)\big)$). The answer is afforded by the proof of 8.9.6, which shows that the circle Γ, consisting of rotations of $\mathbf{R}^3$ around a fixed axis, is a non-trivial loop in $O^+(3)$, since S^3 is simply connected.

Observe also that the loop Γ, described twice, becomes null-homotopic, since $\pi_1\big(O^+(3)\big) = \mathbf{Z}^2$. This can be demonstrated either with a paper band (figure 8.10.3.1) or by the trick of the soup bowl (figure 8.10.3.2). The performer moves his arm so as to rotate a soup bowl by 2π around its axis, always holding it right side up on his hand. His arm is then twisted. The surprising thing is that by repeating the same operation he now untwists his arm, instead of twisting it by another 2π! See [PO, 2].

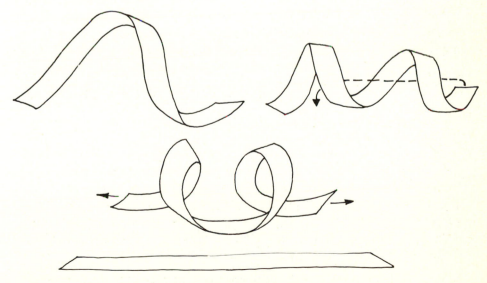

Figure 8.10.3.1

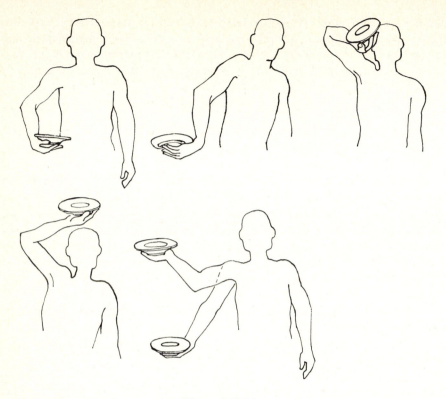

Figure 8.10.3.2

But the fact that $\pi_1\big(O^+(n)\big)$ is non-trivial has a much more important consequence. Indeed, one can lift the group structure of $G = O^+(n)$ to its universal cover $\tilde{G}$, a simply connected topologic space. We obtain an abstract group $\widetilde{O^+}(n)$, and one can ask whether this group has a geometric realization. For $n = 3$, we have realized $\widetilde{O^+}(3)$ as a subset of the quaternions $\mathbf{H}$, the group operation being multiplication in $\mathbf{H}$. What is the analogue of $\mathbf{H}$ for $n > 3$? The answer is Clifford algebras, and $\widetilde{O^+}(n)$ is called the group of spinors. See [PO, chap. 13], [BI2, §9], and the references in 13.7.14.4.

It is galling to realize that spinors themselves did not come out of these lucubrations, but were only discovered by Elie Cartan in 1913, when he classified primitive irreducible linear representations of the Lie algebra of $O^+(n)$. One of these representations, hitherto unknown, had the unexpected dimension 2^n; it has since been known as the spinor representation. Clifford algebras, on the other hand, were discovered in 1876, while Clifford was trying to extend theorems like 8.9.2 and 8.9.8 to higher dimensions.

8.10.4. Other algebraic topological structures associated with $O^+(n)$ (homology groups, cohomology ring, etc.) are all completely known. See, for example, [BO1], and also [HU, 92–95].

8.11. Canonical volume form, mixed product and cross product

Recall that if $n = \dim E$ we have $\dim \Lambda^n E^* = 1$ (cf. 2.7.2.1).

8.11.1. LEMMA. *If E is a Euclidean space, $\Lambda^n E^*$ has a canonical Euclidean structure (here, a norm), defined by*

$$\|\omega\| = |\omega(e_1, \ldots, e_n)|$$

for every $\omega \in \Lambda^n E^*$ and every orthonormal basis (e_i) of E.

Proof. Since $\omega \in \Lambda^n E^*$ is determined by its value on a basis of E, it is enough to show that $|\omega(e_1, \ldots, e_n)| = |(e'_1, \ldots, e'_n)|$ for two arbitrary orthonormal bases. Defining $f \in \mathrm{GL}(E)$ by $f(e_i) = e'_i$ for $i = 1, \ldots, n$, we have $\omega(e'_1, \ldots, e'_n) = (\det f)\omega(e_1, \ldots, e_n)$, and $\det f = \pm 1$ by 8.2.1. $\square$

8.11.2. NOTE. More generally, the exterior, tensor and symmetric algebras $\Lambda^p E$, $\Lambda^p E^*$, $\otimes^p E$, $\otimes^p E^*$, $\bigcirc^p E$, $\bigcirc^p E^*$ have natural Euclidean structures for every p, inherited from the structure of E (cf. [B12, 115]). In this book we only consider the case $\Lambda^n E^*$.

8.11.3. DEFINITION. *Let E be an oriented Euclidean vector space. The canonical volume form on E is the element $\lambda_E \in \Lambda^n E^*$ characterized by $\|\lambda_E\| = 1$ (cf. 8.11.1) and by assigning a positive value to any positively oriented basis of E (cf. 2.7.2.2). For $n = 3$ we also call λ_E the mixed product, and sometimes write $\lambda_E(\cdot, \cdot, \cdot) = (\cdot, \cdot, \cdot)$.*

Let E be a Euclidean vector space. The canonical density on E is the map $\delta_E : E^n \to \mathbf{R}_+$ defined by $(e_1, \ldots, e_n) \to |\lambda_n(e_1, \ldots, e_n)|$ where λ_E is the canonical volume form on E for an arbitrary orientation.

In other words, λ_E and δ_E are characterized by

8.11.4. $\delta_E(e_1, \ldots, e_n) = 1$ (resp. $\lambda_E(e_1, \ldots, e_n) = 1$) for any orthonormal basis (resp. positively oriented orthonormal basis) (e_i).

8.11.5. In order to calculate $\delta_E(x_1, \ldots, x_n)$, we introduce Gram determinants. If E is a Euclidean space and $(x_i)_{i=1,\ldots,p}$ is a subset of E with p elements, the *Gram determinant* of (x_i) is the scalar

$$\mathrm{Gram}(x_1, \ldots, x_p) = \det((x_i \mid x_j)) = \begin{vmatrix} \|x_1\|^2 & (x_1 \mid x_2) & \cdots & (x_1 \mid x_p) \\ (x_2 \mid x_1) & \|x_2\|^2 & \cdots & (x_2 \mid x_p) \\ \vdots & \vdots & \ddots & \vdots \\ (x_p \mid x_1) & (x_p \mid x_2) & \cdots & \|x_p\|^2 \end{vmatrix}.$$

8.11.6. PROPOSITION. *For every finite subset $(x_i)_{i=1,\ldots,n}$ of E one has*
$$\delta_E(x_1, \ldots, x_n) = \left(\mathrm{Gram}(x_1, \ldots, x_n)\right)^{1/2}.$$

Proof. Let (e_i) be an orthonormal basis, and choose $f \in \mathrm{GL}(E)$ such that $f(e_i) = x_i$ for all $i = 1, \ldots, n$. Let $A = M(f)$ be the matrix of f with respect to the basis (e_i). By Lemma 8.8.2,

$$\mathrm{Gram}(x_1, \ldots, x_n) = \det({}^t A A) = (\det A)^2 = (\det f)^2,$$

whence

$$\delta_E(x_1, \ldots, x_n) = |\det f|\, \delta_E(e_1, \ldots, e_n) = |\det f|. \qquad \square$$

8.11.7. EXAMPLES. Take $x, y \in E$ and apply 8.11.6 to the Euclidean space $V = \mathbf{R}x + \mathbf{R}y$. We have $\delta_V(x, y) = \big(\mathrm{Gram}(x, y)\big)^{1/2}$, so $\mathrm{Gram}(x, y) \geq 0$; but

$$\mathrm{Gram}(x, y) = \begin{vmatrix} \|x\|^2 & (x \mid y) \\ (y \mid x) & \|y\|^2 \end{vmatrix} = \|x\|^2 \|y\|^2 - (x \mid y)^2,$$

proving the Schwarz inequality (cf. 8.1.3). Moreover, by 8.6.3 we see that $\delta_V(x, y) = \sin(\overline{\mathbf{R}x, \mathbf{R}y}) \|x\| \|y\|$, where the angle between lines is non-oriented.

In the case of three vectors, the reader can try to prove directly that the Gram determinant $\mathrm{Gram}(x, y, z)$ is always positive.

The rest of this section is devoted to cross products, an operation used all the time in everything three-dimensional that has to do with our physical universe.

$$\boxed{\text{From now on } E \text{ is oriented.}}$$

8.11.8. PROPOSITION. *Let E have dimension $n \geq 3$. Given $n - 1$ vectors $(x_i)_{i=1,\ldots,n-1}$ of E, there exists a unique vector $x_1 \times \cdots \times x_{n-1}$ of E satisfying*

$$(x_1 \times \cdots \times x_{n-1} \mid y) = \lambda_E(x_1, \ldots, x_{n-1}, y)$$

for every $y \in E$. This vector is called the cross product of the x_i. The cross product possesses the following properties:

i) *the map $E^{n-1} \ni (x_1, \ldots, x_{n-1}) \mapsto x_1 \times \cdots \times x_{n-1} \in E$ is a multilinear alternating form;*

ii) *$x_1 \times \cdots \times x_{n-1} = 0$ if and only if the x_i are linearly dependent;*

iii) *$x_1 \times \cdots \times x_{n-1} \in (\mathbf{R}x_1 + \cdots + \mathbf{R}x_{n-1})^{\perp}$;*

iv) *if the vectors (x_i) are linearly independent, the set $\{x_1, \ldots, x_{n-1}, x_1 \times \cdots \times x_{n-1}\}$ is a positively oriented basis for E;*

v) *$\|x_1 \times \cdots \times x_{n-1}\| = \big(\mathrm{Gram}(x_1, \ldots, x_{n-1})\big)^{1/2}$*

$$= \delta_{\mathbf{R}x_1 + \cdots + \mathbf{R}x_{n-1}}(x_1, \ldots, x_{n-1});$$

vi) *$\|x_1 \times \cdots \times x_{n-1}\|$ is uniquely characterized by (iii), (iv) and (v) together.*

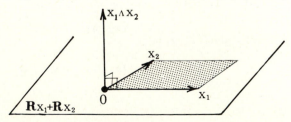

Figure 8.11.8

Proof. Existence and uniqueness follow from 8.1.8.1. Part (i) comes from the fact that λ_E is multilinear and alternating. One direction of (ii) is trivial; to prove the converse, assume that the x_i are linearly independent, and adjoin to them another vector x_n to form a basis. We have

$$(x_1 \times \cdots \times x_{n-1} \mid x_n) = \lambda_E(x_1, \ldots, x_{n-1}, x_n) \neq 0,$$

so the cross product cannot be zero. For (iv), apply 2.7.2.2 to the equation

$$\lambda_E(x_1, \ldots, x_{n-1}, x_1 \times \cdots \times x_{n-1}) = \|x_1 \times \cdots \times x_{n-1}\|^2 > 0.$$

For (v), put $z = x_1 \times \cdots \times x_{n-1}$, and observe that $(z \mid x_i) = 0$ for every $i = 1, \ldots, n-1$, and that, by the definition of Gram determinants,

$$\mathrm{Gram}(x_1, \ldots, x_{n-1}, z) = \mathrm{Gram}(x_1, \ldots, x_{n-1})\|z\|^2,$$

so that

$$\left(\lambda_E(x_1, \ldots, x_{n-1}, z)\right)^2 = \|z\|^4 = \mathrm{Gram}(x_1, \ldots, x_{n-1}, z)$$
$$= \mathrm{Gram}(x_1, \ldots, x_{n-1})\|z\|^2$$

(cf. 8.11.6). Finally, to show that (iii), (iv) and (v) characterize the cross product, one only has to consider the case when the x_i are linearly independent (otherwise the product is zero); but then $\dim(\mathbf{R}x_1 + \cdots + \mathbf{R}x_{n-1}) = 1$, and (iv) and (v) do indeed determine a well-defined vector on this line. $\square$

8.11.9. REMARKS. For $n = 2$, the "cross product" would simply be the map $E \ni x \mapsto \partial(x) \in E$, a rotation by a right angle (cf. 8.3.12 and 8.7.3.5).

The cross product affords an easy method to complete an orthonormal set $\{e_i\}_{i=1,\ldots,n-1}$ into a positive orthonormal basis. The practical calculation comes in 8.11.11, 8.11.12.

8.11.10. THE CROSS PRODUCT IN COORDINATES. Let (e_i) be an orthonormal basis for E, and let $A = (x_1, \ldots, x_{n-1}) = (x_{ij})$ $(i = 1, \ldots, n-1; j = 1, \ldots, n)$ be the matrix whose columns are the vectors x_i in the basis (e_i). Then

8.11.11 $\qquad$ i-th coordinate of $x_1 \times \cdots \times x_{n-1} = (-1)^{n-1} \det A_i$,

where A_i denotes the square matrix obtained from A by deleting the i-th line. To see this, put $z = x_1 \times \cdots \times x_{n-1} = (z_1, \ldots, z_n)$, and expand the determinant giving $(z \mid y)$ with respect to its last column:

$$(z \mid y) = \lambda_E(x_1, \ldots, x_{n-1}, y) = \begin{vmatrix} x_{11} & \cdots & x_{n-1.1} & y_1 \\ \vdots & \ddots & \vdots & \vdots \\ x_{1n} & \cdots & x_{n-1.n} & y_n \end{vmatrix}$$
$$= \sum_i (-1)^{n-i} y_i \det A_i = \sum_i y_i z_i.$$

This completes the proof, since equality holds for any y_i. $\square$

If $n = 3$, for example, we find

8.11.12
$$x = \begin{pmatrix} a \\ b \\ c \end{pmatrix}, \quad y = \begin{pmatrix} a' \\ b' \\ c' \end{pmatrix} \quad \Rightarrow \quad x \times y = \begin{pmatrix} bc' - b'c \\ ca' - c'a \\ ab' - a'b \end{pmatrix}.$$

8.11.13. QUATERNIONS. Consider $\mathbf{R}^3$ with its canonical Euclidean structure and orientation. The cross product affords an easy way to calculate the product of two *pure* quaternions $x, y \in \mathbf{R}^3 \subset \mathbf{H}$ (and, by extension, the product of any two quaternions). By 8.9.1 and 8.11.12, $xy = -(x \mid y) + x \times y$, with $(x \mid y) \in \mathbf{R} \subset \mathbf{H}$.

8.12. Exercises

* **8.12.1.** Let E be a finite-dimensional real vector space, ϕ and ψ two Euclidean structures on E, and $G \subset GL(E)$ a subgroup of the linear group of E; we assume G is irreducible (cf. 8.2.6). Show that if $G \subset O(E, \phi) \cap O(E, \psi)$ (i.e., if every element of G leaves ϕ and ψ invariant, cf. 13.5), then ϕ and ψ are proportional.

* **8.12.2.** A GEOMETRIC PROOF FOR 8.2.15. To prove that $f \in O(E)$ always leaves some line or plane invariant, consider some $x \in S(E)$ such that $\|f(x) - x\|$ is minimal, and show that x, $f(x)$ and $f^2(x)$ lie in the same plane (cf. 9.3.2; use 18.4).

8.12.3. A CALCULUS PROOF FOR 8.2.15. The data being the same as for 8.12.2, prove that the plane determined by x and $f(x)$ is invariant under f by showing that the derivative with respect to x of the map

$$S(E) \ni y \mapsto \|f(y) - y\| \in \mathbf{R}$$

is zero (cf. 13.5.7.2).

8.12.4. Prove directly that a circle is path-connected.

8.12.5. NORMAL ENDOMORPHISMS. An endomorphism f of a Euclidean space E is called *normal* if it commutes with its adjoint: ${}^t f \circ f = f \circ {}^t f$ (cf. 8.1.8.6). Show that 8.2.15 holds for any normal endomorphism. (See also 13.5.7.) What are the normal endomorphisms when $\dim E = 2$?

8.12.6. Determine all continuous homomorphisms $\mathbf{R} \to GO(E)$, where $\dim E = 2$. Draw the associated orbits (cf. 9.6.9). Use 8.3.13.

* **8.12.7.** Show that, for every $n \in \mathbf{N}^*$ and every $a \in \tilde{A}(E)$, the equation $nx = a$ has exactly n solutions in $\tilde{A}(E)$. Draw the solutions on a circle for a few values of a, with $n = 2, 3, 4, 5$.

8.12.8. TABLE OF TRIGONOMETRIC FORMULAS. Prove the formulas below, where $n \in \mathbf{N}$ and $a, b, c \in \mathbf{R}$ are arbitrary. As usual, we define $\tan x = \dfrac{\sin x}{\cos x}$.

$$\tan na = \frac{\binom{n}{1}\tan a - \binom{n}{3}\tan^3 a + \cdots + (-1)^p \binom{n}{2p+1}\tan^{2p+1} a + \cdots}{1 - \binom{n}{2}\tan^2 a + \cdots + (-1)^p \binom{n}{2p}\tan^{2p} a + \cdots}$$

$$\cos a + \cos b = 2\cos\frac{a+b}{2}\cos\frac{a-b}{2}$$

$$\cos a - \cos b = -2\sin\frac{a+b}{2}\sin\frac{a-b}{2}$$

(state similar formulas for the sum and difference of sines);

$$\tan a + \tan b = \frac{\sin(a+b)}{\cos a \cos b}.$$

Show that the maximum value of $\alpha\cos t + \beta\cos t$, for $t \in \mathbf{R}$, is $\sqrt{\alpha^2 + \beta^2}$. In the next four formulas, assume a, b, c positive and $a + b + c = \pi$:

$$\sin a + \sin b + \sin c = 4\cos\frac{a}{2}\cos\frac{b}{2}\cos\frac{c}{2}$$

$$\cos a + \cos b + \cos c = 4\sin\frac{a}{2}\sin\frac{b}{2}\sin\frac{c}{2}$$

$$\tan a + \tan b + \tan c = \tan a \tan b \tan c$$

$$\tan\frac{b}{2}\tan\frac{c}{2} + \tan\frac{c}{2}\tan\frac{a}{2} + \tan\frac{a}{2}\tan\frac{b}{2} = 1$$

$$\cos\frac{\pi}{10} = \frac{\sqrt{5 + 2\sqrt{5}}}{4}; \quad \cos\frac{\pi}{48} = \frac{1}{2}\sqrt{2 + \sqrt{2 + \sqrt{2 + \sqrt{3}}}}$$

$$1 + \cos a + \cdots + \cos na = \frac{\sin(n+1)\frac{a}{2}}{\sin\frac{a}{2}}\cos\frac{na}{2}$$

$$\sin a + \cdots + \sin na = \frac{\sin(n+1)\frac{a}{2}}{\sin\frac{a}{2}}\sin\frac{na}{2}$$

$$\sin\frac{\pi}{n}\sin\frac{2\pi}{n}\cdots\sin\frac{(n-1)\pi}{n} = n2^{1-n}$$

$$\cos a + \cos 3a + \cdots + \cos(2n-1)a = \frac{1}{2}\frac{\sin 2na}{\sin a}$$

$$\sin a + \sin 3a + \cdots + \sin(2n-1)a = \frac{\sin^2 na}{\sin a}$$

$$\lim_{n \to \infty} \cos\frac{a}{2}\cos\frac{a}{4}\cdots\cos\frac{a}{2^n} = \frac{\sin a}{a}.$$

* **8.12.9.** CROSS PRODUCTS IN $\mathbf{R}^3$. For every $a, b, c \in \mathbf{R}^3$, prove the following formulas:

$$a \times (b \times c) = (a \mid c)b - (a \mid b)c,$$
$$(a \times b, a \times c, b \times c) = (a, b, c)^2,$$
$$(a \times b) \times (a \times c) = (a, b, c)a.$$

Show that $\mathbf{R}^3$, endowed with the operations of addition and vector product, is an anticommutative algebra which, instead of being associative, satisfies the *Jacobi identity*

$$a \times (b \times c) + b \times (c \times a) + c \times (a \times b) = 0.$$

Such an algebra is called a *Lie algebra*.

If p, q, r denote the projections from $\mathbf{R}^3$ onto the three coordinate planes, show that $\|a \times b\|^2 = \mathrm{Gram}(p(a), p(b)) + \mathrm{Gram}(q(a), q(b)) + \mathrm{Gram}(r(a), r(b))$. Find a geometrical interpretation for this result (see figure 8.12.9). Generalize this result for $\mathrm{Gram}(a_1, \ldots, a_p)$ in a space of dimension n, where n and p are arbitrary.

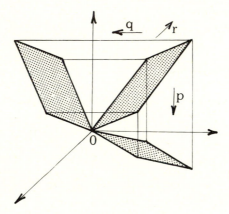

Figure 8.12.9

Study the equation $x \times a = b$ (for a and b given); find whether there is a solution and whether it is unique.

* **8.12.10.** We call a *bisector* of two half-lines A, B in a Euclidean vector space the unique half-line that is a bisector of A and B in the plane generated by A and B, and lies between the two half-lines (cf. 8.7.5.2 and 8.7.3.3). Let S, T, U be three half-lines in a 3-dimensional Euclidean vector space. Find three half-lines A, B, C such that S is a bisector of A, B, T is a bisector of B, C, and U is a bisector of C, A. Study possible generalizations of this problem: replacing half-lines by lines, considering more than three lines, or considering higher-dimensional spaces.

* **8.12.11.** AUTOMORPHISMS OF $\mathbf{H}$. Show that every automorphism of $\mathbf{H}$ is of the form $a \rightarrow \mathcal{R}(a) + \rho(\mathcal{P}(a))$, where $\rho \in O^+(3)$.

8.12.12. Compute some explicit compositions of rotations. In particular, calculate the composition of the rotations of order three and five of the icosahedron given in coordinates in 12.5.5.3.

8.12.13. Show that every finite-dimensional linear representation of a compact group is *semisimple*. This means that, given a compact group G and a homomorphism $f : G \to \mathrm{GL}(V)$ from G into the linear group $\mathrm{GL}(V)$ of a finite-dimensional real vector space (a *representation*), any vector subspace $W \subset V$ such that $f(g)(V) = V$ for all $g \in G$ admits a direct sum complement W such that $f(g)(W) = W$ for all $g \in G$.

8.12.14. Let $\{e_i\}_{i=1,\ldots,n}$ be an orthonormal basis for an n-dimensional Euclidean vector space. Determine explicitly the reduced form (cf. 8.4.1) of the element f of $O(n)$ defined by $f(e_i) = e_{i+1}$ for every $i = 1,\ldots,n-1$ and $f(e_n) = e_1$.

Chapter 9
Euclidean affine spaces

Euclidean affine spaces, in two and three dimensions, are the objects of classical geometry and correspond to our physical universe. Their structure is rich and has been studied for thousands of years, starting with the Greeks; as a consequence, there is a wealth of results and we've had to choose from among them. The choice is consistent with the rest of the book: we start with a quick discussion of the foundations, then offer results that are simple and nice, but at the same time not trivial to prove; and finally we give an overview of related questions, more difficult and perhaps still under investigation.

The first three sections include, aside from basic definitions and results, formulas for the explicit calculation of distances (section 9.2), in which Gram determinants play an essential role (they will be encountered again in section 9.7). Section 9.4 applies the results from section 9.3, on the structure of plane isometries, to the problem of polygons of minimal perimeter inscribed in a convex polygon, which is relevant to the study of trajectories of light rays or billiards, among others. This problem is in some sense opposite to the question of ergodicity of billiards.

Section 9.5 studies similarities and gives non-trivial characterizations for them, in particular Liouville's theorem. Section 9.6 uses plane similarities in the solution of several problems: similar divisions, the double pedal curve of two circles, and logarithmic spirals.

The last seven sections are quite varied and are partially meant to prepare the material for the remainder of the book, but they also introduce fundamental notions. Section 9.7 studies the relationship among distances between several points, and mentions the problem of giving a purely metric characterization of affine spaces, in which the Cayley–Menger determinant play an essential role. Section 9.8 studies subgroups leaving invariant a fixed subset, developing the connection between compactness of the subgroup and the existence of a common fixed point. Section 9.9 discusses the length of a curve and the shortest path between two points. In section 9.10 we use differential calculus to derive the distance as a function of its endpoints (first variation formula), and we deduce some applications. In 9.11 we extend the metric to the so-called Hausdorff distance, defined on the set of compact subsets of the space; this will be used in chapter 12. Section 9.12 defines the canonical measure of a Euclidean affine space, and the derived notion of the volume of a compact set (which is really our everyday notion of volume); we meet them again in chapter 12. Finally, section 9.13 discusses Steiner symmetrization, which has played an important historical role and is still used in the proof of the isoperimetric inequality (chapter 12); in this chapter we use Steiner symmetrization to prove Bierberbach's isodiametric inequality.

In addition to the exercises given in 9.14, we recommend that the reader leaf through the numerous ones given in [HD], [R–C], [FL], [PE], [CR1]. Some of these have a great graphic interest.

> Unless we state otherwise, X stands for a Euclidean affine space of finite dimension n.

9.1. Definitions. Isometries and rigid motions

9.1.1. DEFINITION. *A Euclidean affine space is an affine space* $(X, \vec{X})$, *where* $\vec{X}$ *is a Euclidean vector space. An affine frame* $\{x_i\}_{i=0,\dots,n}$ *of* X *is called orthonormal if* $\{\overrightarrow{x_0 x_1}\}_{i=1,\dots,n}$ *is an orthonormal basis for* $\vec{X}$. *We endow* X *with the metric given by* $d(x, y) = xy = \|\overrightarrow{xy}\|$.

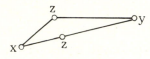

Figure 9.1.1

9.1.1.1. The strict triangle inequality. As can be immediately seen from 8.1.3, the equality $xz = xy + yz$ holds if and only if $z \in [x, y]$, the ixrmsegment joining x and y (cf. 3.4.3).

9.1.2. REMARKS. The topology generated by this metric is the canonical topology of X (2.7.1.1). An (affine) subspace $S \subset X$ inherits from X a natural Euclidean affine space structure.

The standard n-dimensional Euclidean affine space is $X = \mathbf{R}^n$, where $\vec{X} = \mathbf{R}^n$ is the standard n-dimensional Euclidean vector space (8.1.2.2). An *isometry* between two Euclidean affine spaces X, Y is a bijection $f : X \to Y$ that preserves distances: $d(f(x), f(y)) = d(x, y)$ for all $x, y \in X$. Every n-dimensional Euclidean affine space is isometric to the standard space $\mathbf{R}^n$. To see this, take an orthonormal frame $\{x_i\}$ for X and define $f : X \to \mathbf{R}^n$ by

$$f\left(x_0 + \sum_i \lambda_i \overrightarrow{x_0 x_i}\right) = (\lambda_1, \dots, \lambda_n).$$

This justifies the expression "n-dimensional Euclidean space", but we shall not use it too often.

The set of isometries of X is denoted by $\mathrm{Is}(X)$ (cf. 0.3). Translations and, more generally, any $f \in \mathrm{GA}(X)$ such that $\vec{f} \in O(\vec{X})$, are isometries: we have $d(f(x), f(y)) = \|\overrightarrow{f(x)f(y)}\| = \|\vec{f}(\overrightarrow{xy})\| = \|\overrightarrow{xy}\| = d(x, y)$, by 2.3.2 and 8.1.5. All isometries are of this type:

9.1.3. PROPOSITION. *A map* $f : X \to X$ *lies in* $\mathrm{Is}(X)$ *if and only if* $f \in \mathrm{GA}(X)$ *and* $\vec{f} \in O(\vec{X})$.

Proof. Pick an arbitrary $a \in X$, and let $t = t_{\overrightarrow{f(a)a}}$ be the translation by the vector $\overrightarrow{f(a)a}$. Then $g = t \circ F \in \mathrm{Is}_a(X)$, because $T(X) \subset \mathrm{Is}(X)$, where $T(X)$ is the set of translations of X. It suffices to show that $g \in O(X_a)$. Calculating in X_a, and applying the definition of an isometry and 8.1.2.4, we get $(g(x) \,|\, g(y)) = (x \,|\, y)$ for every $x, y \in X$. This implies $g \in O(X_a)$ by 8.1.5. $\qquad \square$

9.1.4. PROPOSITION AND DEFINITION. *Put* $\mathrm{Is}^{\pm}(X) = \{\, f \in \mathrm{Is}(X) \mid \vec{f} \in O^{\pm}(X) \,\}$. *The elements of* $\mathrm{Is}^+(X)$ *are called (proper) motions (and those of* $\mathrm{Is}^-(X)$ *are sometimes called "improper motions"). For every* $a \in X$, *we have the semidirect products*

$$\mathrm{Is}(X) = T(X)\,\mathrm{Is}_a(X), \qquad \mathrm{Is}^+(X) = T(X)\,\mathrm{Is}_a^+(X),$$

and the set-theoretical product

$$\mathrm{Is}^-(X) = T(X) \times \mathrm{Is}_a^-(X).$$

Moreover, the correspondence $f \mapsto \vec{f}$ *induces group isomorphisms* $\mathrm{Is}_a(X) \cong O(\vec{X})$ *and* $\mathrm{Is}_a^+(X) \cong O^+(\vec{X})$. *In the topology induced by the topology of* $GA(X)$ *(cf. 2.7.1.4), the space* $\mathrm{Is}(X)$ *has exactly two connected components* $\mathrm{Is}^+(X)$ *and* $\mathrm{Is}^-(X)$, *which are path-connected.* $\qquad \square$

It is the path-connectedness of $\mathrm{Is}^+(X)$ which justifies the term "motion" for $f \in \mathrm{Is}^+(X)$, because there exists a continuous homotopy $F : [0,1] \to \mathrm{Is}^+(X)$ joining $F(0) = \mathrm{Id}_X$ and $F(1) = f$. It is this F which, properly speaking, is a motion of X. This is more intuitive if we apply F to a fixed subset C of X and draw the images $C(t) = F(t)(C)$.

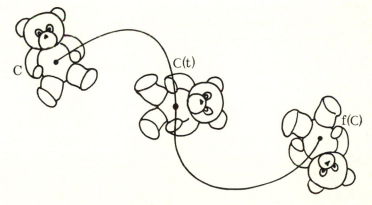

Figure 9.1.4

9.1.5. We can thus write X as a homogeneous space $X = \mathrm{Is}(X)/\mathrm{Is}_a(X)$, where a is given. This is often more interesting than the other homogeneous space decompositions $X = X_a$ and $X = GA(X)/GA_a(X)$, because in the first case, the definition of an affine space, the stabilizer is trivial, and in the second the stabilizer $\text{GA}_a(X)$ is too big, whereas $\text{Is}_a(X)$ is compact. But $\text{Is}(X)$ is big enough in the following sense:

9.1.6. PROPOSITION. *Is(X) acts simply transitively on orthonormal frames of X. Its action is 2-transitive (cf. 1.4.5.) in the following sense: for any four points $a, b, a', b' \in X$ such that $a'b' = ab$, there exists $f \in \text{Is}(X)$ such that $f(a) = a'$ and $f(b) = b'$. If, moreover, $\dim X = 2$, there is exactly one such f in $\text{Is}^+(X)$.*

Proof. This follows from 8.2.7. $\square$

9.1.7. NOTE. We have singled out the fact that $\text{Is}(X)$ is 2-transitive because this apparently unremarkable property is in fact extremely strong. Under certain additional regularity assumptions, metric spaces whose isometry group is 2-transitive can be classified. Besides affine spaces, we shall encounter three other examples in the book: spheres (chapter 18), real projective spaces and hyperbolic spaces (chapter 19). The other possibilities are the complex and quaternionic projective spaces, the octonionic projective plane, and their non-compact counterparts. See [BU2, 95], a recent reference, and also [B–K, 117] and [PV2].

9.2. Orthogonal subspaces. Distances between subspaces

9.2.1. METADEFINITION. *The notions of orthogonality, angle and oriented angle between (affine) subspaces and oriented lines of X are extended from the vector case by considering the directions in $\vec{X}$ of the relevant objects. The same notations are used as in the vector case.*

For example, if D, D' are two lines of X, their angle (in $[0, \pi/2]$) will be denoted by $\overline{DD'} = \overline{\vec{D}\vec{D'}}$, and orthogonality between two lines D and D' is indicated by $D \perp D'$.

9.2.2. PROPOSITION. *Let S be a subspace of X and $x \in S$ a point. There exists a unique subspace T containing x and such that $\vec{X} = \vec{S} \oplus^\perp \vec{T}$. The point $S \cap T$ (cf. 2.4.9.4) is characterized by $xy = \inf \{ xz \mid z \in S \}$. This number, denoted by $d(x, S)$, is called the distance from x to S (cf. 0.3). The distance xz is a strictly increasing function of the distance yz.*

Proof. Since $(\overrightarrow{xy} \mid \overrightarrow{yz}) = 0$, we have $xz^2 = xy^2 + yz^2$ by 8.1.2.4, and the function $t \mapsto \sqrt{1 + t^2}$ is strictly increasing. $\square$

9.2.3. REMARK. We have observed *en passant* that if x, y, z are distinct points of X the condition $xy^2 + yz^2 = xz^2$ is equivalent to $\langle x, y \rangle \perp \langle y, z \rangle$. This is the so-called Pythagorean theorem, which gives rise, in axiomatizations of Euclidean geometry, to some very involved demonstrations.

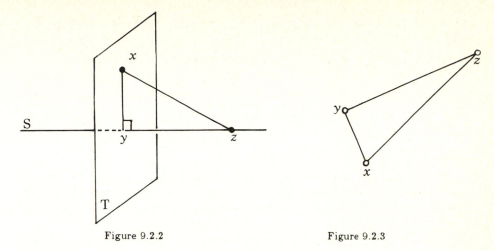

Figure 9.2.2   Figure 9.2.3

9.2.4. PROJECTIONS AND REFLECTIONS. Fix a subspace S of X. Proposition 9.2.2 yields a map $\pi_S : X \ni x \mapsto y \in S$, called the *orthogonal projection* onto S. The *reflection* through S is the map $\sigma_S : X \to X$ defined by $\sigma_S(x) = x + 2\overrightarrow{x\pi_S(x)}$ (or by the condition that $\pi_S(x)$ is the midpoint of x and $\sigma_S(x)$). We have $\sigma_S \in \mathrm{Is}(X)$; conversely, every involutive isometry f is a reflection through an appropriate subspace, as can be seen by noting that f leaves the point $(x + f(x))/2$ invariant, for arbitrary x and applying 8.2.9. If S is a hyperplane, σ_S is called a *hyperplane reflection*.

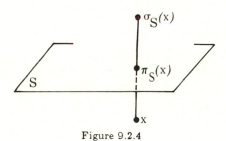

Figure 9.2.4

9.2.5. DISTANCE BETWEEN SUBSPACES. Let S and T be subspaces. We define the *distance* between S and T (cf. 0.3) as

$$d(S,T) = \inf \{\, st \mid s \in S, t \in T \,\}.$$

The infimum is always achieved for some $s \in S$ and $t \in T$; two such points are characterized by the condition $\overrightarrow{st} \in (\vec{S})^\perp \cap (\vec{T})^\perp$ (if $S \cap T = \emptyset$). The pair (s,t) is unique if and only if $\vec{S} \cap \vec{T} = \{0\}$.

The reader can supply an algebraic proof of the existence of the pair (s,t), or use a compactness argument. As for the condition $\overrightarrow{st} \in (\vec{S})^\perp \cap (\vec{T})^\perp$, it is necessary by 9.2.2. Conversely, if it is satisfied, consider the subspace S' parallel to S and containing t (cf. 2.4.9.2). Take $x \in S$, $y \in T$, and

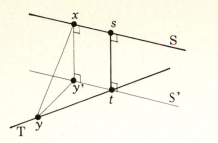
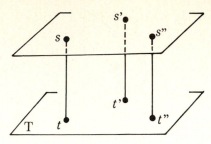

Figure 9.2.5

$y' = \pi_{S'}(x)$. We have $(\overrightarrow{xy'} \mid \overrightarrow{y'y}) = 0$, so $xy \geq xy' = st$. Uniqueness follows from the criterion $\overrightarrow{st} \in (\vec{S})^{\perp} \cap (\vec{T})^{\perp}$, for if $st = s't' = d(S, T)$, we must have $\overrightarrow{ss'} = \overrightarrow{tt'} \in \vec{S} \cap \vec{T}$.

9.2.6. COMPUTATIONS. We have collected below some formulas that allow the solution of a good number of computational problems, including three-dimensional ones. We have not included formulas involving orthonormal bases and coordinates, as they follow from the ones given below and 8.1.2.6, 8.11.5.8 and 8.11.11. We assume X to be oriented so as to be able to use all the results from 8.11.

9.2.6.1. Let S be a subspace of X, $x \in X$ a point and $\{x_i\}_{i=0.1.....k}$ a (not necessarily orthonormal) affine frame of S. Then

9.2.6.2
$$d^2(x, S) = \frac{\mathrm{Gram}(\overrightarrow{x_0 x}, \overrightarrow{x_0 x_1}, \ldots, \overrightarrow{x_0 x_k})}{\mathrm{Gram}(\overrightarrow{x_0 x_1}, \ldots, \overrightarrow{x_0 x_k})}.$$

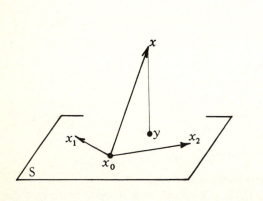
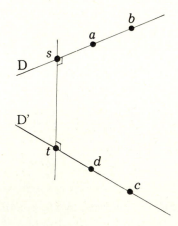

Figure 9.2.6.2

Proof. Put $y = \pi_S(x)$. By 8.11.6, Gram determinants do not change when we add a multiple of one column to another, so we have

$$\mathrm{Gram}(\overrightarrow{x_0 x}, \overrightarrow{x_0 x_1}, \ldots, \overrightarrow{x_0 x_k}) = \mathrm{Gram}(\overrightarrow{xy}, \overrightarrow{x_0 x_1}, \ldots, \overrightarrow{x_0 x_k}).$$

But the right-hand side is equal to $\|\overrightarrow{xy}\| \operatorname{Gram}(\overrightarrow{x_0 x_1}, \dots, \overrightarrow{x_0 x_k})$, as observed in the proof of 8.11.8 (v).   □

9.2.6.3. Let $H = f^{-1}(0)$ be a hyperplane of X, where $f \in A(X; \mathbf{R})$ (cf. 2.7.3.1). If $x \in X$ is a point, we have

9.2.6.4
$$d(x, H) = \frac{|f(x)|}{\|\vec{f}\|},$$

where $\| \cdot \|$ denotes the canonical norm in $(\vec{X})^*$ (cf. 8.1.8.2).

Proof. Put $y = \pi_H(x)$. We have $f(y) = 0$, so $f(x) - f(y) = f(x) = \vec{f}(\overrightarrow{yx})$. But $|\vec{f}(\overrightarrow{yx})| = |(\vec{f}^\sharp \mid \overrightarrow{yx})| = \|\vec{f}^\sharp\| \|\overrightarrow{yx}\| = \|\vec{f}\| \cdot xy$, since $\overrightarrow{yx} \in H^\perp$, so the vectors $\vec{f}^\sharp$ and $\overrightarrow{yx}$ are proportional. We then apply 8.1.8.1, 8.1.8.2 and 8.1.3.   □

9.2.6.5. Let D, D' be lines of X and a, b (resp. a', b') points in D (resp. D'). Then

$$d^2(D, D') = \frac{\operatorname{Gram}(\overrightarrow{aa'}, \overrightarrow{ab}, \overrightarrow{a'b'})}{\operatorname{Gram}(\overrightarrow{ab}, \overrightarrow{a'b'})}.$$

This is proved just like 9.2.6.1, by introducing s, t such that $d(D, D')$. The line $\langle s, t \rangle$, which is in general unique, is called the *common perpendicular* to D and D'.

9.3. Structure of isometries. Generators of $\operatorname{Is}(X)$ and $\operatorname{Is}^+(X)$

A good number of questions that can be asked about $f \in \operatorname{Is}(X)$, concerning fixed points, globally invariant subspaces, and so on, are answered by the following structure theorem:

9.3.1. THEOREM. *Take $f \in \operatorname{Is}(X)$. There exists a unique decomposition $f = t_{\vec{\xi}} \circ g$ satisfying the following conditions: g is an isometry of X, the set G of fixed points of g is non-empty, and the translation vector $\vec{\xi}$ lies in $\vec{G}$. Under these conditions we also have $\vec{G} = \operatorname{Ker}(\vec{f} - \operatorname{Id}_{\vec{X}})$ and $t_{\vec{\xi}} \circ g = g \circ t_{\vec{\xi}}$.*

Proof. We fist observe that $\vec{X} = \operatorname{Ker}(s - \operatorname{Id}_{\vec{X}}) \oplus^\perp \operatorname{Im}(s - \operatorname{Id}_{\vec{X}})$ for every $s \in O(\vec{X})$. In fact, the dimensions add up, and orthogonality follows from the definitions:

$$\big(u \mid s(v) - v\big) = \big(u \mid s(v)\big) - \big(u \mid v\big) = \big(s(u) \mid s(v)\big) - \big(u \mid v\big) = 0.$$

Take an arbitrary $a \in X$ and write the vector $\overrightarrow{af(a)}$ as a sum $\vec{\xi} + \vec{h}$, with $\vec{h} = \vec{f}(\vec{t}) - \vec{t}$ and $f(\vec{\xi}) = \vec{\xi}$. Put $x = a - \vec{t}$. By construction,

$$\overrightarrow{xf(x)} = \overrightarrow{xa} + \overrightarrow{af(a)} + \overrightarrow{f(a)f(x)} = \vec{t} + \vec{\xi} + \vec{f}(\vec{t}) - \vec{t} - \vec{f}(\vec{t}) = \vec{\xi}.$$

By putting $g = t_{\vec{\xi}}^{-1} \circ f$, we do indeed have $g(x) = x$. The remaining assertions are trivial.   □

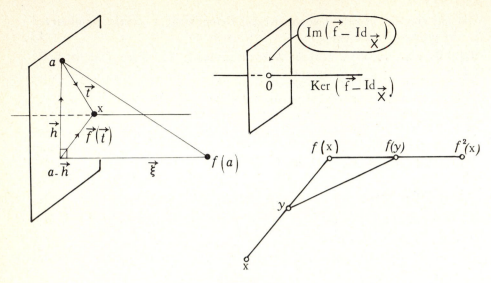

Figure 9.3.1

For more details and the proof of Corollary 9.3.3, see [FL, 194–197].

9.3.2. REMARK. It was the existence of a line globally invariant under f (and necessarily having direction $\mathrm{Ker}(\vec{f} - \mathrm{Id}_{\vec{X}})$) that allowed us to find the decomposition 9.3.1. Such a line can be found by a purely metric and topological reasoning, and is in fact given by $\langle x, f(x)\rangle$, where $x \in X$ is such that $xf(x) = \inf\{\, yf(y) \mid y \in X \,\}$. Proof: let y be the midpoint of x and $f(x)$. Since f is an isometry (and *a fortiori* an affine morphism), $f(y)$ is the midpoint of $f(x)$ and $f^2(x)$ and we have, by assumption,

$$d\big(y, f(y)\big) \geq d\big(x, f(x)\big) = d\big(f(x), f^2(x)\big) = d\big(x, f(x)\big).$$

But

$$d\big(y, f(y)\big) \leq d\big(y, f(x)\big) + d\big(f(x), f(y)\big) = 2 \times \frac{1}{2} d\big(x, f(x)\big).$$

Thus equality obtains everywhere, as this can only happen (cf. 9.1.1.1, the strict triangle inequality) if $f(x) \in \langle y, f(y)\rangle$. In other words, the three points $x, f(x), f^2(x)$ are collinear.

Conversely, every $x \in \vec{G}$ in 9.3.1 satisfies $xf(x) = \inf\{\, yf(y) \mid y \in X \,\}$, as can be seen by applying 9.2.3 to the triangle $\{a, a - \vec{h}, f(a)\}$.

9.3.3. COROLLARY. *Set* $s = \dim \vec{X} - \dim\big(\mathrm{Ker}(\vec{f} - \mathrm{Id}_{\vec{X}})\big)$.

i) *If* $\mathrm{Ker}(\vec{f} - \mathrm{Id}_{\vec{X}}) = \{0\}$, *the map* f *has a unique fixed point, called its center.*

ii) *Every* $f \in \mathrm{Is}(X)$ *is the product of at most* $n + 1$ *hyperplane reflections. At least* s *are necessary if* f *has a fixed point, and at least* $s + 2$ *if* f *has no fixed point.*

iii) *Every $f \in \text{Is}^+(X)$ is the product of s codimension-two reflections. Every*
$f \in T(X)$ is the product of 2 codimension-two reflections.   □

The following classifications also derive from 9.3.1:

9.3.4. STRUCTURE OF PLANE ISOMETRIES. If $n = 2$, every $f \in \text{Is}^+(X) \setminus$
$T(X)$ has a unique fixed point a, called its *center*. We say that f is a *rotation*
with center a, and f is characterized by its center and its *angle*, defined as
$\Phi^{-1}(\vec{f}) \in \tilde{A}(\vec{X})$. Every $f \in \text{Is}^-(X)$ possesses a unique globally invariant line
D and can be written in a unique way as $f = t_{\vec{\xi}} \circ \sigma_D$, where $\vec{\xi} \in \vec{D}$. The line
D is called the *reflection axis of D.*

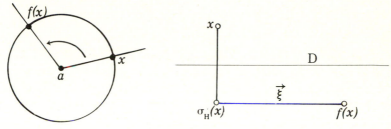

Figure 9.3.4

9.3.5. STRUCTURE OF THREE-DIMENSIONAL ISOMETRIES. Let $n = 3$. A
rotation is any element f of $\text{Is}^+(X) \setminus \text{Id}_X$ having a fixed point. By 8.4.7.1, a
rotation leaves pointwise invariant a unique line D, called its *axis*; the *angle*
of f is the constant $\overline{\Delta f(\Delta)} \in \,]0, \pi]$, where Δ is an oriented line orthogonal
to D (figure 9.3.5.1). The axis and the angle, however, are not sufficient to
determine f wholly; there are still two possibilities. See [DE2, 132–135], or
9.14.5, for more details.

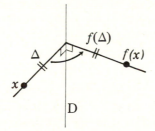

Figure 9.3.5.1

A map $f \in \text{Is}^+(X) \setminus T(X)$ is called a *screw motion*; a screw motion leaves
globally invariant a unique line D, called its *axis*, and can be written in a
unique way as $t_{\vec{\xi}} \circ g$, where g is a rotation with axis D and $\vec{\xi} \in \vec{D}$ (figure
9.3.5.2).

There are two kinds of elements in $\text{Is}^-(X)$. The first kind (figure 9.3.5.3)
leaves at least one point invariant, and can be written as $r \circ \sigma_H$, where H is a
uniquely determined hyperplane and r is a rotation with axis perpendicular

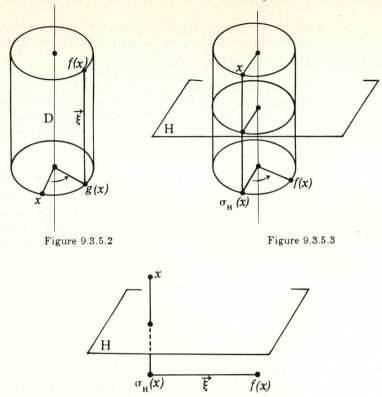

Figure 9.3.5.2   Figure 9.3.5.3

Figure 9.3.5.4

to H (or $r = \mathrm{Id}_X$). The second kind has no fixed point and can be written in a unique way as $t_{\vec{\xi}} \circ \sigma_H$, where H is a hyperplane and $\vec{\xi} \in \vec{H}$ (figure 9.3.5.4).

9.3.6. EXAMPLES IN DIMENSION 2. Let $n = 2$, and take a, b, a', b' such that $ab = a'b'$. By 9.1.6, we know there exists a unique $f \in \mathrm{Is}^+(X)$ (and a unique $f \in \mathrm{Is}^-(X)$) such that $f(a) = a'$ and $f(b) = b'$. It is natural to try to find an explicit geometric construction for this f. This is not hard.

If $f \in \mathrm{Is}^+(X)$ (and excepting the obvious case $f \in T(X)$), we find the center ω of f (cf. 9.3.4) by taking the intersection of the bisectors of the segments $[a, a']$ and $[b, b']$. Another construction, also good for the case of orientation-preserving similarities (cf. 9.7.5), uses the fact that ω is the intersection of the circles defined by $a, a', \langle a, b \rangle \cap \langle a', b' \rangle$ and $b, b', \langle a, b \rangle \cap \langle a', b' \rangle$. (See figure 9.3.6.1.)

If $f \in \mathrm{Is}^-(X)$, it is enough to remark that the axis of f (cf. 9.3.4) contains the midpoint of x and $f(x)$ for every $x \in X$ (figure 9.3.6.2).

It is also useful to be able to construct compositions explicitly (figure 9.3.6.3). In 1.7.5.3 we made essential use of the following result: if r and s are rotations with center a and b, respectively, and $tsr = \mathrm{Id}_X$, the center c of

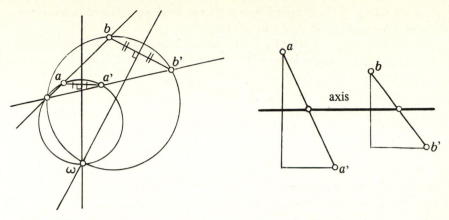

Figure 9.3.6.1   Figure 9.3.6.2

t and the angles of r, s, t in $\mathcal{A}(\vec{X})$ satisfy the equalities

$$2\,\widehat{\overrightarrow{ac},\,\overrightarrow{ab}} = \text{angle of } r, \quad 2\,\widehat{\overrightarrow{ba},\,\overrightarrow{bc}} = \text{angle of } s, \quad 2\,\widehat{\overrightarrow{cb},\,\overrightarrow{ca}} = \text{angle of } t.$$

To prove this result, apply 8.3.5 to write $r = \sigma_{\overrightarrow{ab}}\sigma_D$ and $s = \sigma_E\sigma_{\overrightarrow{ab}}$, where D and E are appropriately chosen lines. Then $sr = \sigma_E\sigma_D$, so the center of sr, and consequently the center of $t = (sr)^{-1}$, is in the intersection $D \cap E$. We can thus write $c = D \cap E$. But then $r = \sigma_{\overrightarrow{ab}}\sigma_{\overrightarrow{ac}}$ and $s = \sigma_{\overrightarrow{bc}}\sigma_{\overrightarrow{ab}}$, and it suffices to apply 8.7.7.7.

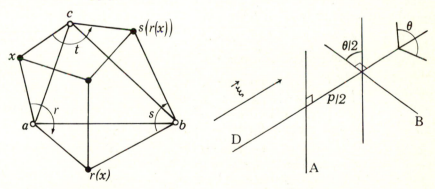

Figure 9.3.6.3   Figure 9.3.7

9.3.7. EXAMPLES IN DIMENSION 3. In the case $n = 3$, corollary 9.3.3 (iii) takes a simple and nice form, and can be independently proved using elementary means. A screw motion f with axis D, translation vector $\vec{\xi}$ and angle $\theta \in \,]0, \pi]$ can be decomposed as the product of two codimension-two reflections σ_A and σ_B. Further, A and B are arbitrary lines orthogonal to D and intersecting D, satisfying the conditions $d(A, B) = p/2$ (where $p = \|\vec{\xi}\|$ is called the *pitch* of the screw) and $\widehat{AB} = \theta/2$. This decomposition makes for an easy study of the composition of screw motions, and yields numerous applications; see [FL, 338–339] and 9.14.38.

Screw motions are naturally found in mechanics and in differential geometry (for example as the first-order approximation to a differentiable curve in Is(X)); see 9.14.7 or [SY, 282].

9.4. Polygonal billiards and the structure of plane isometries

In this section X is always a Euclidean affine plane.

In this section we utilize our knowledge of the structure of plane isometries to study a more elaborate example, involving two related problems: polygonal billiards and least-perimeter polygons inscribed in a given polygon. For the sake of readability, we deal first with the case of triangles, and then extend the discussion to polygons with arbitrarily many sides. The interesting thing is that the solution is radically different depending on whether the number of sides is even or odd.

9.4.1. THE CASE OF A TRIANGLE. Let a, b, c be an arbitrary triangle in X. A triangle *inscribed* in $\{a, b, c\}$ is a triple of points α, β, γ, where $\alpha \in [b, c]$, $\beta \in [c, a]$, $\gamma \in [a, b]$ (cf. 3.4.3). The *perimeter* of $\{\alpha, \beta, \gamma\}$ is $\alpha\beta + \beta\gamma + \gamma\alpha$ (cf. 10.3 or 12.3.1). The triangle $\{\alpha, \beta, \gamma\}$ is called a *billiard trajectory*, or a *light polygon*, if $\alpha \in]b, c[$, $\beta \in]c, a[$, $\gamma \in]a, b[$ and each side of $\{a, b, c\}$ is an exterior bisector of $\{\alpha, \beta, \gamma\}$ (this means that, in the terminology of 8.7.3.2, $\overrightarrow{bc}$ is the bisector of $\overrightarrow{\gamma\alpha}$ and $\overrightarrow{\alpha\beta}$ and so on). The connection between billiards and least perimeters is afforded by the following

9.4.1.1. Lemma. *Consider, in a Euclidean plane, a line D and two points a and b lying in the same open half-plane determined by D (cf. 2.7.3). There exists a unique $x \in D$ such that $ax + bx$ is minimal, and x is characterized by the fact that D supports the bisectors of $\overrightarrow{ax}$ and $\overrightarrow{xb}$.*

Proof. The classical trick is to introduce $a' = \sigma_D(a)$. Then $\langle a', b \rangle$ meets D in a unique point x, and we have, for every $y \in D$,

$$ay + by = a'y + by \geq a'b = ax + bx. \qquad \square$$

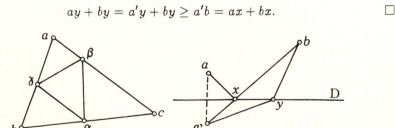

Figure 9.4.1.0   Figure 9.4.1.1

9.4.1.2. Note. The fact that the minimum satisfies the bisector condition also follows from the first variation formula (9.10.5).

9.4.1.3. Proposition. *If the triangle $\{a, b, c\}$ is acute (i.e., its angles lie in $[0, \pi/2]$), it possesses a unique billiard trajectory, which is also the least-perimeter triangle inscribed in $\{a, b, c\}$, and is formed by the feet of the altitudes of $\{a, b, c\}$.*

If $\{a, b, c\}$ has an angle $\geq \pi/2$ at a, it possesses no billiard trajectory, but has a unique inscribed least-perimeter triangle $\{\alpha, \beta, \gamma\}$, where α is the foot of the altitude from a and $\beta = \gamma = a$.

9.4.1.4. Proof. Least-perimeter polygons always exist by compactness, since the perimeter function

$$[b, c] \times [c, a] \times [a, b] \ni (\alpha, \beta, \gamma) \mapsto \alpha\beta + \beta\gamma + \gamma\alpha \in \mathbf{R}$$

is continuous. Suppose the minimum is achieved for $\alpha \in \,]b, c[$, $\beta \in \,]c, a[$ and $\gamma \in \,]a, b[$. Then $\{\alpha, \beta, \gamma\}$ is a billiard trajectory by 9.4.1.1. Let $A = \langle b, c \rangle$, $B = \langle c, a \rangle$, $C = \langle a, b \rangle$ be the lines supporting the sides of $\{a, b, c\}$, and introduce the reflections σ_A, σ_B, σ_C. If $\{\alpha, \beta, \gamma\}$ is a billiard trajectory, the composition $f = \sigma_C \sigma_B \sigma_A$ satisfies $f(\langle \alpha, \gamma \rangle) = \langle \alpha, \gamma \rangle$. But $f \in \mathrm{Is}^-(X)$, so $\langle \alpha, \gamma \rangle$ is well-determined, being the axis of f (cf. 9.3.4). Thus a billiard trajectory, if it exists, is unique.

Now the axis of f is determined by the feet of the altitudes (check this), and the three feet lie in the interior of the corresponding sides if and only if the triangle is acute. There remains to show that in the acute case the billiard trajectory obtained has minimal perimeter, and that in the obtuse case the least-perimeter polygon is indeed the stated one. The latter is done by hand, observing that at least one of α, β, γ must coincide with a vertex of $\{a, b, c\}$ by the paragraph above.

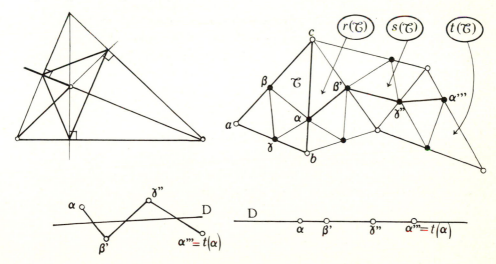

Figure 9.4.1.4

To treat the acute case, we unfold an arbitrary inscribed triangle α, β, γ out on the plane as follows: consider the three isometries $r = \sigma_A$, $s = \sigma_{B'} r$, where $B' = r(B)$, and $t = \sigma_{C''} s$, where $C'' = s(C)$. We find

$$\sigma_{B'} = \sigma_{r(B)} = \sigma_A \sigma_B \sigma_A^{-1} = \sigma_A \sigma_B \sigma_A,$$

so $s = \sigma_A \sigma_B$, and similarly $t = \sigma_A \sigma_B \sigma_C$ (so that $t = f^{-1}$!). Set $\beta' = r(\beta)$, $\gamma'' = s(\gamma)$, $\alpha''' = t(\alpha)$. Then

$$\alpha\beta + \beta\gamma + \gamma\alpha = \alpha\beta' + \beta'\gamma'' + \gamma''\alpha''' \geq \alpha t(\alpha)$$

by the triangle inequality. Applying 9.3.2, we see that the function $x \mapsto xt(x)$ has its minimum on the axis D of t, which is the same as the axis of f. Conversely, if α belongs to this axis (and to $[a, b]$) we easily see that α, β', γ'' and α''' all lie on this axis, in the order named, so that $\alpha\beta' + \beta'\gamma'' + \gamma''\alpha''' = \alpha t(\alpha)$. $\square$

9.4.2. ARBITRARY POLYGONS. We consider an n-sided convex polygon P (cf. 12.1) with vertices $(a_i)_{i=1,\ldots,n}$ and sides $[a_i, a_{i+1}]$ supported by lines $D_i = \langle a_i, a_{i+1} \rangle$ (where, by convention, $n + 1 = 1$). We generalize the notions introduced in 9.4.1: a polygon *inscribed* in P is an n-tuple of points $(\alpha_i)_{i=1,\ldots,n}$ such that $\alpha_i \in [a_i, a_{i+1}]$ for $i = 1, \ldots, n$. The inscribed polygon $(\alpha_i)_{i=1,\ldots,n}$ is a *light polygon* if $\alpha_i \in]a_i, a_{i+1}[$ and $\sigma_{D_i}(\langle \alpha_i, \alpha_{i-1} \rangle) = \langle \alpha_i, \alpha_{i+1} \rangle$ for all $i = 1, \ldots, n$. The *perimeter* of (α_i) is defined as

$$p((\alpha_i)) = \sum_{i=1}^{n} \alpha_i \alpha_{i+1}.$$

9.4.2.1. Theorem. *Any convex polygon has least-perimeter inscribed polygons. Any strictly inscribed least-perimeter polygon is a light polygon. Conversely, any light polygon has minimal perimeter.*

Assume there exists a light polygon. If n is odd, the polygon is unique. If n is even, there are infinitely many light polygons.

If n is even, a necessary condition for the existence of light polygons is that the relation $\sum_{i=1}^{n/2} \widehat{D_{2i-1}D_{2i}} = 0$ hold in the set $\mathcal{A}(\vec{X})$ of oriented angles between lines; in particular, if $n = 4$, the vertices must lie on the same circle. A sufficient condition for the existence of a light polygon for n arbitrary is that there be a line D such that $f(D) = D$ and $D \cap g_i(]a_{i+1}, a_{i+2}[) \neq \emptyset$ for all $i = 1, \ldots, n$, where f and g_i are defined below.

Proof. The proof involves no new ideas; we just adapt the ones used in 9.4.1.4.

9.4.2.2. Least-perimeter inscribed polygons exist by compactness. From 9.4.1.1 we see that strictly inscribed least-perimeter polygons are light polygons. The converse follows from the second part of the theorem.

9.4.2.3. We generalize the constructions in 9.4.1.4 by putting $f = \sigma_{D_n} \cdots \sigma_{D_1}$ and defining g_i by recurrence:

9.4.2.4
$$g_1 = \sigma_{D_1}, \quad g_2 = \sigma_{g_1(D_2)} g_1, \quad \ldots, \quad g_{i+1} = \sigma_{g_i(D_{i+1})} g_i \quad (i = 1, \ldots, n).$$

Since we have $\sigma_{h(C)} = h \circ \sigma_C h^{-1}$, we see that

9.4.2.5

$$g_i = \sigma_{D_1} \cdots \sigma_{D_i} \text{ for all } i; \quad \text{in particular } g = g_n = \sigma_{D_1} \cdots \sigma_{D_n} = f^{-1}.$$

If (α_i) is a polygon inscribed in (a_i), we define points β_i by recurrence, as follows:

9.4.2.6 $\quad \beta_1 = \alpha_1, \quad \beta_2 = g_1(\alpha_2), \quad \ldots, \quad \beta_{i+1} = g_i(\alpha_{i+1}) \quad (i = 1, \ldots, n).$

The union $\bigcup_{i=1}^n [\beta_i \beta_{i+1}]$ is a broken line obtained by unfolding the inscribed polygon (α_i). This broken line is a line segment if and only if (α_i) is a light polygon.

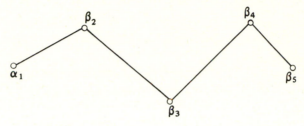

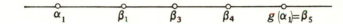

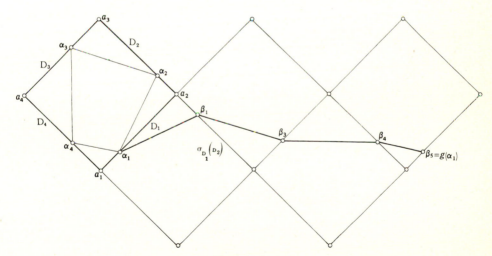

Figure 9.4.2.6

By construction we have $p((\alpha_i)) = \sum_{i=1}^n \beta_i \beta_{i+1}$. By the strict triangle inequality we also have

9.4.2.7 $$p((\alpha_i)) \geq \alpha_i g(\alpha_i),$$

and $p((\alpha_i)) = \alpha_i g(\alpha_i)$ if and only if the β_i lie on the same line, in the right order.

9.4.2.8. Assume that (α_i) is a light polygon. The line $D = \langle \alpha_n, \alpha_1 \rangle$ is invariant under f. Thus, if n is odd, $f \in \mathrm{Is}^-(X)$ and D is the axis of f, well-determined by 9.3.4. This shows that light polygons are unique when n is odd. (Non-uniqueness for n even will be shown at the end of the proof.) When n is even the condition $f(D) = D$ implies, by 9.3.4, that f is a translation, or, equivalently, $\underline{\Phi}^{-1}(f) = 0$ in $\tilde{A}(\vec{X})$. Writing f as a product of pairs of reflections,

$$f = (\sigma_{D_n} \sigma_{D_{n-1}}) \cdots (\sigma_{D_2} \sigma_{D_1})$$

and applying 8.7.7.8, we obtain the desired result $\sum_{i=1}^{n/2} \widehat{D_{2i-1} D_{2i}} = 0$. Co-cyclicity for $n = 4$ follows from 10.9.5.

9.4.2.9. Now assume that the conditions in the last assertion of the theorem are satisfied. From the convexity of (a_i) we conclude that the points $\beta_i = D \cap g_i(]a_{i+1}, a_{i+2}[)$ are disposed on D in the right order, so the inscribed polygon (α_i) defined by $\alpha_i = g_{i-1}^{-1}(\beta_i)$ for $i = 1, \ldots, n$ has perimeter $p((a_i)) = \beta_1 g(\beta_1)$ (cf. 9.4.2.7). Conversely, the perimeter of an arbitrary polygon (a_i'), again by 9.4.2.7, is equal to $\alpha_1' g(\alpha_1')$; but 9.3.2 and 9.3.4 show that $p((\alpha_i')) \geq p((\alpha_i))$ always, whether n is odd (in which case f belongs to $\mathrm{Is}^-(X)$ and has a unique axis) or even (and f is a translation and leaves invariant all lines parallel to

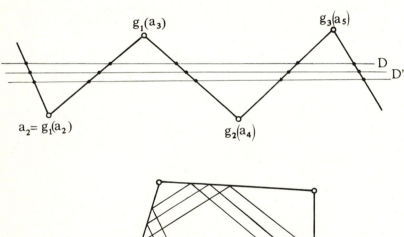

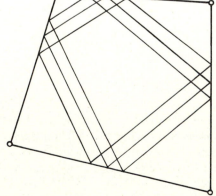

Figure 9.4.2.9

the direction of translation). We conclude that the polygon (α_i) has minimal perimeter and is a light polygon. This also shows that every light polygon has minimal perimeter, since the line $D = \langle \alpha_1, \alpha_n \rangle$ satisfies the conditions at the end of the theorem, by definition of the g_i.

9.4.2.10. To conclude, assume that n is even and (α_i) is a light polygon. The line $D = \langle \alpha_1, \alpha_n \rangle$ enjoys the right intersection properties with the segments $g_i(]a_{i+1}, a_{i+2}[)$; by continuity, all lines D' parallel to D and sufficiently close to it enjoy the same properties and satisfy $f(D') = D$, since $f \in T(X)$. $\square$

9.4.3. REMARKS. The somewhat annoying fact that light polygons are not always to be found (cf. 9.4.1.3) occurs for every value of n; see 9.14.10 and 9.14.33. When interested in finding a least-perimeter polygon for an actual problem, one should first construct f and apply 9.2.4.1 to check for the existence of light polygons. If they do not exist, one of the α_i will be a vertex of P, and one has to carry out a case analysis by hand, applying 9.4.1.1 systematically.

One can extend the above study to billiard trajectories that close only after two or more turns (cf. 9.14.9). Another possible direction is to look for light polygons in arbitrary convex compact sets of the plane. The situation is nicer if the set is strictly convex (cf. 11.6.4); then there exist light polygons with n vertices for every $n \geq 2$ (see 9.14.33 for a proof). The case of an ellipse is particularly impressive: we shall see in 17.6.6 that not only do light polygons exist for every n, but one of their vertices can actually be chosen arbitrarily on the ellipse. On the other hand, if the set is not strictly convex, the trajectory of a light ray may leave the set.

Figure 9.4.3

9.4.4. ERGODICITY. An interesting problem, currently under intensive research, is the ergodicity of polygonal or compact convex billiards. A billiard trajectory is determined by an initial point and direction: the ball follows this direction until it hits the boundary, then bounces off in the direction prescribed by the equality between the incidence and reflection angles, and so on forever. Some trajectories may hit the boundary at a vertex, where the reflection direction is not well-determined, but this does not matter, as ergodic theory proposes to study the set of trajectories only up to a set of measure zero. A compact convex set is called *weakly ergodic* if (almost) all

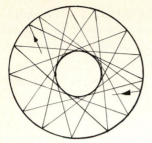
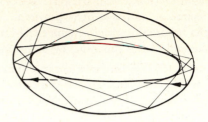

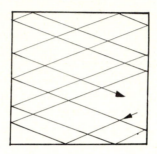

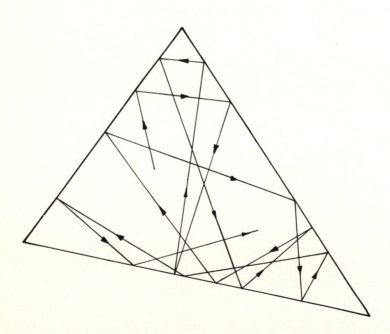

Figure 9.4.4

trajectories are dense in the set, and *strongly ergodic* if all trajectories are dense in the product of the set and S^1 (where the second component gives the direction at each point of the trajectory). This problem is in some sense opposite to the search for closed trajectories, undertaken in 9.4.2.

Some convex billiards are completely understood: circles and ellipses are never weakly ergodic (trivial for the circle, a consequence of 17.6.6 for the ellipse). Some triangles are not strongly ergodic (9.14.11). On the other hand, a square is certainly weakly ergodic, since trajectories whose directions have irrational slopes are dense, but not strongly ergodic, because each trajectory has only two slopes, opposite to each other.

In other cases the situation is much more complicated. Even for arbitrary triangles with given angles, for example, it is an open question whether ergodicity holds. In order to glimpse the complexity of the problem, the reader can just try to follow a trajectory; after a number of bounces, the situation quickly seems to get out of hand. Recently Lazutkin [LZ] showed that strict convex sets with C^2 boundary are never ergodic; this is a consequence of the existence of caustics, curves that generalize the homofocal ellipses tangent to trajectories in elliptical billiards.

For references on the ergodicity of billiards, see [A–A], [CZ], [ME], [SI].

9.5. Similarities

Consider $f \in \mathrm{GA}(X)$ such that $\vec{f} \in \mathrm{GO}(\vec{X})$ (cf. 8.8.2). Writing μ for the ratio of $\vec{f}$, we have $f(x)f(y) = \mu xy$, and $f(x')f(y') = \mu x'y'$, whence

$$\frac{f(x')f(y')}{f(x)f(y)} = \frac{x'y'}{xy}$$

whenever $x \neq y$. In other words, f preserves ratios between distances. The converse also holds:

9.5.1. PROPOSITION AND DEFINITION. *Let $f : X \to X$ be a non-constant set-theoretical map such that $\dfrac{f(x')f(y')}{f(x)f(y)} = \dfrac{x'y'}{xy}$ whenever $x \neq y$ and $f(x) \neq f(y)$. Then $f \in \mathrm{GA}(X)$ and $\vec{f} \in \mathrm{GO}(\vec{X})$. Such maps are called similarities of X, and they are orientation-preserving or reversing according to whether $\vec{f} \in \mathrm{GO}^+(X)$ or $\vec{f} \in \mathrm{GO}^-(X)$. The set of similarities (resp. orientation-preserving, orientation-reversing) similarities of X is denoted by $\mathrm{Sim}(X)$ (resp. $\mathrm{Sim}^+(X)$, $\mathrm{Sim}^-(X)$). The ratio of f is defined as the ratio of $\vec{f}$.*

Proof. Take distinct points x_0, y_0 such that $f(x_0) \neq f(y_0)$. The hypothesis implies that f is bijective and $f(x)f(y) = \mu xy$ for every $x, y \in X$, where μ is defined by $\mu = \dfrac{f(x_0)f(y_0)}{x_0 y_0}$. If we now take an arbitrary homothety h of ratio μ^{-1}, the composition $h \circ f$ is an isometry of X, whence, by 9.1.3, $h \circ f \in \mathrm{GA}(X)$ and $\overrightarrow{h \circ f} \in O(\vec{X})$. The conclusion follows by composing again with h^{-1}. □

9.5.2. PROPOSITION. *Consider* $f \in \mathrm{Sim}(X) \setminus \mathrm{Is}(X)$. *There exists a unique* $\omega \in X$ *such that* $f(\omega) = \omega$; *this point is called the center of the similarity* f. *We have a decomposition* $f = h \circ g = g \circ h$, *where* $h \in H_{\omega,\mu}$ *and* $g \in \mathrm{Is}_\omega(X)$.

Proof. This follows from 9.3.3. There is also a topological proof, using contracting maps—obviously either f or its inverse are contracting, depending on whether the ratio of f is greater or less than one. □

9.5.3. CHARACTERIZATION OF SIMILARITIES

9.5.3.1. Let f be a similarity. From 9.2.1 and 8.8.5.1 it follows that f preserves angles between lines and half-lines and that, in the two-dimensional case, f preserves or reverses oriented angles depending on whether it preserves or reverses orientation. In particular, f preserves orthogonality between lines:

$$D \perp D' \Rightarrow f(D) \perp f(D').$$

Furthermore, the image under f of a sphere S of X (cf. 10.7 if necessary) is also a sphere (with radius μ times bigger). Either of these two properties characterizes similarities:

9.5.3.2. **Theorem.** *Let* f *be a set-theoretical bijection of* X, *where* $\dim X \geq 2$. *The following conditions are equivalent:*
  i) f *is a similarity;*
  ii) *if* a, b, c, d *are points of* X *satisfying* $a \neq b$, $c \neq d$ *and* $\langle a, b \rangle \perp \langle c, d \rangle$, *the images satisfy* $\langle f(a), f(b) \rangle \perp \langle f(c), f(d) \rangle$;
  iii) *the image* $f(S)$ *of any sphere* S *of* X *is also a sphere.*

9.5.3.3. *Proof.* Assume first that (ii) is satisfied. In view of 2.6.5 and 8.8.5.1, it is enough to show that f takes collinear points a, b, c into collinear points $f(a)$, $f(b)$, $f(c)$. To do so, take vectors $a_2, \ldots, a_n$ such that $\{a, b, a_2, \ldots, a_n\}$ is an orthogonal affine frame; by assumption and 8.1.2.5, $\{f(a), f(b), f(a_2), \ldots, f(a_n)\}$ is an orthogonal affine frame. But

$$\langle f(a), f(c) \rangle \perp \langle f(a), f(a_i) \rangle$$

for all $i = 2, \ldots, n$, showing that $\langle f(a), f(b) \rangle = \langle f(a), f(c) \rangle$.

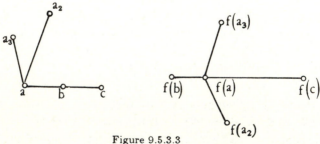

Figure 9.5.3.3

9.5.3.4. Now assume that (iii) is satisfied; we start by showing that f takes collinear points into collinear points. This is easy in the opposite direction: if a', b', c' are distinct collinear points, the points $a = f^{-1}(a')$, $b = f^{-1}(b')$ and

$c = f^{-1}(c')$ are collinear, else we could consider a sphere S containing a, b, c and its image $f(S)$ would contain the collinear points a', b', c', a contradiction (see 10.7.2 if necessary). In the forward direction, take distinct collinear points a, b, c and assume that their images $a' = f(a)$, $b' = f(b)$ and $c' = f(c)$ do not satisfy $c' \in \langle a', b' \rangle$. By the previous argument $f^{-1}(\langle a', b' \rangle) \subset \langle a, b \rangle$. Put $P' = \langle a', b', c' \rangle$ and let d' be an arbitrary point in $\langle a', b' \rangle$; then $f^{-1}(\langle c', d' \rangle) \subset \langle a, c \rangle = \langle a, b \rangle$, so $f^{-1}(P') \subset \langle a, b \rangle$. Now let S be a sphere containing a and b; we have $S \cap \langle a, b \rangle = \{a, b\}$, and $C' = f(S) \cap P'$ is a circle because $f(S)$ is, by assumption, a sphere. But this would imply $f^{-1}(C') \subset \langle a, b \rangle$, which is absurd, since a circle must contain at least three points.

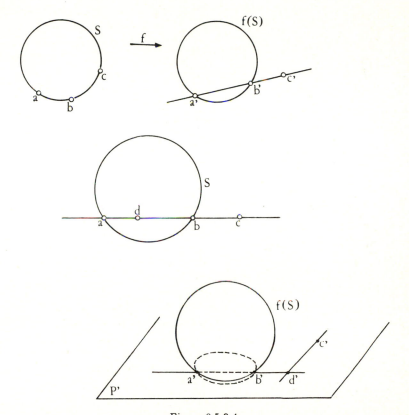

Figure 9.5.3.4

9.5.3.5. After an application of 2.6.5, there remains to show that a map $f \in \mathrm{GA}(X)$ that takes spheres into spheres is a similarity; in fact, it is enough to assume that there is one sphere whose image is also a sphere. After composing f with a dilatation, we can assume that there is a sphere S such that $f(S) = S$. We first show that f fixes the center ω of S. This is done by considering two diametrically opposed points a, b on S, and the hyperplanes H_a and H_b, tangent to S at a and b. Hyperplanes tangent to a sphere are characterized by intersecting it in a single point, and this condition is preserved by f; thus $f(H_a)$ and $f(H_b)$ are the hyperplanes tangent to S at $f(a)$ and $f(b)$. Now two points on S are diametrically opposed if and only if the tangent hyperplanes at these points are parallel; since parallelism is also preserved by f, we conclude that $f(H_a)$ and $f(H_b)$ are parallel and $f(a)$, $f(b)$ are diametrically opposed. Using the fact that f is affine, we obtain

$$\omega = \frac{f(a) + f(b)}{2} = f\left(\frac{a+b}{2}\right) = f(\omega),$$

as desired.

This brings us to the vector case, $f \in \mathrm{GL}(X_\omega)$; but then saying that $f(S) = S$ is equivalent to saying that f preserves the norm of X_ω, whence $f \in O(X_\omega)$, by 8.1.5.   □

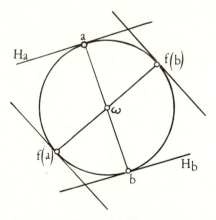

Figure 9.5.3.5

9.5.3.6. Note. If one weakens the assumptions in 9.5.3.2, requiring that f be a bijection between two parts of X only, condition (iii) is not sufficient to guarantee that f is the restriction of a similarity. For example, f can be an inversion (10.8.2) or a composition of inversions (18.10.4). For a discussion of bijective maps f between subsets of affine spaces that take spheres into spheres, see [CD] and [G–W].

9.5.4. Liouville's theorem

9.5.4.1. Liouville's theorem provides a characterization of similarities in terms of differential geometry, but its discussion is relevant to other parts of this book as well: 10.8.5, 18.10, 20.6.

We start from a similarity f of X, considered as a C^∞ map from X into itself. Every morphism $f \in A(X;X)$ is of class C^∞ and its derivative $f' : X \to L(X;X)$ is just the constant map $x \mapsto \vec{f}$, that is, $f'(x) = \vec{f}$ for all $x \in X$ (this follows from 2.7.7). In the particular case $f \in \mathrm{Sim}(X)$, we have $f'(x) = \vec{f} \in \mathrm{GO}(\vec{X})$ for all $x \in X$; this can be expressed by saying that f preserves infinitesimal angles. The question arises whether there are other maps with the same property. In dimension one the question is trivial, since every map with nowhere vanishing derivative satisfies the property; we thus restrict our analysis to the case $n > 1$.

9.5.4.2. Definition. *Let U and V be open subsets of a Euclidean affine space X. A map $f \in C^1(U; V)$ (that is, a C^1 map $f : U \to V$) is said to be conformal if f is bijective and $f'(x) \in \mathrm{GO}(\vec{X})$ for every $x \in U$. A conformal map f is orientation-preserving (resp. reversing) if $f'(X) \in \mathrm{GO}^+(\vec{X})$ (resp. $\mathrm{GO}^-(\vec{X})$). The corresponding sets are denoted by $\mathrm{Conf}(U; V)$ and $\mathrm{Conf}^\pm(U; V)$; we also write $\mathrm{Conf}(U)$ and $\mathrm{Conf}^\pm(U)$ for $\mathrm{Conf}(U; U)$ and $\mathrm{Conf}^\pm(U; U)$.*

It is easy to see, using injectivity, the theorem of invariance of domains and the fact that $\mathrm{GO}(\vec{X}) \subset \mathrm{Isom}(\vec{X}; \vec{X})$, that a map $f \in \mathrm{Conf}(U; V)$ is necessarily a diffeomorphism onto its image.

9.5.4.3. First example: holomorphic functions

Assume that X has dimension $n = 2$, and identify it with an affine complex line (cf. 8.3.12, 9.6.4). A function $f : U \to X$, where U is an open subset of X, is called *holomorphic* if it has a complex derivative at every point; this notion is well-defined because, even though we have to orient X to make sense of it, the choice of orientation does not affect the outcome. A classical result states that $f \in \mathrm{Conf}^+(U; f(U))$ if and only if f is injective, holomorphic, and its derivative is never zero. The proof follows from 8.8.4.1 (see [CH2, 67] if necessary).

We thus obtain a great number of infinitesimal similarities which are not global similarities, because injective holomorphic maps on open sets U with nowhere vanishing derivative are very abundant. The space of such maps, in fact, cannot be parametrized in any reasonable way by finitely many parameters.

9.5.4.4. Second example: inversions

We shall see in 10.8.5 that an inversion f with center a and arbitrary power belongs to $\mathrm{Conf}(X \setminus a)$. The composition of a finite number of inversions is thus an element of $\mathrm{Conf}(U; f(U))$, where U is equal to X minus a finite number of points. Again, these maps are not global similarities.

9.5.4.5. The number of degrees of freedom in the choice of inversions is finite, and we shall see in 18.10.4 that it is equal to $((n + 2)(n + 1))/2$.

The next theorem shows that, for infinitesimal similarities, there is a fundamental difference between the cases $n = 2$ and $n \geq 3$; the theorem also shows that there is a radical difference between the "local" case (the open set U is a proper subset of X) and the "global" case ($U = X$). See 12.8, 16.4 and 18.3.8.6 for other examples of passing from local to global.

9.5.4.6. Theorem. *For $n = 2$ we have $\mathrm{Conf}(X) = \mathrm{Sim}(X)$. For arbitrary n, a C^4 map $f \in \mathrm{Conf}(X)$ belongs to $\mathrm{Sim}(X)$. If $n \geq 3$, U is an open set of X and $f \in \mathrm{Conf}(U; f(U))$ is of class C^4, then f is the restriction to U of a product of inversions of X (and so must be C^∞).*

9.5.4.7. Note. The equality $\mathrm{Conf}(X) = \mathrm{Sim}(X)$ actually holds for all n (that is, the C^4 condition in the theorem could be replaced by C^1), but the proof is much more difficult, resorting to fine analytical techniques (cf. [HM]). The proof given below for the case $n \geq 3$ is due to R. Nevanlinna [NA]; see 9.5.4.21. See also 9.14.45 and [LF3, 59, exercise 12].

9.5.4.8. The case $n = 2$. After vectorializing and orienting X and identifying it with $\mathbf{C}$, we have a map $f : \mathbf{C} \to \mathbf{C}$, injective and of class C^1. Since $\mathbf{C}$ is connected, we have either $f'(z) \in \mathrm{GO}^+(\mathbf{C})$ or $f'(z) \in \mathrm{GO}^-(\mathbf{C})$ for all $z \in \mathbf{C}$. After composing f with a conjugation $z \to \bar{z}$, we can assume we are in the first case. By 9.5.4.3 the map $f : \mathbf{C} \to \mathbf{C}$ is holomorphic and injective, and, by a classical result, must be of the form $z \to az + b$, for some $a, b \in \mathbf{C}$, and hence a similarity (see 9.6.4 if necessary). (The result quoted can be found in [CH2, 181–182]; the proof is nice, but it uses the notions of essential singularities and meromorphic functions.)

The proof for $n \geq 3$ uses, among other thing, the following famous lemma:

9.5.4.9. Braid lemma (see also 1.9.14). *Let V and W be vector spaces over a field of characteristic $\neq 2$, and $k : V \times V \times V \to W$ a trilinear map, symmetric on the first two variables and skew-symmetric on the last two. Then $k = 0$.*

This lemma is often pivotal in differential geometry, as exemplified by this case; see also [SB, 333] and [KO–NO, volume 1, p. 160]. The name comes from the fact that its proof mimics the first six steps in making a braid: after six crossings, three of the first two strands and three of the last two, we're back in the original situation; but swapping the last two variables three times switches the sign of the result, leading to the desired conclusion. □

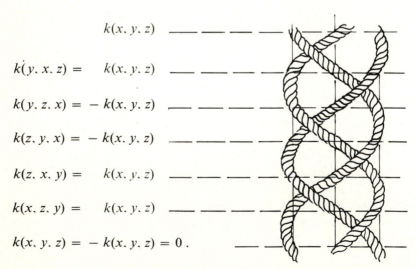

$$k(x, y, z)$$

$$k(y, x, z) = \quad k(x, y, z)$$

$$k(y, z, x) = -\,k(x, y, z)$$

$$k(z, y, x) = -\,k(x, y, z)$$

$$k(z, x, y) = \quad k(x, y, z)$$

$$k(x, z, y) = \quad k(x, y, z)$$

$$k(x, y, z) = -\,k(x, y, z) = 0\,.$$

Figure 9.5.4.9

9.5.4.10. By assumption, $f'(x) \in \mathrm{GO}(\vec{X})$ for all $x \in U$. Denoting by $\mu(x)$ the corresponding similarity ratio, we have

9.5.4.11 $\qquad \big(f'(x)(u) \mid f'(x)(v)\big) = \mu^2(x)(u \mid v) \quad$ for all $u, v \in \vec{X}$.

Let u, v be fixed orthogonal vectors; we have $\big(f'(x)(u) \mid f'(x)(v)\big) = 0$, and differentiation gives, for all $w \in \vec{X}$,

$$\big(f''(x)(u, w) \mid f'(x)(v)\big) + \big(f'(x)(u) \mid f''(x)(v, w)\big) = 0.$$

The braid argument and the symmetry of f'' show that $\big(f''(x)(u, v) \mid f'(x)(w)\big) = 0$ for any w orthogonal to u and v. But for such a w the vectors $f'(x)(u)$, $f'(x)(v)$ and $f'(x)(w)$ are orthogonal, and consideration of a basis of the orthogonal complement of $\mathbf{R}u + \mathbf{R}v$ leads to the conclusion that $f''(u, v) \in \mathbf{R}f'(x)u + \mathbf{R}f'(x)v$. Thus there exist two functions $\alpha, \beta : U \to \mathbf{R}$ such that

9.5.4.12 $\qquad f''(x)(u, v) = \alpha(x)f'(x)(u) + \beta(x)f'(x)(v)$

for every $x \in U$ and fixed orthogonal vectors u, v.

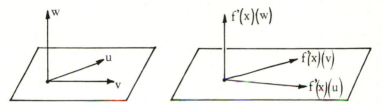

Figure 9.5.4.12

Upon differentiation of $\|f'(x)(u)\|^2 = \mu^2(x)\|u\|^2$, we obtain

$$\big(f''(x)(u, v) \mid f'(x)(u)\big) = \mu'(x)(v)\mu(x)\|u\|^2,$$

and substituting this into 9.5.4.12 gives

9.5.4.13 $\qquad \alpha(x) = \dfrac{\mu'(x)(v)}{\mu(x)}, \qquad \beta(x) = \dfrac{\mu'(x)(u)}{\mu(x)}.$

We now put $\rho = \mu^{-1}$ and combine 9.5.4.12 and 9.5.4.13 into

9.5.4.14 $\qquad \rho'(x)(v)f'(x)(u) + \rho'(x)(u)f'(x)(v) + \rho(x)f''(x)(u, v) = 0$

for every $x \in U$ and fixed orthogonal vectors u, v.

9.5.4.15. After catching our breath, we differentiate again, this time formula 9.5.4.14, obtaining

$$\rho''(x)(v, w)f'(x)(u) + \rho'(x)(v)f''(x)(u, w) + \rho''(u, w)f'(x)(v)$$
$$+ \rho'(x)(u)f''(x)(v, w) + \rho'(w)f''(x)(u, v) + \rho(x)f'''(x)(u, v, w) = 0.$$

We verify that the sum of the last five terms is symmetric in u and w, so the same must hold for the first term:

$$\rho''(x)(v, w)f'(x)(u) = \rho''(x)(v, u)f'(x)(w)$$

for all orthogonal vectors u, v, w (since ρ itself does not depend on u and v). But the two vectors $f'(x)(u)$, $f'(x)(w)$ are linearly independent, so that $\rho''(x)(u, v) = 0$ for all orthogonal pairs u, v because $n \geq 3$.

By the proof of 8.8.5.1 (and here infinitesimal arguments do correspond to reality), $\rho''(x)$ must be proportional to the Euclidean structure. In other words, there exists $\sigma : U \to \mathbf{R}$ such that

9.5.4.16 $$\rho''(x)(u, v) = \sigma(x)(u \mid v)$$

for all $x \in U$, $u, v \in \vec{X}$. The factor σ is in fact constant; to see this, we differentiate 9.5.4.16 (which we can do because f is C^4) to get

$$\rho'''(x)(u, v, w) = \sigma'(x)(w)(u \mid v).$$

The second term is symmetric in v and w because ρ''' is symmetric; we thus have $\big(\sigma'(x)(w)v - \sigma'(x)(v)w \mid u\big) = 0$ for all u, so $\sigma'(x)(w)v = \sigma'(x)(v)w$ and $\sigma = 0$ because v and w are linearly independent.

9.5.4.17. The equation $\rho''(u, v) = \sigma(u \mid v)$ can be integrated by inspection, yielding

9.5.4.18 $$\rho(x) = a\|\overrightarrow{x_0 x}\|^2 + b,$$

where a, b are constants and $x_0 \in X$. If $a = 0$, μ is a constant and differentiating 9.5.4.11 gives $f'' = 0$, so f is affine and must be the restriction to U of a similarity. If b is zero, we observe that $x_0 \neq U$ and compose f with an inversion i of pole x_0. The composition $i \circ f$ is still an infinitesimal similarity, and we have $\rho_{i \circ f}(x) = \rho_i\big(f(x)\big)\rho_f(x)$. But the factor ρ_i for an inversion with pole x_0 is $\|\overrightarrow{x_0 x}\|^2$ up to a scalar multiple (cf. 10.8.5.1, the power of an inversion), so the factor $\rho_{i \circ f}$ is constant, and $i \circ f$ is a similarity restricted to U. Notice that the case $b = 0$ is excluded if $U = X$, so an infinitesimal similarity $f : X \to X$ of class C^4 in dimension ≥ 3 is a global similarity.

9.5.4.19. To conclude the proof, we will show that the case $a \neq 0$, $b \neq 0$ cannot occur. Let f be the inverse of $f : U \to f(U)$. This diffeomorphism is still an infinitesimal similarity and, as observed in 9.5.4.18, we have $\mu_g\big(f(x)\big)\mu_f(x) = \mu_{g \circ f} = 1$ for any $x \in U$. By 9.5.4.18 there exist constants c, d such that

9.5.4.20 $$\big(a\|\overrightarrow{x_0 x}\|^2 + b\big)\big(c\|\overrightarrow{f(x_0)f(x)}\|^2 + d\big) = 1$$

for all $x \in U$. This shows, to begin with, that f maps subsets of spheres of center x_0 contained in U into subsets of spheres of center $f(x_0)$; next, that f maps a segment of U whose support contains x_0 into a segment of $f(U)$ whose support contains $f(x_0)$. Fixing $u \in \vec{X}$ with $\|u\| = 1$, we obtain a real-valued function ϕ defined by the equality

$$f(x_0 + tu) = f(x_0) + \phi(t)v,$$

where the unit vector v and the interval of variation of t are appropriately chosen. By 9.5.4.11 we have $\phi'(t) = (at^2 + b)^{-1}$, and also $(at^2 + b)\big(c\phi^2(t) + d\big) = 1$, and the two relations can only coexist if either a or b vanish. $\quad\square$

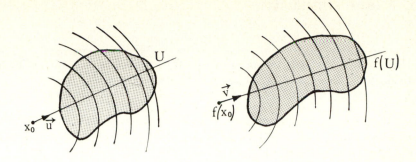

Figure 9.5.4.20

9.5.4.21. Note. Liouville's original demonstration was quite unexpected, using facts apparently unrelated to the theorem, namely: a system of three mutually orthogonal surfaces in dimension three intersects along curvature lines (Dupin's theorem); and spheres are characterized by the fact that all their points are umbilical, that is, they have many more curvature lines than ordinary surfaces. Liouville finished off with 9.5.3.2 (iii).

9.5.5. The umbilical locus and the cyclic points. We study here the affine analogue of 8.8.6. Using the notation of 7.6, we denote by $\tilde{X}^C$ the complexified projective completion of our Euclidean affine space X. We have $\tilde{X}^C = X^C \cup \infty_{X^C}$, with $\infty_{X^C} = P(\vec{X}^C)$. As usual, $p : \tilde{X}^C \to P(\vec{X}^C)$ denotes the projection from a vector space onto the associated projective space. By 8.8.6 and using the terminology of chapter 14, the quadratic form N^C defined on $\infty_{X^C} = P(\vec{X}^C)$ is a quadric with image $\Omega = p((N^C)^{-1}(0))$. If $n = 2$, Ω consists of two points, and if $n = 3$, Ω is a conic (cf. 14.1.3.7).

9.5.5.1. Definition. *The subset Ω of ∞_{X^C} is called the umbilical (locus) of the the affine Euclidean space X. For $n = 3$, Ω is a conic, and for $n = 2$, Ω consists of two points, called the cyclic points of X and denoted by $\{I, J\}$. Choosing one of the two cyclical points is equivalent to orienting X, according to the convention stated in 8.8.6.2.*

There is no abuse in identifying the isotropic lines I, J with the points to which they give rise in the projective space.

We now combine 8.8.6.4 and 5.2.2:

9.5.5.2. Proposition. *A necessary and sufficient condition for $f \in GA(X)$ to be a similarity is that $\tilde{f}^C(\Omega) = \Omega$. If $f = 2$, $f \in \text{Sim}^+(X)$ (resp. $f \in \text{Sim}^-(X)$) if, in addition, $\tilde{f}^C$ fixes (resp. switches) the two cyclic points.* $\square$

9.5.5.3. Example. Orthogonality between two lines D, D' of X can be translated by

$$D \perp D' \iff [\infty_{D^C}, \infty_{D'^C}, I, J] = -1,$$

for switching I and J does not alter the cross-ratio in the case of a harmonic division (cf. 8.8.7.4).

9.6. Plane similarities

> In this section X stands for an oriented affine plane. If X is not oriented
> to begin with, we choose an arbitrary orientation for it.

This section consists mostly of elementary examples with visual appeal, after
the first five paragraphs, which contain a summary of properties and classical
expressions for plane similarities.

9.6.1. STRUCTURE. A map $f \in \mathrm{Sim}^+(X)$ is either a translation or the
product, taken in either order, of a rotation and a homothety with same
center ω, where ω, the center of f, is the unique fixed point of f. A map
$f \in \mathrm{Sim}^-(X)$ is either an isometry, in which case it is a reflection through
a line followed by a translation, or the product, taken in either order, of a
homothety of center ω and a reflection through a line containing ω, where ω
is the unique fixed point of f.

   This follows immediately from 9.3.4 and 9.5.2.

9.6.2. SIMPLE TRANSITIVENESS. For any four points a, b, a', b' of X with
$a \neq b$ and $a' \neq b'$, there exists a unique $f \in \mathrm{Sim}^+(X)$ such that $f(a) = a'$,
$f(b) = b'$.

   To see this we use a homothety to reduce the problem to the case of an
isometry, $ab = a'b'$, and then apply 9.1.6.

   Given a, b, a', b', it is interesting to try to construct geometrically the
center ω of the unique f taking a to a' and b to b'. The solution consists in
finding the intersection point $e = \langle a, b \rangle \cap \langle a', b' \rangle$, drawing the circles containing
a, a', e and b, b', e, and taking the intersection ω of the two circles other than
e (unless the circles are tangent at e). This construction is a consequence of
9.5.3.1 and 10.9.4. See also 9.14.14.

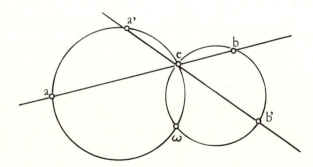

Figure 9.6.2

9.6.3. ANGLE OF AN ORIENTATION-PRESERVING SIMILARITY. We decompose $f \in \mathrm{Sim}^+(X)$ as in 8.8.3 and assign to it the angle of the associated rotation. Denoting this angle by α, we have $\widehat{\Delta f(\Delta)} = \alpha$ for every oriented line Δ.

9.6.4. PLANE SIMILARITIES AND COMPLEX NUMBERS. The fundamental remark here is the affine counterpart of 8.3.12 and 8.8.4. Let X be an oriented affine plane; then X has a natural complex line structure. For this natural structure, we have $\mathrm{GA}(X) = \mathrm{Sim}^+(X)$. In the coordinates associated with an arbitrary complex affine frame of X, every $f \in \mathrm{Sim}^+(X)$ is of the form $z \mapsto az + b$ and every $f \in \mathrm{Sim}^-(X)$ is of the form $z \mapsto a\bar{z} + b$. The angle of $f \in \mathrm{Sim}^+(X)$ (cf. 9.6.3) is simply the argument of the complex number a (cf. 8.7.8.4), and the modulus of f is the absolute value $|a|$ of a.

The discussion about simply transitiveness (9.6.2) and all practical calculations in $\mathrm{Sim}^+(X)$ are facilitated by writing $f : z \mapsto az + b$.

9.6.5. CROSS-RATIO OF FOUR POINTS. Since X is, in a natural way, a complex affine line $\underline{X}$, we can complete it into the complex projective line $\tilde{X} = \underline{X} \cup \infty_X$ (cf. 5.1.3). (Notice that $\tilde{X}$ should not be confused with $\tilde{X}^C$, introduced in 9.5.5. The latter is a complex projective plane, with a line at infinity, while the former is a complex projective line, with only one point at infinity.) The cross-ratio of four points of X is their cross-ratio as points of $\tilde{X}$; it is a number in $\mathbf{C} \cup \infty$, and is finite if the four points are distinct. The connection between this complex cross-ratio and the Euclidean structure of X is very nice:

9.6.5.1. Proposition. *If $z = [a, b, c, d]$, where $a, b, c, d \in X$, the absolute value of z is determined by the distances: $|z| = \dfrac{ac/bc}{ad/bd}$, and the argument of z is determined by the oriented angles: $arg(z) = \widehat{\overrightarrow{ca}, \overrightarrow{cb}} - \widehat{\overrightarrow{da}, \overrightarrow{db}} \, in \tilde{\mathcal{A}}(\vec{X})$.*

9.6.5.2. This result has many elementary consequences, which we do not enumerate exhaustively; we mention but two of them. The first refers to harmonic quadrilaterals, i.e., quadruples of points of X whose cross-ratio is -1 (cf. 9.14.15). The second is that $z = [a, b, c, d]$ is real if and only if its argument is 0 or ϖ; this is equivalent, by 8.7.7, to the condition $\langle c, a \rangle, \langle c, b \rangle = \langle d, a \rangle, \langle d, b \rangle$ in $\mathcal{A}(\vec{X})$; but we shall see that this is exactly the condition that a, b, c, d be cocyclic or collinear. It follows, for example, that homographies $z \mapsto \dfrac{az + b}{cz + d}$ of $\tilde{X}$ map circles into circles (or lines). This result will follow again in a more general context (chapter 20; see 18.10.4 and 20.6).

9.6.6. AN ELEMENTARY CONSTRUCTION. The problem is to find a circle through a given point x and tangent to two given lines D, D'. The case $D \parallel D'$ is left to the reader. We put $a = D \cap D'$, and draw any circle C tangent to D and D' and contained in the convex set defined by D, D' and x. The idea, developed in figure 9.6.6, is to find the homothety of center x which maps C into a circle containing x.

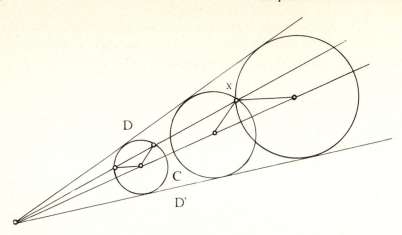

Figure 9.6.6

9.6.7. THE PRINCIPLE OF SIMILAR FIGURES. We give as motivation for this principle the following question: Given two distinct lines D, D' of X and points m, m' on D, D', respectively, each of which moves with constant speed, what is the envelope of the line mm'?

9.6.7.1. Let f be the orientation-preserving similarity such that $f(m) = m'$ and $f(n) = n'$, where m and n are two positions of the point that describes D and m' and n' are two positions of the point describing D' (cf. 9.6.2). Then, if p and p' are two further positions of the points on D and D', we will still have $f(p) = p'$, since the condition of constant speed and the fact that $f \in \mathrm{Sim}^+(X)$ are both equivalent to $\dfrac{\overrightarrow{m'p'}}{\overrightarrow{m'n'}} = \dfrac{\overrightarrow{mp}}{\overrightarrow{mn}}$ (cf. 2.4.6). What we need now is to show that every point linked to m and m' by an orientation-preserving similarity also describes a line. This is where the principle comes in:

9.6.7.2. Principle. *Let f be an orientation-preserving similarity of X and m'' a point linked to the pair (m, m'), where $m' = f(m)$, by an orientation-preserving similarity (this expression means that there exists a similarity $\vec{g} \in \mathrm{GO}^+(\vec{X})$ such that $\overrightarrow{mm''} = g'(\overrightarrow{mm'})$ for every $m \in X$). Then $m \mapsto m''$ is an orientation-preserving similarity. In particular, if m describes a line, a circle, etc., so does m''.*

Proof. We use 9.6.4. Working in an arbitrary frame, we have $f(z) = az + b$ and $\vec{g}(\vec{u}) = c\vec{u}$, where $a, b, c \in \mathbf{C}$. Then

$$m'' = m + mm'' = z + c(f(z) - z) = z + c(az + b - z) = (ac - c + 1)z + bc.$$

But the map $z \mapsto (ac - c + 1)z + bc$ is obviously an orientation-preserving similarity. $\square$

Returning now to the original problem, we see that the foot h of the perpendicular dropped from a to the line $\langle m, m' \rangle$ is linked to (m, m') by an

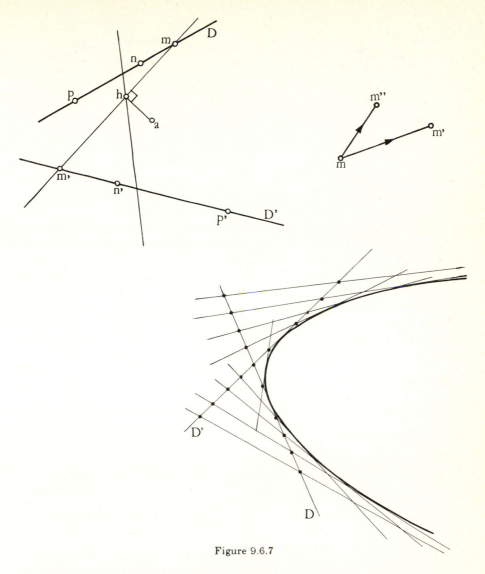

Figure 9.6.7

orientation-preserving similarity, since a is the center of the similarity $m \mapsto m'$; thus m describes a line E and the envelope of the line $\langle m, m' \rangle$ is the parabola with focus a and tangent at the vertex equal to E (cf. 17.2.2.6).

For other applications of the principle, see 10.13.18.

9.6.8. DOUBLE PEDAL CURVE OF TWO CIRCLES. The *pedal (curve)* of a curve C, relative to the point a, is the set of points $m \in X$ for which there exists a tangent to C containing m and orthogonal to the line $\langle a, m \rangle$ (figure 9.6.8.0.1). The *double pedal* of two curves C, C' is the set of points $m \in X$ for which there exists a tangent to C and a tangent to C' intersecting orthogonally at m (figure 9.6.8.0.2).

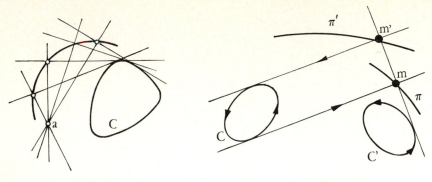

Figure 9.6.8.0.1   Figure 9.6.8.0.2

Pedals of the circle are called *Pascal limaçons* (their eponym is the father of the better-known Blaise Pascal) and occur in many situations, if for no other reason because they are among the simplest curves of fourth degree (bicircular quartics). Figure 9.6.8.0.3 shows their possible forms. The one curve displaying a cusp is called a *cardioid*; there is a discussion of it in 9.14.33 for those who find it attractive. In 9.14.22 we study the general case. The particular case where the tangents at the intersection point form an angle of $2\pi/3$ can be found in [DQ, 169, exercise 77].

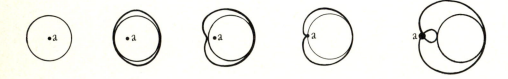

Figure 9.6.8.0.3

In the case of double pedals, it is easy to see that if the two curves are closed, regular and of class C^1, the double pedal can be naturally decomposed into two curves π, π', according to the orientation of the tangents deriving from a fixed orientation of C, C'. But this geometric decomposition does not, in general, imply an algebraic decomposition. We shall see that if C, C' are circles there is really an algebraic decomposition, and π, π' are the pedals of C, C' with respect to appropriately chosen points (figure 9.6.8.0.4).

9.6.8.1. We assume that the circles C, C' are not concentric, otherwise the problem is trivial. Observe first that there exist exactly two orientation-preserving similarities f_1 and f_2 that rotate by a right angle and take C into C' (cf. 8.7.3.5 and 9.6.3). To see this, take $m \in C$ and diametrically opposite points $m'_1, m'_2 \in C'$ such that, denoting by a, a' the centers of C, C', we have

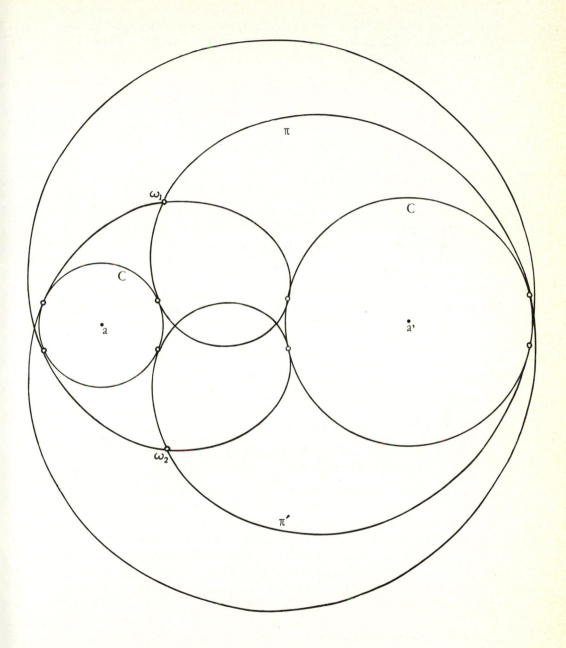

Figure 9.6.8.0.4

$am \perp a'm'_1$. For f_1 and f_2 take the orientation-preserving similarities such that $f_i(a) = a'$ and $f_i(m) = m'_i$ for $i = 1, 2$; such similarities exist and are well-defined by 9.6.2. It is clear that $f_i(C) = C'$ because $f_i(C)$ is the circle of center $f(a) = a'$ and radius equal to the radius of C'.

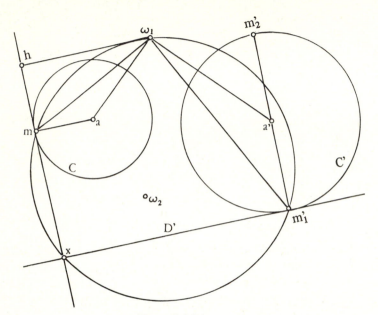

Figure 9.6.8.1

Now let $D = T_m C$, $D' = T_{m'_1} C'$ be the tangents to C, C' at m, m'_1, respectively, set $x = D \cap D'$, and let h be the projection of ω_1, the center of f_1, onto D. We shall use the following trivial lemma (figure 9.6.8.2):

9.6.8.2. Lemma. *Let $\{a, b, c\}$ and $\{a', b', c'\}$ be two right triangles, (that is, such that $\langle a, b \rangle \perp \langle a, c \rangle$ and $\langle a', b' \rangle \perp \langle a', c' \rangle$). There exists an orientation-preserving similarity f satisfying $f(a) = a'$, $f(b) = b'$, $f(c) = c'$ if and only if the following equality holds in $A(\vec{X})$:*

$$\langle b, a \rangle, \widehat{\langle b, c \rangle} = \langle b', a' \rangle, \widehat{\langle b', c' \rangle}.$$

Furthermore, $\vec{f}$ is uniquely determined by the value of this oriented angle between lines. $\square$

9.6.8.3. Applying this lemma, we see that the angle $\langle m'_1, \omega_1 \rangle, \widehat{\langle m'_1, m \rangle}$ is well-determined by $\vec{f_1}$; we call α this element of $A(\vec{X})$, which depends only on C, C' and the choice of f_1. By 10.9.5, we have $\alpha = \langle xm'_1, \omega_1 \rangle, \widehat{\langle m'_1, m \rangle} = \langle x, m \rangle, \widehat{\langle x, \omega_1 \rangle}$ because the four points ω_1, m, m'_1, x are cocyclic, since

$$\langle \omega_1, m \rangle, \widehat{\langle \omega_1, m'_1 \rangle} = \langle x, m \rangle, \widehat{\langle x, m'_1 \rangle} = \delta$$

(cf. 8.7.7.4 and 10.9.5). Thus $\langle x, h \rangle, \widehat{\langle x, \omega_1 \rangle} = \alpha$, and another application of the lemma shows that x is deduced from h by a fixed similarity g_1. Since h describes the pedal π_1 of C with respect to ω_1, the point x must describe $g_1(\pi_1)$; and the conclusion is that the double pedal of C and C' is made up of $g_1(\pi_1)$ and the analogous curve $g_2(\pi_2)$.

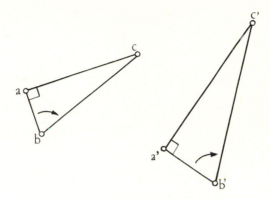

Figure 9.6.8.2

9.6.9. EADEM MUTATA RESURGO. We discuss here a property of logarithmic spirals that impressed Jacob Bernoulli to the extent that he requested that such a spiral, together with the above Latin quotation (loosely translated as "Here am I again, transformed"), be engraved on his tombstone in Basel (see 9.14.32). The property is that a logarithmic spiral is globally preserved by certainly similarities. This can be easily explained by the fact that $\mathrm{Sim}_\omega^+(X)$, for a fixed $\omega \in X$ (or, alternatively, $\mathrm{Sim}^+(\vec{X})$) contains non-trivial subgroups.

9.6.9.1. For fixed $k \in \mathbf{R}_+^*$ and $\omega \in X$, introduce the set

$$G = \{ f(t) = H_{\omega.kt} \circ \underline{\Theta}^{-1}(\Lambda(t)) \mid t \in \mathbf{R} \} \subset \mathrm{Sim}_\omega^+(X),$$

where $H_{\omega.kt}$ is the homothety of center ω and ratio kt, and $\underline{\Theta}^{-1}(\Lambda(t))$ denotes the rotation of center ω and angle $\Lambda(t)$. We see that G is a subgroup of $\mathrm{Sim}_\omega^+(X)$; it is homeomorphic and isomorphic to the additive group $\mathbf{R}$, and we call it a one-parameter subgroup of $\mathrm{Sim}_\omega^+(X)$. A *logarithmic spiral* of pole ω is an orbit of G (distinct from $\{\omega\}$), for an arbitrary k.

By definition, logarithmic spirals are invariant under G. The image of a logarithmic spiral under a similarity is still a logarithmic spiral. Every curve linked to a logarithmic spiral C and its pole ω by an orientation-preserving similarity is still a logarithmic spiral. An example is the pedal of C with respect to ω, as can be seen by noticing that the tangent $T_m C$ to C at m makes an constant angle $\widehat{\langle m, \omega \rangle, T_m C} \in \mathcal{A}(\vec{X})$ with the line $\langle m, \omega \rangle$. This is shown in the following way: since G acts transitively on C, by definition, there exists $f \in G$ such that $f(m) = n$ for any $m, n \in C$; in particular,

$$\widehat{\langle m, \omega \rangle, T_m C} = \widehat{\langle f(m), f(\omega) \rangle}, f(T_m C) = \widehat{\langle n, \omega \rangle, T_n C}.$$

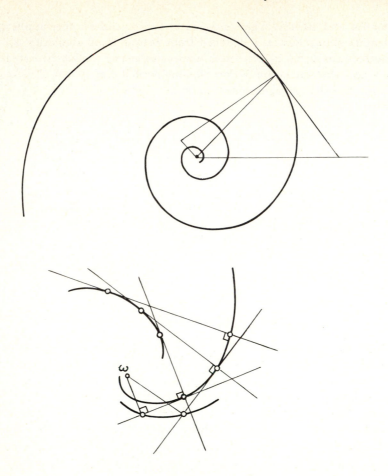

Figure 9.6.9.1

For a converse and other properties of logarithmic spirals, see exercise 9.14.21 (or Bernoulli's tomb). For examples of logarithmic spirals in nature, see [WL, 69–72].

9.7. Distances between several points

9.7.1. PROPOSITION. *Let* $(x_i)_{i=0.1,\ldots,n}$ *be an affine frame of* X. *If two points* $x, y \in X$ *satisfy* $x_i x = x_i y$ *for all* $i = 0, 1, \ldots, n$, *we have* $x = y$. *Let* k *be an integer, and* $(x_i), (y_i)$ $(i = 1, \ldots, k)$ *subsets of* X *such that* $x_i x_j = y_i y_j$ *for all* i *and* j; *there exists* $f \in \mathrm{Is}(X)$ *such that* $f(x_i) = y_i$ *for all* i.

Proof. Vectorialize X at x_0. By 8.1.2.4, we have $(x_i \mid x) = (x_i \mid y)$ for all i. But the linear forms $(x_i \mid \cdot)$ are linearly independent for $i = 1, \ldots, n$ because (x_i) is an affine frame; this shows that $x = y$.

By the first part, it is enough to show the second for $k \leq n+1$ and affinely independent subsets (x_i), (y_i). In fact, we can assume $k = n + 1$, since every isometry $f \in \mathrm{Is}(Y, Y')$ between subspaces of X can be trivially extended to an isometry $\tilde{f}$ of X. We will argue by induction on $n = k - 1$. For $n = 1$ we apply 9.1.6 (or we can start from $n = 0$, which is trivial). Suppose our conclusion holds for n, and let $Y = \langle x_0, x_1, \ldots, x_n \rangle$, $Y' = \langle y_0, y_1, \ldots, y_n \rangle$ be the subspaces spanned by each set of points. There exists $g \in \mathrm{Is}(X)$ such that $g(Y') = Y$. Applying the induction assumption to the subsets $(x_i)_{i=0.1.....n}$ and $\bigl(g(y_i)\bigr)_{i=0.1.....n}$ of the n-dimensional space Y, we see that it is enough to prove the result for $n+1$ when (x_i) and (y_i) satisfy $x_i = y_i$ for $i = 0, 1, \ldots, n$.

Vectorialize X at x_0. By 8.1.2.4 we have $(x_i \mid x_{n+1}) = (x_i \mid y_{n+1})$ for $i = 1, \ldots, n$, so x_{n+1} and y_{n+1} belong to the same affine line D of X, orthogonal to Y. But $x_0 x_{n+1} = x_0 y_{n+1}$, so 9.2.3 shows that $d(x_{n+1}, Y) = d(y_{n+1}, Y)$, whence $x_{n+1} = y_{n+1}$ or $y_{n+1} = \sigma_Y(x_{n+1})$. $\qquad\square$

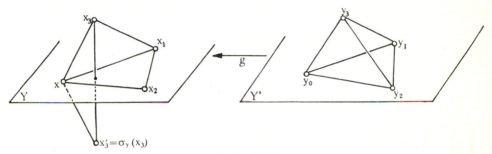

Figure 9.7.1

9.7.2. REMARK. Proposition 9.7.1 expresses the uniqueness (up to isometry) of subsets $(x_i)_{i=0.1.....k}$ for which the distance $d_{ij} = x_i x_j$ are given. The existence of such sets will be discussed in 9.7.3.4.

9.7.3. RELATION BETWEEN DISTANCES

9.7.3.1. It is reasonable to suspect that the distances $d_{ij} = x_i x_j$ between $n+1$ points $(x_i)_{i=0.1.....n}$ in a space X of dimension $n-1$ satisfy some universal relation. For $n = 1$, since one of the three points in necessarily between the other two, the product $(d_{01} + d_{02} - d_{12})(d_{01} + d_{12} - d_{02})(d_{02} + d_{12} - d_{01})$ must be zero. In arbitrary dimension n, the argument used in proposition 9.7.1 also leads to suspecting the existence of such a universal relation: taking the case $n = 2$, for example, and giving x_0, x_1, x_2 and d_{13}, d_{23}, there are only two possible choices for x_3, symmetrically placed with respect to the line $\langle x_1, x_2 \rangle$.

We will in fact find this universal relation, by expressing the volume of the parallelepiped built on the points $(x_i)_{i=0.1.....n}$ in n dimensions as a function of the distances $d_{ij} = x_i x_j$. If the points x_i actually belong to an $(n-1)$-dimensional subspace, the n-dimensional volume will be zero, and this will give the desired relation.

The volume is derived from 8.11.5 and 8.11.6, after vectorializing X at x_0 and using again equality 8.1.2.4, in the form $(x_i \mid x_j) = \frac{1}{2}(d_{0i}^2 + d_{0j}^2 - d_{ij}^2)$. The desired relation is thus

$$
\begin{vmatrix}
d_{01}^2 & \cdots & \frac{1}{2}(d_{01}^2 + d_{0n}^2 - d_{1n}^2) \\
\vdots & \ddots & \vdots \\
\frac{1}{2}(d_{10}^2 + d_{n0}^2 - d_{n1}^2) & \cdots & d_{0n}^2
\end{vmatrix} = 0.
$$

A rigorous proof would involve embedding X in a space X' of dimension n and applying 8.11.8 (ii) and (v), for example, to the vectorialization of X' at x_0.

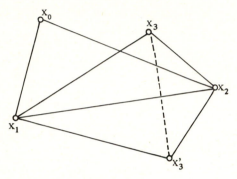

Figure 9.7.3

The relation above has the drawback of not being symmetric with respect to the indices; the index 0 is special. To remedy this, we define the *Cayley–Menger determinant* of $(x_i)_{0,1,\dots,k}$ as

9.7.3.2
$$
\Gamma(x_0, x_1, \dots, x_k) =
\begin{vmatrix}
0 & 1 & 1 & \cdots & 1 \\
1 & 0 & d_{01}^2 & \cdots & d_{0k}^2 \\
1 & d_{10}^2 & 0 & \cdots & d_{1k}^2 \\
\vdots & \vdots & \vdots & \ddots & \vdots \\
1 & d_{k0}^2 & d_{k1}^2 & \cdots & 0
\end{vmatrix}.
$$

9.7.3.3. Lemma. *We have* (cf. 8.11.5)

$$
\mathrm{Gram}(\overrightarrow{x_0 x_1}, \cdots, \overrightarrow{x_0 x_k}) = \frac{(-1)^{k+1}}{2^k} \Gamma(x_0, x_1, \dots, x_k).
$$

Proof. A refinement of the argument used in 8.2.2 and 8.11.6. Take an arbitrary orthonormal frame, and denote by x_i^j the coordinates of x_i in this frame. Using standard operations on determinants, we obtain

$$\lambda(\overrightarrow{x_0 x_1}, \dots, \overrightarrow{x_k x_1}) = \begin{vmatrix} x_1^1 - x_0^1 & \cdots & x_1^k - x_0^k \\ \vdots & \ddots & \vdots \\ x_k^1 - x_0^1 & \cdots & x_k^k - x_0^k \end{vmatrix}$$

$$= \begin{vmatrix} x_0^1 & \cdots & x_0^k & 1 \\ x_1^1 & \cdots & x_1^k & 1 \\ \vdots & \ddots & \vdots & \vdots \\ x_k^1 & \cdots & x_k^k & 1 \end{vmatrix} = \begin{vmatrix} x_0^1 & \cdots & x_0^k & 1 & 0 \\ x_1^1 & \cdots & x_1^k & 1 & 0 \\ \vdots & \ddots & \vdots & \vdots & \vdots \\ x_k^1 & \cdots & x_k^k & 1 & 0 \\ 0 & \cdots & 0 & 0 & 1 \end{vmatrix}.$$

The determinant of this last matrix D is then multiplied by the transpose of the determinant of the matrix obtained from D by switching its last two rows and its last two columns, giving

$$\lambda(\overrightarrow{x_0 x_1}, \dots, \overrightarrow{x_k x_1}) = \begin{vmatrix} (x_0 \mid x_0) & (x_0 \mid x_1) & \cdots & (x_0 \mid x_k) & 1 \\ (x_1 \mid x_0) & (x_1 \mid x_1) & \cdots & (x_1 \mid x_k) & 1 \\ \vdots & \vdots & \ddots & \vdots & \vdots \\ (x_k \mid x_0) & (x_k \mid x_1) & \cdots & (x_k \mid x_k) & 1 \\ 1 & 1 & \cdots & 1 & 0 \end{vmatrix},$$

where the "vectors" x_i are to be understood as the n-tuples of coordinates of x_i in the frame being used. We next replace $(x_i \mid x_j)$ by $\frac{1}{2}(\|x_i\|^2 + \|x_j\|^2 - d_{ij}^2)$, and eliminate all the $\|x_i\|^2$ by subtracting the appropriate multiples of the last column and line from the others. We finally obtain $\dfrac{(-1)^{k+1}}{2^k} \Gamma(x_0, x_1, \dots, x_n)$.

$\square$

9.7.3.4. Theorem. *Let $(x_i)_{i=0,1,\dots,n}$ be arbitrary points in an $(n-1)$-dimensional Euclidean affine space X. Then $\Gamma(x_0, x_1, \dots, x_n) = 0$. A necessary and sufficient condition for $(x_i)_{i=0,1,\dots,n-1}$ to be a simplex of X is that $\Gamma(x_0, x_1, \dots, x_{n-1}) \neq 0$. Given $k(k+1)/2$ real numbers d_{ij} $(i, j = 0, 1, \dots, k)$, a necessary and sufficient condition for the existence of a simplex $(x_i)_{i=0,1,\dots,n}$ satisfying $d_{ij} = x_i x_j$ is that for every $h = 2, \dots, k$ and every h-element subset of $\{0, 1, \dots, k\}$ the corresponding Cayley–Menger determinant be nonzero and its sign be $(-1)^{k+1}$.*

Proof. The first two assertions follow from 8.11.6 and also 9.7.6.6. The last one is shown by induction on k, it being obvious for $k = 1$. By the induction assumption, we can construct in the $(k-2)$-dimensional space Z a simplex $(x_i)_{i=2,\dots,k}$ such that $x_i x_j = d_{ij}$ $(i, j = 2, \dots, k)$. Next we find in the $k-1$-dimensional space Y containing Z two points x_0' and x_1 satisfying $x_0', x_1 \in Y \setminus Z$, $x_i x_i = d_{1i}$ and $x_0' x_i = d_{0i}$ for all $i = 2, \dots, k$. Call h the projection of x_0' onto Z; embed Y as a hyperplane of a k-dimensional space X and let W be the codimension-two subspace of X that contains h and is orthogonal to Z. Consider a point x describing the circle C of center h in W and passing through x_0'. This circle intersects Y at x_0''; as x describes C, the distance $x x_1$ takes on all values in the interval $[x_1 x_0', x_1 x_0'']$.

Now consider the determinant in 9.7.3.2, with the distance d_{01}^2 replaced by an arbitrary real number ξ; it is clear that the determinant $\Gamma(\xi)$ obtained is of the form $\Gamma(\xi) = -\xi^2 \Gamma(x_2, \ldots, x_k) + \alpha\xi + \beta$, with $\alpha, \beta \in \mathbf{R}$. We investigate the change in $\Gamma(\xi)$ as ξ takes values in $\mathbf{R}_+$; we know that $\Gamma(\xi)$ is a quadratic trinomial whose term in ξ^2 has sign $(-1)^{k-1}$, by assumption. We also know, by the first part of the theorem, that $\Gamma(\xi)$ vanishes for two distinct values $\xi' = x_1 x_0'$ and $\xi'' = x_1 x_0''$; it follows that $\Gamma(\xi)$ has sign $(-1)^{k-1}$ exactly for $\xi \in [x_1 x_0', x_1 x_0'']$. But $\Gamma(d_{01})$ is non-zero, and its sign is $(-1)^{k+1} = (-1)^{k-1}$, by assumption; we conclude that there exists $x_0 \in C \setminus \{x_0', x_0''\}$ such that $x_0 x_1 = d_{01}$. $\qquad\square$

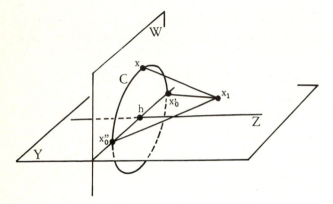

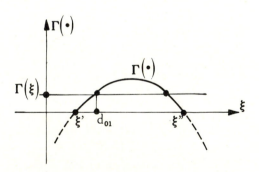

Figure 9.7.3.4

9.7.3.5. The reader may have remarked two facts in the proof of 9.7.3.4. The first is that not all hypotheses were used; in particular, if we know a priori that the x_i exist and form a simplex, every subset made up of these points is also a simplex. This leads to the suspicion that the algebraic properties of Γ are such that, if $\Gamma(x_0, x_1, \ldots, x_n)$ is non-zero and has the right sign, the Γ of any subset will have the same properties. Such is indeed the case, cf. 9.14.23. Thus the necessary and sufficient condition in 9.7.3.4 could be weakened to read that just the determinant Γ of the d_{ij} $(i, j = 0, \ldots, k)$ has to be non-zero and have sign $(-1)^{k+1}$.

The second observation is that 9.7.3.4 does not answer the question posed in 9.7.2 in the case when the points do not form a simplex. The solution in this case is a bit longer, though not more difficult; it is stated in 9.14.23.

The determinants Γ also give a criterion to determine if $n+2$ points of the n-dimensional space X lie on a sphere (cf. 9.7.5), and solve the related problem of determining the radius of the sphere circumscribed around a simplex. For this we introduce a new determinant Δ as follows:

9.7.3.6
$$\Delta(x_1, \ldots, x_k) = \begin{vmatrix} 0 & d_{12}^2 & \cdots & d_{1k}^2 \\ d_{21}^2 & 0 & \cdots & d_{2k}^2 \\ \vdots & \vdots & \ddots & \vdots \\ d_{k1}^2 & d_{k2}^2 & \cdots & 0 \end{vmatrix}.$$

9.7.3.7. Proposition. *If the points $(x_i)_{i=1,\ldots,n}$ form a simplex in an $(n-1)$-dimensional space X, the radius R of the sphere circumscribed around this simplex is given by*

$$R^2 = -\frac{1}{2} \frac{\Delta(x_1, \ldots, x_n)}{\Gamma(x_1, \ldots, x_n)};$$

in particular, $\Delta(x_1, \ldots, x_n) \neq 0$ for a simplex.

In order that $n + 2$ points $(x_i)_{i=1,\ldots,n+2}$ in an n-dimensional space X belong to the same sphere or the same hyperplane, it is necessary and sufficient that

$$\Delta(x_1, \ldots, x_{n+2}) = 0.$$

Proof. An easy calculation with determinants shows that, if $d_{0i} = R$ for $i = 1, \ldots, n$, we have

$$\Gamma(x_0, x_1, \ldots, x_n) = -2R^2 \Gamma(x_1, \ldots, x_n) - \Delta(x_1, \ldots, x_n);$$

the desired formula follows from taking x_0 as the center of the sphere circumscribed around $(x_i)_{i=1,\ldots,n}$, for then $\Gamma(x_0, x_1, \ldots, x_n) = 0$ by 9.7.3.4.

The same method would also show that $\Delta(x_1, \ldots, x_{n+2}) = 0$ if the points x_i are cospheric, but it wouldn't work for the converse. For this reason we prove the equivalence of the two conditions directly. Choose an arbitrary orthonormal frame, and let x_i^j $(i = 1, \ldots, n+2; j = 1, \ldots, n)$ be the coordinates of the points x_i. If the points belong to the same hyperplane or sphere, there exists scalars a, b, c_j, not all of which vanish, satisfying

$$a\|x_i\|^2 + b + \sum_{i=1}^{n} c_i x_i^j = 0$$

for all $i = 1, \ldots, n + 2$ (cf. 10.7.6, for example). It follows that the two determinants below are zero:

$$\begin{vmatrix} \|x_1\|^2 & 1 & x_1^1 & \cdots & x_1^n \\ \vdots & \vdots & \vdots & \ddots & \vdots \\ \|x_{n+2}\|^2 & 1 & x_{n+2}^1 & \cdots & x_{n+2}^n \end{vmatrix} = 0,$$

$$\begin{vmatrix} 1 & \|x_1\|^2 & -2x_1^1 & \cdots & -2x_1^n \\ \vdots & \vdots & \vdots & \ddots & \vdots \\ 1 & \|x_{n+2}\|^2 & -2x_{n+2}^1 & \cdots & -2x_{n+2}^n \end{vmatrix} = 0.$$

Thus the determinant of the matrix obtained by multiplying one of these by the transpose of the other is also zero; but this product is none other than $\Delta(x_1, \ldots, x_{n+2})$!

Conversely, if $\Delta(x_1, \ldots, x_{n+2}) = 0$, the first determinant above is zero, and there must exist numbers a, b, c_j, not all zero, satisfying the condition above. By 10.7.6, the points x_i belong to the sphere or hyperplane given by the equation

$$a\| \cdot \|^2 + b + \sum_{i=1}^{n} c_j \cdot = 0. \qquad \square$$

9.7.3.8. Examples. Put $a = d_{12}$, $b = d_{23}$, $c = d_{31}$. We find

$$\Gamma(x_1, x_2, x_3) = -(a + b + c)(a + b - c)(a - b + c)(-a + b + c);$$

this is the area of a triangle of sides a, b, c (see 10.3.3).

Put $\alpha = d_{12}d_{34}$, $\beta = d_{13}d_{24}$, $\gamma = d_{14}d_{23}$. We find

$$\Delta(x_1, x_2, x_3, x_4) = -(\alpha + \beta + \gamma)(\alpha + \beta - \gamma)(\alpha - \beta + \gamma)(-\alpha + \beta + \gamma).$$

This is Ptolemy's theorem (see 10.9.2).

9.7.4. A FUNDAMENTAL PROBLEM ABOUT METRIC SPACES. The important role played by Euclidean affine spaces leads to the question of whether they can be characterized in metric terms alone. This fundamental problem was solved by K. Menger in 1928; Cayley–Menger determinants are the key to the solution, which is not significantly more difficult than the proof of 9.7.3.4. A complete solution is to be found in [BL, chapter IV].

The analogous problem can also be posed for other classical metric spaces, like spheres, elliptic and hyperbolic spaces (cf. chapters 18 and 19). The solution to this is also in [BL], which, by the way, is an excellent reference for the systematic study of metric spaces with no additional structure. See also [BER2].

9.7.5. EQUIDISTANT HYPERPLANES

9.7.5.1. Proposition. *Given distinct points x, y of X, the set $\{z \in X \mid zx = zy\}$ is a hyperplane, said to be equidistant from x and y. (For $n = 2$, this set is also known as the perpendicular bisector of x and y.)*

More generally, if $(x_i)_{i=0,1,...,k}$ are affine independent, the set

$$\{ z \in X \mid zx_0 = zx_1 = \cdots = zx_k \}$$

is a subspace of dimension $n + 1 - k$. In particular, if $k = n$, that is, if (x_i) is a simplex, there exists a unique point whose distance to all the vertices of the simplex is the same; in other words (cf. 10.7), there exists a unique sphere circumscribed around this simplex.

Proof. Vectorialize X at x_0 and observe that, as often before, the required relations between distances yield scalar products. Then apply 2.4.8. $\qquad\square$

9.7.6. APPOLONIUS' FORMULA, BARYCENTERS AND DISTANCES. Using the notation and the nomenclature of 3.4.5, we recover a well-known result:

9.7.6.1. Appolonius' formula. *Let $\{(\lambda_i, x_i)\}$ be a finite family of punctual masses of X, and let $(\sum_i \lambda_i, g)$, or $(0, \vec{\xi})$, be their barycenter. Then, for any $z \in X$, we have*

$$\sum_i \lambda_i \, zx_i^2 = \sum_i \lambda_i \, gx_i^2 + \left(\sum_i \lambda_i \right) zg^2,$$

or

$$\sum_i \lambda_i \, zx_i^2 = \sum_i \lambda_i \, yx_i^2 + 2(\overrightarrow{zy} \mid \vec{\xi}).$$

Proof. It suffices to write $zx_i^2 = zg^2 + gx_i^2 + 2(\overrightarrow{zg} \mid \overrightarrow{gx_i})$, and to use 3.4.6.5. $\qquad\square$

9.7.6.2. Corollary. *With the same data as in 9.7.6.1 and k a real number, put*

$$L = \left\{ z \in X \;\middle|\; \sum_i \lambda_i \, zx_i^2 = k \right\}.$$

Then

i) *if $\sum_i \lambda_i = 0$, L is an affine hyperplane for every k, and $(\vec{\xi})^\perp$ is its direction;*

ii) *if $\sum_i \lambda_i > 0$, we have $L = \emptyset$ if $k < \sum_i \lambda_i \, gx_i^2$, and $L =$ sphere of center g and radius $\left(\dfrac{k - \sum_i \lambda_i \, gx_i^2}{\sum_i \lambda_i} \right)^{\frac{1}{2}}$ otherwise.* $\qquad\square$

The reader can supply the case $\sum_i \lambda_i < 0$.

9.7.6.3. Corollary. *With the same data as in 9.7.6.1 and the condition $\sum_i \lambda_i > 0$, the function $z \mapsto \sum_i \lambda_i \, zx_i^2$ achieves its minimum at g and nowhere else.* $\qquad\square$

9.7.6.4. Corollary 9.7.6.3 shows that the center of mass of a finite set of points is invariant under any isometry that leaves this set globally invariant; but this also follows, and more simply, from 3.7.3. and 9.1.3.

9.7.6.5. Corollary 9.7.6.2 allows one to describe an impressive number of loci. We start by recovering 9.7.5.1, upon taking the family $\{(1, x), (-1, y)\}$.

More generally, the locus of points z such that the distance from z to two fixed points x, y is a constant $k \neq 1$ is a sphere of center

$$\frac{x}{1-k} - \frac{ky}{1-k}.$$

Similarly, the set $\{ z \in X \mid zx^2 + zy^2 = k \}$ is a sphere, and $\{ z \in X \mid zx^2 - zy^2 = k \}$ is a hyperplane; in the first case we have the "formula of the median":

$$zx^2 + zy^2 = 2\, zg^2 + 2\, gx^2 = 2\, zg^2 + \frac{1}{2}\, xy^2.$$

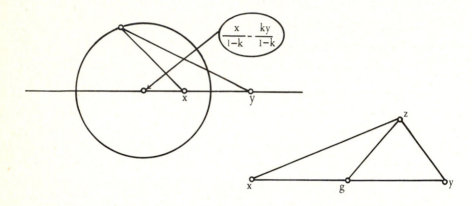

Figure 9.7.6.5

Finally, the set $\{ z \mid zx^2 - zy^2 = k \}$ is a hyperplane, orthogonal to $\overrightarrow{xy}$.

9.7.6.6. Formula 9.7.6.1 gives a way of proving the first statement in 9.7.3.4 without having to fiddle around with determinants (cf. 9.7.3.3). It goes like this: by 3.6, at least one of the points, say x_n, can be written as a barycenter of the others, $x_n = \sum_{i=0}^{n-1} \lambda_i x_i$ with $\sum_{i=0}^{n-1} \lambda_i = 1$. This means the barycenter of the family

$$\{ (\lambda_0, x_0), \ldots, (\lambda_{i-1}, x_{i-1}), (-1, x_n) \}$$

is $(0, \vec{0})$! By 9.7.6.1, the function $\sum_{i=0}^{n} \lambda_i\, zx_i^2$ does not depend on z; let k be its value. Substituting $x_0, x_1, \ldots, x_n$ for z we get

$$\lambda_1 d_{01}^2 + \lambda_2 d_{02}^2 + \cdots + \lambda_{n-1} d_{0.n-1}^2 - d_{0n}^2 = k$$
$$\lambda_0 d_{10}^2 \qquad\quad + \lambda_2 d_{12}^2 + \cdots + \lambda_{n-1} d_{1.n-1}^2 - d_{1n}^2 = k$$

$$\cdot \quad\cdot\quad\cdot \qquad\qquad \cdot \qquad\qquad\qquad \cdot \qquad\qquad \cdot$$

$$\lambda_0 d_{n0}^2 + \lambda_1 d_{n1}^2 + \lambda_2 d_{n2}^2 + \cdots + \lambda_{n-1} d_{n.n-1}^2 \qquad\qquad = k$$
$$\lambda_0 \;+\; \lambda_1 \;+\; \lambda_2 \;+ \cdots + \quad \lambda_{n-1} \qquad\quad - 1 = 0,$$

whence $\Gamma(x_0, x_1, \ldots, x_n) = 0$.

9.8. Stabilizers of subsets

9.8.1. In the study of a subset $A \subset X$, it is natural to introduce its *stabilizer* or *isotropy group* $\mathrm{Is}_A(X) = \{\, g \in \mathrm{Is}(X) \mid g(A) = A \,\}$ in $\mathrm{Is}(X)$. The bigger $\mathrm{Is}_A(X)$, the more "symmetries" A has: $\mathrm{Is}_A(X)$ is intimately associated with A. We have already met the subgroups $\mathrm{Is}_A(X)$ in chapter 1, and they will be essential in 12.5. The map $\chi : A \mapsto \mathrm{Is}_A(X)$ from the set of subsets of X into the set of subgroups of $\mathrm{Is}(X)$ is a unifying thread in this section.

9.8.2. REMARKS. If A is compact and $g \in \mathrm{Is}(X)$, the condition $g(A) \subset A$ implies $g(A) = A$, but this no longer holds if A is just closed or bounded (cf. 9.14.26).

In general, $\mathrm{Is}_A(X) \subset \mathrm{Is}_{\overline{A}}(X)$, where $\overline{A}$ is the closure of A, but the inclusion may be strict, cf. 9.14.27.

With the metric induced from X, the subset A has a group of isometries $\mathrm{Is}(A)$, whence a restriction map $\rho : \mathrm{Is}_A(X) \to \mathrm{Is}(A)$. In general, ρ is not injective; for instance, if $Y = A$ is a proper subspace of X, the reflection σ_Y induces the identity on X. Injectivity holds if $\langle A \rangle = X$, where $\langle A \rangle$ is the affine subspace generated by the set A (see 2.4.2.5). Surjectivity always holds, by 9.7.1.

The map χ itself is neither surjective nor injective. The two sets in figure 9.8.2, for example, have the same group of symmetries; on the other hand, given $\omega \in X$, the subgroup $G = \mathrm{Is}_\omega^+(X)$ cannot be the stabilizer of any subset A, for any orbit of G is the union of spheres of center ω, and thus must be invariant under $\mathrm{Is}_\omega(X)$. However, if G is a finite subgroup of $\mathrm{Is}(X)$, there always exists $A \subset X$ such that $G = \mathrm{Is}_A(X)$ (see 9.14.36).

Figure 9.8.2

9.8.3. CLOSEDNESS. Unlike $\mathrm{Is}_a(X)$, where $a \in X$ is a point, $\mathrm{Is}_A(X)$ is not closed in $\mathrm{Is}(X)$: just take A to be an orbit of G, where G is the subgroup of $\mathrm{Is}(X)$ generated by an irrational plane rotation. But if A is closed in X, so is $\mathrm{Is}_A(X)$ in $\mathrm{Is}(X)$.

9.8.4. BOUNDEDNESS. If A is bounded, so is $\mathrm{Is}_A(X)$. For, as we shall see below, $\mathrm{Is}_{\overline{A}}(X)$ is bounded, and $\mathrm{Is}_A(X) \subset \mathrm{Is}_{\overline{A}}(X)$ (cf. 9.8.2). On the other hand, it may happen that $\mathrm{Is}_A(X)$ is bounded but A is not: see 9.14.27.

9.8.5. FINITENESS. It may happen that $\mathrm{Is}_A(X)$ is finite and A is not (see 9.14.27), and that A is finite but $\mathrm{Is}_A(X)$ is not—for example, when A is contained in a subspace of codimension ≥ 2. On the other hand, if $\dim\langle A\rangle \geq \dim X - 1$, the finiteness of A implies that of $\mathrm{Is}_A(X)$. This is clear, since a map $g \in \mathrm{Is}(X)$ that induces the identity on A also induces the identity on $\langle A\rangle$; thus, if $\langle A\rangle = X$ we must have $g = \mathrm{Id}_X$, and if $\langle A\rangle$ is a hyperplane we have $g = \mathrm{Id}_X$ or $\sigma_{\langle A\rangle}$.

9.8.6. COMPACTNESS AND FIXED POINTS.

9.8.6.1. Proposition. *If A is a compact subset of X, there exists $x \in X$ such that $\mathrm{Is}_A(X) \subset \mathrm{Is}_x(X)$. In particular, $\mathrm{Is}_A(X)$ is compact in $\mathrm{Is}(X)$. If $G \subset \mathrm{Is}(X)$ is compact, there exists $x \in X$ such that $G \subset \mathrm{Is}_x(X)$.*

Proof. The second part follows from the first, by considering any orbit of G. The "in particular" follows from 9.8.3 and the compactness of $\mathrm{Is}_x(X)$ (cf. 8.2.3.3). The first assertion can be shown in at least three ways:

9.8.6.2. First proof. This is the same as the proof of 2.7.5.9; here the condition that the interior be non-empty is not necessary, since $\mathrm{Is}_x(X)$ is always compact, contrary to $\mathrm{GA}_x(X) \cong \mathrm{GL}(X)$.

9.8.6.3. Second proof. This follows from 11.5.8. In fact, denoting by B the ball of least radius in X containing A, we have $g(B) = B$ whenever $g(A) = A$, since B is unique and is defined in purely metrical terms. The desired fixed point x is the center of B.

9.8.6.4. Third proof (Bruhat–Tits lemma). This lemma guarantees the existence of a fixed point for the group of isometries of a compact subset of any metric space in a certain class. This class encompasses Euclidean affine spaces and many others, including hyperbolic spaces, cf. 19.4.7.

Observe that 9.8.6.1 cannot hold for an arbitrary metric space; for example, the isometry group $\mathrm{Is}(S)$ of the sphere S centered at x does not leave any point of S fixed, and yet is compact, $\mathrm{Is}(S) \cong \mathrm{Is}_x(X)$.

The class of metric spaces X under advisement is characterized by the following property:

(CN) *For every $x, y \in X$ there exists $m \in X$ such that*

$$d^2(x, z) + d^2(y, z) \geq 2d^2(m, z) + \frac{1}{2}d^2(x, y)$$

for all $z \in X$.

To see that this condition is satisfied by Euclidean affine spaces, take m to be the midpoint of x and y; the formula of the median (9.7.6.5) shows that equality holds in (CN). The midpoint of a segment also works in hyperbolic space (19.4.7). On the other hand, spheres do not satisfy (CN).

9.8.6.5. Lemma (Bruhat–Tits) (cf. [BR–TI, 63]). *Let X be a complete metric space satisfying (CN), and let A be a bounded subset of X. There exists a point x such that, for every $g \in \mathrm{Is}(X)$ satisfying $g(A) = A$, we have $g(x) = x$.*

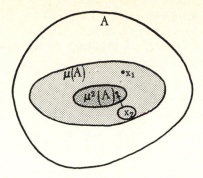

Figure 9.8.6.5

Proof. First, an idea of the proof, assuming A compact. Consider the set $\mu(A)$ of points m obtained by applying (CN) to pairs of points $x, y \in A$ satisfying $d(x, y) = \text{diam}(A)$. Condition (CN) implies that $\text{diam}(\mu(A)) \leq \lambda \, \text{diam}(A)$, where $\lambda < 1$ is a universal constant; we thus obtain a decreasing sequence of compact sets whose diameter approaches zero, and the intersection of these sets is the desired point.

Now for a detailed proof in the general case where A is a bounded subset of X. Fix $k \in [0, 1]$ and, for every $Y \subset X$, denote by $\mu(Y)$ the set of points m obtained by applying (CN) to pairs $x, y \in Y$ such that $d(x, y) \geq k \, \text{diam}(Y)$. Condition (CN) gives

$$\sup \big\{ \, d(x, y) \mid x \in Y, y \in \mu(Y) \, \big\} \leq k_1 \, \text{diam}(Y),$$
$$\text{diam}(\mu(Y)) \leq k_2 \, \text{diam}(Y),$$

where $k_1 = \sqrt{1 - k^2/4}$ and $k_2 = \sqrt{1 - k^2/2}$. Define, by induction,

$$\mu^n(Y) = \mu\big(\mu^{n-1}(Y)\big).$$

Let A be bounded; from the inequalities above, we deduce that

$$\text{diam}\big(\mu^n(A)\big) \leq k_2^n \, \text{diam}(A)$$

for all $n \in \mathbf{N}^*$, so the intersection of the $\mu^n(A)$ has at most one point. Let $(x_n)_{n \in \mathbf{N}}$ be an arbitrary sequence of points $x_n \in \mu^n(A)$, $n \in \mathbf{N}$. By the preceding inequalities, $d(x_n, x_{n+1}) \leq k_1 k_2^n \, \text{diam}(A)$, so (x_n) is a Cauchy sequence, and converges to a point x by completeness. Thus $\bigcap_{n \in \mathbf{N}} \mu^n(A) = \{x\}$. Since the sets $\mu^n(A)$ are obtained from A by a purely metric procedure, they must be invariant under $\text{Is}_A(X)$, and the same holds for x. $\qquad\square$

9.8.7. REMARK. As in 2.7.5.11, one sees that all maximal compact subgroups of $\text{Is}(X)$ are conjugate.

9.8.8. EXAMPLES. For examples, see the proof of 1.7.5.1 and [BR–TI, 64].

9.9. Length of curves

In this section M denotes an arbitrary metric space, whose distance is denoted by $d(\cdot, \cdot)$.

9.9.1. DEFINITIONS. *A curve or path in M is a pair $([a, b], f)$ $(a < b)$ consisting of a closed interval of $\mathbf{R}$ and a continuous map $f : [a, b] \rightarrow M$. The endpoints of $([a, b], f)$ are the points $(f(a), f(b))$; they are said to be joined by f. (Sometimes $f(a)$ is called the starting point and $f(b)$ the endpoint). The set of curves joining $f(a)$ to $f(b)$ is denoted by $C(x, y)$. The length of f is the scalar $\operatorname{leng}(f) \in \mathbf{R}_+ \cup \infty$ given by*

$$\sup \left\{ \sum_{i=0}^{n-1} d\big(f(t_i), f(t_{i+1})\big) \,\middle|\, a = t_0 < t_1 < \cdots < t_n = b \text{ and } n \in \mathbf{N}^* \right\}.$$

The curve f is said to be rectifiable if $\operatorname{leng}(f) \in \mathbf{R}_+$.

9.9.2. REMARKS. Saying that $C(x, y) \neq \emptyset$ for every $x, y \in M$ is the same as saying that M is path-connected. If $\theta : [c, d] \rightarrow [a, b]$ is a homeomorphism and $f : [a, b] \rightarrow M$ is a curve in M, we have $\operatorname{leng}(f \circ \theta) = \operatorname{leng}(f)$. From the point of view of the length, we can then take the quotient of the set of curves by the equivalence relation thus obtained. The quotient classes are called *geometric arcs*; this notion shall not be needed in this book.

Observe that a drawing like 9.9.2 assumes the existence of "segments" connecting each pair of consecutive points $\big(f(t_i), f(t_{i+1})\big)$; we shall discuss this matter in 9.9.4.2.

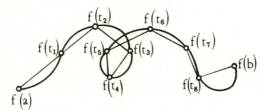

Figure 9.9.2

9.9.3. EXAMPLES.

9.9.3.1. One necessarily has $\operatorname{leng}(f) \geq d(x, y)$ for every $f \in C(x, y)$.

9.9.3.2. If $f : [a, b] \rightarrow M$ and $g : [b, c] \rightarrow M$ are curves such that $f(b) = g(b)$, the curve $f \cup g : [a, c] \rightarrow M$ satisfies

$$\operatorname{leng}(f \cup g) = \operatorname{leng}(f) + \operatorname{leng}(g).$$

9.9.3.3. Even in good spaces M, there exist non-rectifiable curves. For example, one can construct a curve in the plane $\mathbf{R}^2$ by iterating infinitely often the procedure suggested by figure 9.9.3.3. (The reader should verify that the map thus obtained is continuous.) The length must be $\geq (4/3)^n$ for

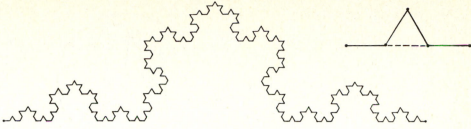

Figure 9.9.3.3

every $n \in \mathbf{N}^*$, so it cannot be finite. Observe also that, for any $t < t' \in [0, 1]$, the length of the restriction $f|_{[t.t']}$ is still infinite.

In spite of their pathological appearance, such curves (called "fractals") occur very naturally: as coastlines, for example (cf. [MB]).

9.9.4. SEGMENTS. INTRINSIC AND EXCELLENT SPACES

9.9.4.1. Definition. *A curve $f : [a, b] \to M$ is called a segment if $d\big(f(t), f(t')\big) = t' - t$ for every $t < t' \in [a, b]$. In particular, $\mathrm{leng}(f) = d\big(f(a), f(b)\big)$.*

9.9.4.2. No conflict arises with the notion, introduced in 3.4.3, of the segment $[x, y]$ with endpoints x, y in a Euclidean affine space X. Indeed, such a segment determines a segment in the new sense, unique up to a translation of its interval of definition:

$$[a, a + d(x, y)] \ni t \mapsto f(t) = x + \frac{t - a}{d(x, y)} \overrightarrow{xy} \in X.$$

x   f(t)   f(t')   y

Figure 9.9.4.2

In general, even if $C(x, y) \neq \emptyset$, there may not be a segment joining x and y. There are reasonably general conditions that guarantee the existence of segments: see [CT, 135] or [BL, 70].

9.9.4.3. Counterexamples are easily manufactured: take $M = \mathbf{R}^2 \setminus (0, 0)$ with the metric induced from $\mathbf{R}^2$. There is no segment joining x and $-x$, as such a segment should have length $2\|x\|$, and the only path in $\mathbf{R}^2$ with this length is the segment $[x, -x]$, which unfortunately contains $(0, 0) \notin M$.

Another example is the sphere $S^n \subset \mathbf{R}^{n+1}$, with the metric induced from $\mathbf{R}^{n+1}$. For any distinct points $x, y \in S^n$, and any curve f joining x and y, we have $\mathrm{leng}(f) > d(x, y)$, because

$$\mathrm{leng}(f) \geq d\big(x, f(t)\big) + d\big(f(t), y\big)$$

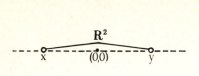

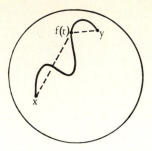

Figure 9.9.4.3

for any $t \in]a, b[$, and

$$d\big(x, f(t)\big) + d\big(f(t), y\big) > d(x, y)$$

by 9.1.1.1.

Such counterexamples lead naturally to the following

9.9.4.4. Definition. *A metric space M is called* intrinsic *if it is path-connected and $d(x, y) = \inf\big\{ \operatorname{leng}(f) \mid f \in C(x, y) \big\}$ for every $x, y \in X$. It is called* excellent *if, for any $x, y \in X$, there exists a segment joining x and y.*

9.9.4.5. Examples. By 9.9.4.1, excellent spaces are always intrinsic. The converse is not true, as can be seen from the first example in 9.9.4.3.

Euclidean affine spaces are excellent, by 9.4.4.2. Moreover, 9.1.1.1 shows that a segment joining two given points is unique (up to a translation of the interval of definition). This is not true for other excellent spaces, as can be seen from the example of the circle (9.9.8).

We shall meet other excellent spaces in 18.4.2, 19.1.2 and 19.3.2. The first of these examples deals with the sphere, which, as seen in 9.9.4.3, is not an intrinsic space with the metric induced from $\mathbf{R}^{n+1}$; this flaw is remedied in 18.4.

9.9.5. SHORTEST PATH. One often hears that, in a Euclidean space, the shortest path between two points is the segment joining them. More precisely, if $f : [a, b] \to X$ is a curve joining x and y and having length $\operatorname{leng}(f) = d(x, y)$, we have, for every $t \le t' \in]a, b[$:

$$f(t') \in [x, y] \quad \text{and} \quad f(t) \in [x, f(t')].$$

This, too, is a consequence of 9.1.1.1 and the following elementary remark: if $\operatorname{leng}(f) = d(x, y)$, then, for every $t \in [a, b]$, $\operatorname{leng}\big(f|_{[a.t]}\big) = d\big(x, f(t)\big)$ and $\operatorname{leng}\big(f|_{[t.b]}\big) = d\big(f(t), y\big)$.

9.9.6. REMARKS. Let X be a Euclidean affine space. Recall (cf., for example, [DR, 314]) that if a curve $f : [a, b] \to X$ is of class C^1, it is rectifiable and has length

$$\operatorname{leng}(f) = \int_a^b \|f'(t)\|\, dt.$$

An analogous formula holds for curves in differentiable submanifolds of Euclidean affine spaces, and even in abstract riemannian manifolds. If the manifold is complete, the metric obtained is excellent, and uniqueness of segments holds locally. These considerations lead to the fundamental notion of *geodesics:* see [BER1], [KG1, 144 ff.] or [MA, 271], for example, for these generalizations.

9.9.7. SYSTEMATIC CONSTRUCTION OF AN INTRINSIC METRIC. If (M, d) is a non-intrinsic metric space, we can almost always derive from it an intrinsic metric $\bar{d}$ via the following natural procedure:

9.9.7.1. Proposition. *Let (M, d) be a metric space such that, for every $x, y \in X$, there exists $f \in C(x, y)$ such that $\mathrm{leng}(f) < \infty$. Then*

$$\bar{d} : M \times M \ni (x, y) \mapsto \bar{d}(x, y) = \inf \left\{ \mathrm{leng}(f) \mid f \in C(x, y) \right\} \in \mathbf{R}_+$$

is an intrinsic metric on M. Furthermore, $\bar{\bar{d}} = \bar{d}$.

Proof. Left to the reader (see 9.14.30). □

The fact that $\bar{\bar{d}} = \bar{d}$ means that the procedure stops after one step.

In the case of (S^n, d), where d is the metric induced from $\mathbf{R}^{n+1}$, the new metric $\bar{d}$ obtained is none other than $\bar{d}(x, y) = \mathrm{Arccos}((x \mid y))$, introduced in 8.6.3. We study the case $n = 1$ in 9.9.8 and the general case in 18.4 (see 18.4.3). Computing $\bar{d}$ with the help of 9.9.6 and the parametrization $t \mapsto (t, \sqrt{1 - t^2})$ of the circle S^1, we recover the measure of angles obtained from the integral $\int_0^t \dfrac{ds}{\sqrt{1 - s^2}}$; this process is somewhat more elementary than the one employing complex exponentials (8.3.13).

9.9.8. THE INTRINSIC METRIC OF THE CIRCLE. Consider the circle $C = \{ x \in X \mid \|x\| = 1 \}$ in the Euclidean vector plane X. This can be identified with the set of half-lines of X (cf. 8.6.1).

9.9.8.1. Theorem. *With the metric $\overline{xy} = \mathrm{Arccos}((x \mid y))$, the circle C is an intrinsic metric space. The relation $\overline{xy} + \overline{yz} = \overline{xz}$ holds if and only if the half-line $\mathbf{R}_+ y$ lies between $\mathbf{R}_+ x$ and $\mathbf{R}_+ z$ (cf. 8.7.5.2). Given $x, y \in C, x \neq -y$, the unique shortest path from x to y (cf. 9.9.5) is the arc of circle from x to y (cf. 8.7.5.4); if $x = -y$, there are two shortest paths, the semicircles with endpoints x and y.*

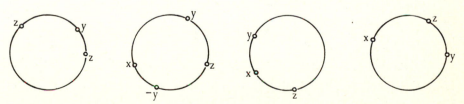

Figure 9.9.8.1.1

Proof. Consider three points $x, y, z \in C$. Leaving aside the cases when two of them are equal or opposite, which are treated directly, we are left with the four possibilities shown in figure 9.9.8.1.1. From 8.7.5.3, we see that in case I we have $\overline{xy} + \overline{yz} = \overline{xz}$; in case II, $\overline{xy} + \overline{yz} + \overline{xz} = 2\pi$; in case III, $\overline{xz} = \overline{yz} - \overline{xy}$; and in case IV, $\overline{xz} = \overline{xy} - \overline{yz}$. Bearing in mind that $\overline{xy} + \overline{yz} > \pi$ in case II, we do obtain $\overline{xy} + \overline{yz} \geq \overline{xz}$ in all cases, and equality only in case I. In particular, $(C, \overline{\cdot\cdot})$ is a metric space.

To see that this metric is intrinsic, we show that it is excellent. Let $x, y \in C$. Following the construction in 8.7.5, let $\overline{xy} = t$, orient X and join x to y by the curve

$$f : [0, t] \ni s \mapsto \cos s \cdot e_1 + \sin s \cdot e_2 \in C.$$

For every $s, s' \in [0, t]$ with $s \leq s'$, we have $\overline{f(s)f(s')} = s' - s$, so f is really a segment. As for shortest paths, we just have to consider the strict triangle inequality as given in the first part of the theorem, and the remark at the end of 9.9.5.

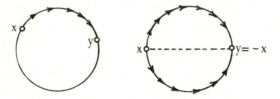

Figure 9.9.8.1.2

9.9.9. NOTE. For results on curves in general metric spaces, and in particular for a definition of curvature and torsion that does not use differential calculus, see [BL, 74 ff.] or [B–M, chapter 10]; see also 9.14.31.

9.10. Distance and differential geometry. The first variation formula

This imposing-sounding formula merely gives the derivative d' of the distance function $d : X \times X \to \mathbf{R}$ on a Euclidean affine space. In spite of its simplicity, it has numerous nice consequences. See also 9.10.7.

9.10.1. THE FIRST VARIATION FORMULA. *For every pair of distinct points $x, y \in X$ and every $\vec{u}, \vec{v} \in \vec{X}$, we have*

$$d'(x, y)(\vec{u}, \vec{v}) = \frac{1}{\|\overrightarrow{xy}\|}(\overrightarrow{xy} \mid \vec{v} - \vec{u}) = \|\vec{v}\| \cos(\overrightarrow{v}, \overrightarrow{xy}) - \|\vec{u}\| \cos(\overrightarrow{u}, \overrightarrow{xy}),$$

where the angles are between oriented lines (cf. 8.6.3).

Proof. Let e be the square of d, so that

$$e(x, y) = d^2(x, y) = \|\overrightarrow{xy}\|^2.$$

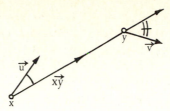

Figure 9.10.1

From the rules of differential calculus,

$$e'(x,y)(\vec{u}, \vec{v}) = 2(\overrightarrow{xy} \mid \vec{v} - \vec{u}),$$

since the scalar product is bilinear (see [CH1, 33], for example). Again by calculus, the derivative of $d = \sqrt{e}$ is given by the desired formula. □

9.10.2. For example, if two curves C and C' are everywhere orthogonal to a family $D(t)$ of lines, we must have $d(C \cap D(t), C' \cap D(t)) = $ constant, because the two angles in 9.10.1 are equal to $\pi/2$ and their cosine is zero. In particular, the tangent at m to a curve drawn on a sphere of center x is orthogonal to the line $\langle x, m \rangle$ (compare 10.7.4).

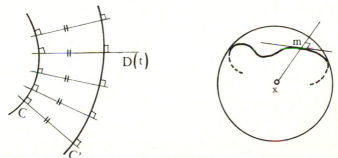

Figure 9.10.2

9.10.3. Let $x(t)$ be a point describing a curve C of class C^1, assumed to be regular (i.e., $x'(t) \neq 0$ for all t). Let $T_{x(t)}C$ be the tangent to C at $x(t)$, and $y(t)$ a point of $T_{x(t)}C$ that describes a curve D everywhere orthogonal to $T_{x(t)}C$ (this is the case, for example, if C is the *evolute* of D, that is, the envelope of the normals to D). Under these conditions we have, for all values of s, t:

$$\left| x(t)y(t) - x(s)y(s) \right| = \operatorname{leng}(C|_{[s,t]}).$$

To see this, take $\vec{u} = x'(t)$ and $\vec{v} = y'(t)$ in 9.10.1. The angle between $\vec{v}$ and $\overrightarrow{xy}$ will be $\pi/2$ and have cosine zero, whereas the angle between $\vec{u}$ and $\overrightarrow{xy}$ will be 0 or π. Thus $d'(x(t), y(t))(x'(t), y'(t)) = \pm \|x'(t)\|$, from which we obtain the desired formula by integrating from s to t and using 9.9.6.

9.10.4. A similar but somewhat more complicated example, to be used in 17.6.4, is the following: two points $x(t)$, $y(t)$ describe the same curve C, assumed regular. The tangents $T_{x(t)}C$, $T_{y(t)}C$ intersect at the point $z(t) = $

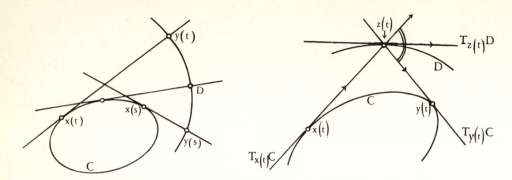

Figure 9.10.3

$(T_{x(t)}C) \cap (T_{y(t)}C)$, which describes a curve D such that the tangent $T_{z(t)}D$ is the exterior bisector of the oriented lines $\overrightarrow{z(t)x(t)}$ and $\overrightarrow{z(t)y(t)}$. Then

$$\boldsymbol{F}(t) = x(t)z(t) + y(t)z(t) - \text{leng}(C|_{\text{from } x(t) \text{ to } y(t)}) = \text{constant}.$$

To see this, differentiate this function using 9.9.6 and 9.10.1:

$$F'(t) = \left\| z'(t) \right\| \cos(\overrightarrow{x(t)z(t)}, z'(t)) - \left\| x'(t) \right\|$$

$$+ \left\| z'(t) \right\| \cos(\overrightarrow{y(t)z(t)}, z'(t)) - \left\| y'(t) \right\| - \left\| y'(t) \right\| + \left\| x'(t) \right\|.$$

The two terms involving $\left\| z'(t) \right\|$ vanish because $T_{z(t)}D$ is assumed to be the exterior bisector.

9.10.5. Formula 9.10.1 could have been used to guess 9.4.1.1. In fact, if $F = ax + bx$ has a minimum at x, we have, for any vector $\vec{\xi} \in \vec{D}$:

$$F'(x)\vec{\xi} = \left\| \vec{\xi} \right\| \left(\cos(\overrightarrow{\xi, xa}) + \cos(\overrightarrow{\xi, xb}) \right) = 0,$$

so D must be the exterior bisector of $\overrightarrow{xa}$ and $\overrightarrow{xb}$ (figure 9.10.5).

9.10.6. Formula 9.10.1 can be used to guess the solution to the following problem of Fermat: given a triangle $\{a, b, c\}$ in the Euclidean plane, find a point x that minimizes $ax + bx + cx$. If x achieves this minimum and is not a vertex, we have, for every $\vec{u} \in \vec{X}$:

$$\left(\vec{u} \middle| \frac{\overrightarrow{ax}}{\left\| \overrightarrow{ax} \right\|} \right) + \left(\vec{u} \middle| \frac{\overrightarrow{bx}}{\left\| \overrightarrow{bx} \right\|} \right) + \left(\vec{u} \middle| \frac{\overrightarrow{cx}}{\left\| \overrightarrow{cx} \right\|} \right) = 0.$$

In other words, the three unit vectors in the direction of $\overrightarrow{ax}$, $\overrightarrow{bx}$ and $\overrightarrow{cx}$ must add up to zero, and this is the case if and only if their angles at x are all equal to $2\pi/3$. This is the starting point for the solution developed in 10.4.3.

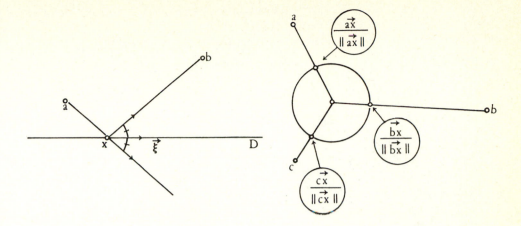

Figure 9.10.5   Figure 9.10.6

9.10.7. NOTE. The name "first variation formula" signifies the fact that we are dealing with the first derivative d'. The computation of d'' does not result in anything new that cannot be done directly (roughly speaking, 9.2.2). For submanifolds of Euclidean spaces or abstract Riemannian manifolds, on the other hand, the "second variation formula", giving d'', is a fundamental tool. It is from this formula that most global results in Riemannian geometry are derived, and it is in the second derivative that one finds the curvature of the manifold (which is zero in the Euclidean case). For recent references, see, for example, [KO–NO, volume 2, chapter VIII] and [G–K–M, 121 ff.].

9.11. The Hausdorff metric

For later use, especially in chapter 12, where it will be essential, we define a metric structure on the set of all compact subsets of a Euclidean affine space. This metric is due to Hausdorff; we start by considering it in a slightly more general setting.

9.11.1. NOTATION. *Let X be a metric space. Denote by $K = K(X)$ the set of all compact sets of X. If F is a subset of X and $\rho \geq 0$ is a real number, put* (cf. 0.3)

$$U(F, \rho) = \{\, x \in X \mid d(x, F) < \rho \,\}, \qquad B(F, \rho) = \{\, x \in X \mid d(x, F) \leq \rho \,\}.$$

If F, G are subsets of X, define the Hausdorff distance between F and G as the real number

$$\delta(F, G) = \inf \{\, \rho \mid F \subset B(G, \rho) \text{ and } G \subset B(F, \rho) \,\}.$$

Watch out for the difference between $\delta(F, G)$ and $d(F, G)$ (cf. 0.3).

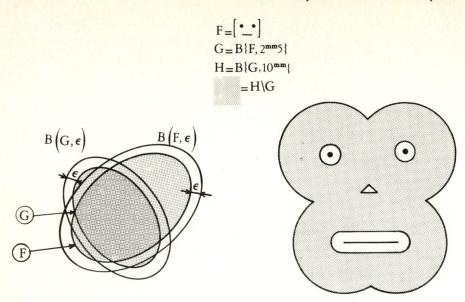

Figure 9.11.1.1   Figure 19.11.1.2 (*Pithecanthropus Geometricus*)

9.11.2. THEOREM. *The Hausdorff distance makes $K(X)$ into a metric space. If X is such that every closed bounded set is compact, then $\big(K(X),\delta\big)$ is complete (resp. compact) if X is.*

Proof. The function $\delta(\cdot,\cdot)$ is obviously symmetric. If $\delta(F,G) = 0$, we have $F \subset B(G,0) = \overline{G} = G$ and $G \subset F$, whence $F = G$. Let F, G, H be arbitrary subsets of X satisfying $G \subset B(H,\sigma)$ and $F \subset B(G,\rho)$; then $F \subset B(G,\rho) \subset B(H, \sigma + \rho)$. Thus, if $\delta(F,G) = \rho$ and $\delta(G,H) = \sigma$, we have $\delta(F,H) \le \rho + \sigma$.

From now on it is understood that $K(X)$, or just K, is endowed with the Hausdorff metric. Let $(F_n)_{n \in \mathbf{N}}$ be a Cauchy sequence in K. For each $n \in \mathbf{N}$, set

$$G_n = \bigcup_{p \in \mathbf{N}} F_{n+p}.$$

Each G_n is bounded because, since (F_n) is a Cauchy sequence, there exists n_0 such that $\delta(F_n, F_{n_0}) \le 1$ for all $n \ge n_0$, and $F_n \subset B(F_{n_0}, 1)$ for $n \ge n_0$. We thus have a decreasing sequence (G_n) of bounded and closed, hence compact, sets; by a classical result of general topology (cf. 0.4), the intersection $F = \bigcap_{n \in \mathbf{N}} G_n$ is non-empty. The set F is in K; there remains to see that $F = \lim_{n \to \infty} F_n$ in K. Given $\epsilon > 0$, let n_0 be such that $\delta(F_n, F_{n_0})$ for all $n \ge n_0$. Then $F \subset G_{n_0} \subset B(F_{n_0}, \epsilon)$. In the opposite sense, there exists n_1 such that $G_n \subset B(F, \epsilon)$ for all $n \ge n_1$; taking $n \ge \sup(n_0, n_1)$ gives $\delta(F, F_n) \le \epsilon$.

To show that K is compact when X is, recall that compactness is equivalent to completeness and the open cover property (for every $\epsilon > 0$ the space has a finite cover by sets of diameter $\le \epsilon$). So assume X is compact and take a finite cover of X by balls $B(x_i, \epsilon)$. Denote by K_0 the set of subsets of $\{x_1, \ldots, x_n\}$; we have $K_0 \subset K$. We show that $K = \bigcup_{K \in K_0} B_\delta(K, \epsilon)$;

this is enough since $\#K_0 = 2^n < \infty$. For $G \in K$ arbitrary, introduce the set $F = \{ x_i \mid d(x_i, G) \leq \epsilon \}$. By construction, $F \subset B(G, \epsilon)$; but since $X = \bigcup_{i=1}^{n} B(x_i, \epsilon)$, there exists for any $y \in G$ an x_i such that $y \in B(x_i, \epsilon)$, showing that $x_i \in F$ and $G \subset B(F, \epsilon)$. Thus $\delta(F, G) \leq \epsilon$, or again $G \subset B_\delta(F, \epsilon)$.

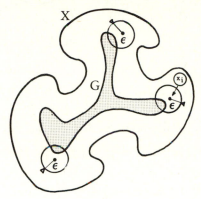

Figure 9.11.2

9.11.3. COROLLARY. *If X is compact, finite subsets form a dense subset of $K(X)$.*

Proof. This follows from the last part of the proof of 9.11.2. □

9.11.4. COROLLARY. *Let X be a Euclidean affine space. Then $K(X)$ is complete. Moreover, for any $a \in X$ and $r \in \mathbf{R}_+$, the set $K_{a,r}(X) = \{ F \in K(X) \mid F \subset B(a, r) \}$ is compact.*

Proof. This result is often called the "Blaschke selection theorem" ([EN, 64]); it implies that every infinite family of sets $K_{a,r}(X)$ has an accumulation point, or, in other words, one can pick compacts subsets of X with certain limit properties. We shall use this in 9.13.8 and 12.11.1.

9.11.5. PROPOSITION. *For every metric space X, the diameter function* diam $: K(X) \mapsto \mathbf{R}$ *(cf. 0.3) is a Lipschitz map with constant 2.*

Proof. Let $F, G \in K$ be such that $\delta(F, G) = \epsilon$, and take $x, y \in F$ satisfying $d(x, y) = \mathrm{diam}(F)$. There exist $z, t \in G$ such that $d(x, z) \leq \epsilon$ and $d(y, t) \leq \epsilon$; this implies

$$\mathrm{diam}(F) = d(x, y) \leq d(x, z) + d(z, t) + d(t, y) \leq 2\epsilon + \mathrm{diam}(G).$$

Switching F and G gives $\left| \mathrm{diam}(F) - \mathrm{diam}(G) \right| \leq 2\epsilon$. □

9.11.6. PROPOSITION. *Let X be a Euclidean affine space and H a hyperplane in X. Then $p : K(X) \ni K \mapsto p(K) \in K(H)$, the orthogonal projection on H (cf. 9.2.4), is Lipschitz with constant 1.*

Proof. It is enough to observe that, for any $K \in K(X)$ and any real number ρ, we have

$$p(B_X(K, \rho)) = B_H(p(K), \rho).$$ □

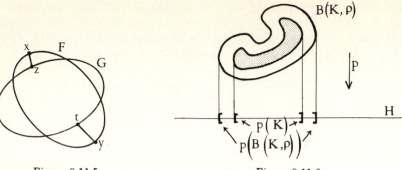

Figure 9.11.5   Figure 9.11.6

9.11.7. From the triangle inequality we deduce that, for every ρ, σ, F and G,

$$\delta\big(B(F,\rho), B(G,\sigma)\big) \leq \delta(F,G) + |\rho - \sigma|.$$

9.12. Canonical measure and volume

9.12.1. Here we combine 2.7.4 and 8.11. In 8.11 we saw that a Euclidean vector space $\vec{X}$ possesses a canonical Lebesgue measure, obtained, for example, as the image of the Lebesgue measure of $\mathbf{R}^n$ under any vector isometry $\mathbf{R}^n \to \vec{X}$. By 2.7.4.3, the canonical measure on $\vec{X}$ gives rise to a canonical measure on X, which we *denote* by μ or μ_X if necessary. By construction, if $\{x_i\}_{i=0.1,\dots,n}$ is an orthonormal frame for X and $f : X \to \mathbf{R}$ is integrable, we have

$$\int_X f\mu = \int_{\mathbf{R}^n} f\mu_0,$$

where μ_0 is the Lebesgue measure on $\mathbf{R}^n$ and f also denotes, by abuse of notation, the function on $\mathbf{R}^n$ obtained from f via the isomorphism $X \to \mathbf{R}^n$ associated with the chosen frame. Naturally, $\mu = \mu_0$ when $X = \mathbf{R}^n$. Clearly μ is invariant under $\mathrm{Is}(X)$, and is the unique measure (up to a scalar) with this property (cf. 2.7.4.4).

9.12.2. An explicit link with the volume form $\lambda_{\vec{X}}$ and the density $\delta_{\vec{X}}$ is provided by the theory of forms and densities on differentiable manifolds. The manifold is X, the form and the density, *denoted* by λ_X and δ_X, are obtained directly from $\lambda_{\vec{X}}$ and $\delta_{\vec{X}}$ because the tangent space to X is canonically identified with $\vec{X}$, and we have, for every function $f : X \to \mathbf{R}$:

$$\int_X f\mu = \int_X f\lambda_X = \int_X f\delta_X.$$

9.12.3. EXAMPLES. In X, every affine subspace $Y \neq X$ has measure zero.

If X and X' are two Euclidean affine spaces of same dimension n and $f : X \to X'$ is a similarity of ratio k (cf. 9.5), the image measure $f(\mu_X)$ is equal to $k^n \mu_X$.

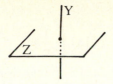

Figure 9.12.3

Let Y and Z be complementary orthogonal affine subspaces of X. The measure μ_X is the product of the measures μ_Y and μ_Z, that is, $\mu_X = \mu_Y \times \mu_Z$, and we can apply Fubini's theorem (cf. 0.6).

9.12.4. VOLUME

9.12.4.1. Definition. *The volume of a compact set K of X, denoted by $\mathcal{L}(K)$, is the integral $\int_X \chi_K \mu$, where χ_K is the characteristic function of K (cf. 0.6). In dimension one (resp. two) we generally talk about length (resp. area) instead of volume.*

These are really our everyday concepts of volume and area. Here we compute only the volume of a box and a simplex; for the volume of common solids, like balls, the reader can consult 9.12.4.7, 9.12.4.8 and 12.12.10.

9.12.4.2. The *box* built on the points $\{x_i\}_{i=0.1.....n}$ is the set

$$P = \left\{ x_0 + \sum_{i=1}^{n} \lambda_{\overrightarrow{x_0 x_i}} \;\middle|\; \lambda_i \in [0, 1] \text{ for } i = 1, \dots, n \right\}.$$

The *solid simplex* built on the same points is the set

$$S = \left\{ \sum_{i=0}^{n} \lambda_i x_i \;\middle|\; \sum_{i=0}^{n} \lambda_i = 1 \text{ and } \lambda_i \geq 0 \text{ for } i = 0, 1, \dots, n \right\}.$$

We have the formulas

9.12.4.3
$$\mathcal{L}(P) = \delta_{\overrightarrow{X}}(\overrightarrow{x_0 x_1}, \dots, \overrightarrow{x_0 x_n}), \qquad \mathcal{L}(S) = \frac{1}{n!}\mathcal{L}(P).$$

The first formula follows from 9.12.3 if $\{x_i\}$ is not an affine frame; otherwise $\{x_i\}$ determines a linear isomorphism $f : X \to \mathbf{R}^n$ (which is generally not an isometry), and the image measure $f(\mu)$ is equal to $\delta_{\overrightarrow{X}}(\overrightarrow{x_0 x_1}, \dots, \overrightarrow{x_0 x_n})\mu_0$ by 2.7.4.3 and 8.11. Now $f(P)$ is the unit cube in $\mathbf{R}^n$, and its volume is 1 (by Fubini, for example); this gives the formula for the box.

The second formula is obtained by induction, using the intermediate formula 9.12.4.4. (We can assume $\{x_i\}$ is an affine frame, otherwise $\mathcal{L}(S) = \mathcal{L}(P) = 0$.) Let η be the distance

$$\eta = d\big(x_0, \langle x_1, \dots, x_n \rangle\big)$$

from x_0 to the hyperplane $Y = \langle x_1, \dots, x_n \rangle$, and σ the volume of the solid simplex S' built on the points $\{x_i\}_{i=1,\dots,n}$ of the $(n-1)$-dimensional) Euclidean space H. Then

9.12.4.4 $$\mathcal{L}(S) = \frac{1}{n}\eta\sigma.$$

For $n = 2$, this says that the area of a triangle is equal to one half the altitude times the base, and for $n = 3$, that the volume of a tetrahedron is one third of the height times the area of the opposite face.

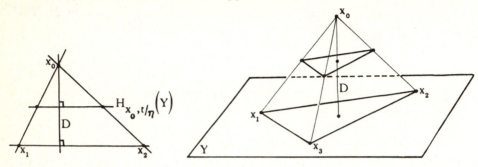

Figure 9.12.4.4

To prove 9.12.4.4, let D be the line orthogonal to Y and containing x_0. We can parametrize D by a real number t so that D is isometric to $\mathbf{R}$ and so that $t = \eta$ at $D \cap Y$ and $t = 0$ at x_0. By 9.12.3, we have

$$\mathcal{L}(S) = \int_0^\eta \sigma(t)\,dt,$$

where $\sigma(t)$ denotes the volume (in the Euclidean space $H_{x_0.t/\eta}(Y)$) of the solid simplex $H_{x_0.t/\eta}(S')$ obtained from S' by a homothety of center x_0 and radius t/η. But this volume is equal to $(t/\eta)^{n-1}$, by 9.12.3; we obtain

$$\mathcal{L}(S) = \int_0^\eta \frac{t^{n-1}}{\eta^{n-1}}\sigma\,dt = \frac{1}{n}\eta\sigma.$$

9.12.4.5. It may seem wasteful to use integration theory to calculate the volume of a box or a simplex; a more elementary theory is developed in 12.2.5.

One may also want to calculate $\mathcal{L}(S)$ as a function of the lengths $d_{ij} = x_i x_j$ of the simplex S only; this follows immediately from combining 8.11.6, 9.7.3.2 and 9.12.4.3.

9.12.4.6. Volume of balls. We denote by $B_n(a, r)$ the ball of center a and radius r in the n-dimensional space X, for any $a \in X$, $r \in \mathbf{R}_+$. Then

$$\mathcal{L}\big(B_{2d}(a,r)\big) = \frac{\pi^d}{d!} r^{2d},$$

$$\mathcal{L}\big(B_{2d+1}(a,r)\big) = \frac{2^{d+1}\pi^d}{1 \cdot 3 \cdot 5 \cdots (2d+1)} r^{2d+1}.$$

By 9.12.3, it is enough to find $\mathcal{L}\big(B_n(0,1)\big)$ in the standard space $\mathbf{R}^n$. This is a classical computation and can be carried out in various ways (cf. [B–G, 6.5.7 and 6.10.9]). In the sequel we shall often need the special value

9.12.4.7 $\beta(n) = \mathcal{L}\big(B_n(0,1)\big)$.

9.12.4.8. Since we're at it, we give the $(n-1)$-dimensional volume of the sphere $S_n(0,1)$ (cf. 9.12.7, 12.10.8, [B–G, 6.5.6.1]). Let $\alpha(n)$ be the $(n-1)$-dimensional volume of the unit sphere S^{n-1} in $\mathbf{R}^n$; we have $\alpha(n) = n\beta(n)$ for every n. Thus

$$\alpha(2d) = \frac{2\pi^d}{(d-1)!},$$

$$\alpha(2d+1) = \frac{2^{d+1}\pi^d}{1 \cdot 3 \cdot \cdots \cdot (2d-1)}.$$

9.12.4.9. Orthogonal projection. Let H and H' be hyperplanes of X, and let $\alpha \in [0, \pi/2]$ be the angle between them: $\alpha = (\vec{H})^\perp, (\vec{H'})^\perp$. Denoting by $p : H' \to H$ the restriction to H' of the orthogonal projection from X onto H, by $\mathcal{L}_H$ (resp. $\mathcal{L}_{H'}$) the volume in H (resp. H') and by K an arbitrary compact set in H', we have

$$\mathcal{L}_H\big(p(K)\big) = \cos\alpha \cdot \mathcal{L}_{H'}(K).$$

To see this, take an orthonormal frame in H' that has $n-1$ vectors in $\vec{H} \cap \vec{H'}$.

9.12.5. VOLUME AND HAUSDORFF DISTANCE. Even for compact sets, it is not to be expected that the volume function $\mathcal{L} : \mathcal{K} \to \mathbf{R}$ is a simple one. For example, there exist compact sets $K \subset \mathbf{R}^2$ whose boundary ∂K not only has positive area, but has positive area everywhere (see [G–O, 135]). The function $\mathcal{L} : \mathcal{K} \to \mathbf{R}$ is certainly not continuous, otherwise (cf. 9.11.3) all compact sets would have volume zero, because finite sets do. But we do have the following

9.12.5.1. Proposition. $\mathcal{L} : \mathcal{K} \to \mathbf{R}$ *is upper semicontinuous.*

Proof. Assume that $F = \lim_{n \to \infty} F_n$ in $\mathcal{K}$; we show that $\limsup_{n \to \infty} \chi_{F_n} \le \chi_F$. Let $x \in X \setminus F$; there exists n_0 such that $\delta(F, F_n) \le d(x, F)$ for all $n \ge n_0$. Thus $\chi_{F_n}(x) = 0$ for all $n \ge n_0$.   □

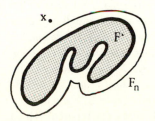

Figure 9.12.5.1

9.12.5.2. Note. We shall see in 12.9.3.4 that $\mathcal{L}$ is continuous when restricted to convex compact sets.

9.12.6. CENTROIDS. Let K be a compact set in a Euclidean affine space X. Assume $\overset{\circ}{K} \neq \emptyset$, and let $g = \text{cent}'(K)$ (cf. 2.7.5.2). A generalization of 9.7.6.1 with all the $\lambda_i = 1$ is the following: for every $x \in X$,

9.12.6.1
$$\int_{a \in X} \chi_K \, xa^2 \mu = \int_{a \in X} \chi_K \, ga^2 \mu + \mathcal{L}(K) \, xg^2.$$

Here again the proof consists in writing

$$xa^2 = xg^2 + 2(\overrightarrow{xg} \mid \overrightarrow{ga}) + ga^2.$$

To generalize 9.7.6.1 with arbitrary $\lambda_i \geq 0$, we choose on K an arbitrary positive measure θ. If the total mass of K is positive, we can define the *barycenter of K for the measure θ* as the vector

$$g = x + \frac{\int_{a \in K} \overrightarrow{xa} \, \theta}{\int_{a \in K} \theta},$$

which does not depend on x. We then have, for every $x \in X$,

9.12.6.2
$$\int_{a \in K} xa^2 \, \theta = \int_{a \in K} ga^2 \theta + \left(\int_{a \in K} \theta \right) xg^2.$$

9.12.6.3. Notes. Such measures θ are naturally encountered when dealing with curves in the Euclidean plane or surfaces in Euclidean three-space, for example. One uses their canonical measure, cf. [B–G, 6.4].

Formulas 9.12.6.1 and 9.12.6.2 show that the function

$$x \mapsto \int_{a \in K} xa^2 \theta$$

has a minimum at g, and only there. In mechanical terms this means that the moment of inertia around the point g is the smallest possible, so if you want to spin something around a point with the least possible effort, you should do it around g. This is clearly useful to know in practice!

For a nice relation between volume and center of mass, see 12.12.20.9. For mechanical methods for the calculation of areas, see [GK, 72 ff.].

9.12.7. THE k-DIMENSIONAL VOLUME. If C is a differentiable curve in the plane its volume $\mathcal{L}(C)$ is zero, and similarly for, say, the sphere S^2 in $\mathbf{R}^3$. This concept of volume is not much good for such objects. For C the appropriate notion would be its length, defined in 9.9 (at least when C is a homeomorphic image of its interval of definition), and for S^2 it would be its area, equal to 4π (cf. 18.3.7, for example). In general, we would like to characterize "k-dimensional subsets" of X, and define for such subsets their "k-dimensional volume" (area if $k = 2$, length if $k = 1$). This is a very difficult problem in general, though certainly natural (think of painting a solid; for an application of this idea, see 12.10.7).

If we restrict ourselves to C^1 submanifolds of X, no difficulty arises: such a submanifold possesses a canonical measure, so if it is compact it has a total mass, and that is its k-volume (k being the dimension of the manifold). See, for example, [B–G, 6.4]. On the other hand, if we want to break free from this very restrictive condition, which precludes even simple bodies as in figure 9.12.7.1, and allow certain kinds of singularity, no one notion of k-dimensional volume ($2 \leq k < n = \dim X$) will do. There are, in fact, many notions, all essentially different. For a complete discussion of this question, see [FR], in particular pages 171–174, where no less than seven k-dimensional measures are introduced. For $k = n = \dim X$, all these measures coincide with the volume, and for compact differentiable manifolds they all coincide with the notion discussed above.

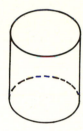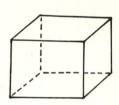

Figure 9.12.7.1

There is, however, one particular case that we discuss in detail, namely, frontiers of compact convex sets. These possess a canonical $(n-1)$-dimensional volume, which we call area for the sake of simplicity. We first define it for polytopes (12.3), then approximate convex sets by polytopes (12.10). See also the case of the sphere in 18.3.7.

The reader can get a feeling for the pitfalls involved in this process by considering the well-known example of the Chinese lantern, where we try to

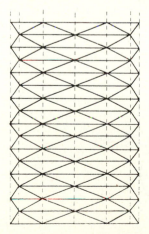

Figure 9.12.7.2 (source: [B–G])

find the area of a cylinder by approximating it by polyhedra, by analogy with
the one-dimensional case (figure 9.12.7.2). Unfortunately, depending on the
ratio between the number of horizontal slices and the number of sides of the
regular polygons as both go to infinity, the surface area of the polyhedra can
be made to approach any limit between the area of the cylinder (the limit we
really want) and infinity.

9.13. Steiner symmetrization

This operation, which transforms a compact subset of X into another,
plays a fundamental role in the proof of certain inequalities about compact
sets: see 9.13.8, but especially the conclusion of chapter 12. Apart from its
inherent beauty, we study its properties here with an eye to applications in
chapter 12.

Let X be an affine space, and fix a hyperplane H of X. Recall that σ_H
stands for the reflection through X (cf. 9.2.4). Let K be a compact subset
of X.

9.13.1. DEFINITION. *The Steiner symmetrization of K with respect to H,
denoted by $\mathrm{st}_H(K)$, is the compact set $K' = \mathrm{st}_H(K)$ uniquely defined by the
following condition:*

9.13.2. *If D is any line orthogonal to H, either $K \cap D = \emptyset$ and $K' \cap D = \emptyset$,
or $K \cap D \neq \emptyset$ and $K' \cap D$ is a line segment contained in D whose midpoint
is $D \cap H$ and whose length is equal to the length of $K \cap D$ (in D).*

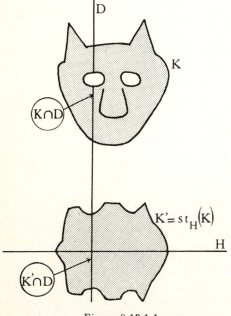

Figure 9.13.1.1

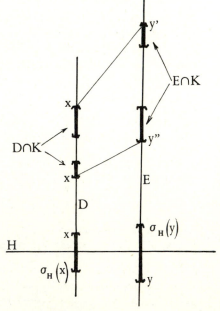

Figure 9.13.1.2

9.13.3. REMARK. Watch out for the fact that $\mathrm{st}_H : K \to K$ is not continuous (with respect to the Hausdorff distance). This is seen in figure 9.13.3,
where the symmetrization of the segment K is a segment (of varying length)
contained in H, except when K reaches the vertical position; then its symmetrization suddenly becomes a vertical segment.

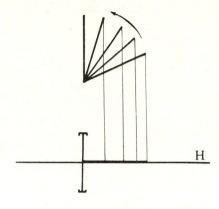

Figure 9.13.3

9.13.4. PROPOSITION. *For every H and K we have*
i) $\sigma_H\big(\mathrm{st}_H(K)\big) = \mathrm{st}_H(K)$;
ii) $\mathcal{L}\big(\mathrm{st}_H(K)\big) = \mathcal{L}(K)$, *that is, Steiner symmetrization preserves volumes;*
iii) $\mathrm{diam}\big(\mathrm{st}_H(K)\big) \le \mathrm{diam}(K)$, *that is, Steiner symmetrization does not in-*
crease diameters.

Proof. Property (i) is immediate from the definition; (ii) follows from 9.12.3
and Fubini's theorem. For (iii), let $x, y \in \mathrm{st}_H(K)$ be such that

$$xy = \mathrm{diam}\big(\mathrm{st}_H(K)\big),$$

and let D, E be lines through x, y and orthogonal to H. By definition, the
compact set $D \cap K$ (resp. $E \cap K$) has endpoints x', x'' (resp. y', y'') such that
$x'x'' \ge x\sigma_H(x)$ (resp. $y'y'' \ge y\sigma_H(y)$); but we have $xy \le \sup(x'x'', y'y'')$. $\square$

The next lemma follows from 9.13.2 and figure 9.13.5:

9.13.5. LEMMA. *Let B be a ball, S its boundary, H a hyperplane containing
the center of B and K a compact set contained in B. Then*

$$\mathrm{st}_H(K) \cap \big((S \setminus K) \cup \sigma_H(S \setminus K)\big) = \emptyset. \qquad\qquad \square$$

The next theorem will be put to essential use in 9.13.8 and 12.11.2; it provides
an explicit way to locate a ball in a family of compacts.

9.13.6. THEOREM (THE BLASCHKE KUGELUNGSATZ). *Let $\mathcal{F}$ be a non-
empty subset of $\mathcal{K} = \mathcal{K}(X)$, closed under the Hausdorff measure and invariant
under Steiner symmetrization with respect to any hyperplane containing a
fixed point a. Then either $\{a\} \in \mathcal{F}$ or there exists $r > 0$ such that $B(a, r) \in \mathcal{F}$.*

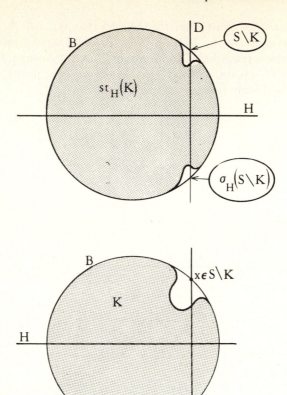

Figure 9.13.5

Observe that when one enlarges a family so as to make it closed and invariant under symmetrization, the one-point set $\{a\}$ can appear in the family, as shown by example 9.13.3.

Proof. Set $r = \inf \{ s \mid$ there exists $F \in \mathcal{F}$ contained in $B(a, s) \}$, and consider the set $\mathcal{F}' = K_{a.r+1}(X) \cap \mathcal{F}$ (cf. 9.11.4). From the assumptions and 9.11.4, this is a compact set, so there exists $F \in \mathcal{F}'$ such that $F \subset B(a, r)$. If $r = 0$, we have $F = \{a\}$; the proof will be completed by showing that we have $F = B(a, r)$ if $r > 0$.

9.13.6.1. First step. We prove by contradiction that $F \supset S = S(a, r)$. Assume there exist $b \in S$ and $\epsilon > 0$ such that $B(b, \epsilon) \cap F = \emptyset$; we shall find an element $F_n \in \mathcal{F}$ such that $F_n \cap S = \emptyset$, contradicting the choice of r. Choose points b_i by induction, so that $b_1 = b$ and $B(b_i, \epsilon) \cap B(b_{i+1}, \epsilon) \cap S \neq \emptyset$ for all $i = 1, 2, \ldots$. A finite number of such balls covers S:

$$S \subset \bigcup_{i=1}^{n} B(b_i, \epsilon).$$

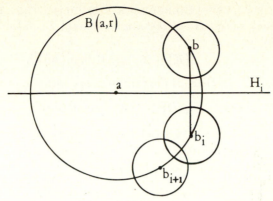

Figure 9.13.6.1

Denote by H_i the plane equidistant from b and b_i, and define F_i by induction as $F_1 = F$, $F_i = \mathrm{st}_{H_i}(F_{i-1})$ $(i = 1, 2, \ldots, n-1)$. Lemma 9.13.5 shows that $F_n \cap S = \emptyset$, and by assumption $F_n \in \mathcal{F}$, as desired.

9.13.6.2. Second step. We show by contradiction by $F = B(a, r)$. Assume $x \in B(a, r) \setminus F$; let D be an arbitrary line containing x, and H the hyperplane orthogonal to D and containing a. Since $F \subset B(a, r)$, the length of $D \cap F$ is strictly smaller than the length of $D \cap B(a, r)$; this would imply that $\mathrm{st}_H(F)$ does not contain S, contradicting the first step applied to $\mathrm{st}_H(F) \in \mathcal{F}$. $\square$

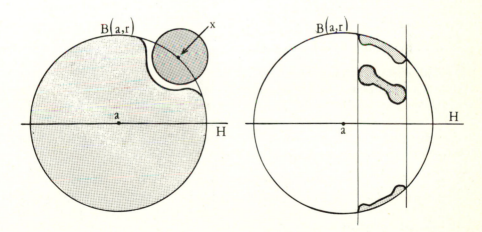

Figure 9.13.6.2

9.13.7. NOTE. There is an older proof, perhaps more intuitive but also harder to formalize, that considers a finite number of linearly independent hyperplanes H_i passing through a and meeting at irrational angles. The subgroup of $\mathrm{Is}_a(X)$ generated by the reflections σ_{H_i} is dense in $\mathrm{Is}_a(X)$, so $\mathcal{F}$ contains a ball centered at a.

9.13.8. COROLLARY (BIEBERBACH'S ISODIAMETRIC INEQUALITY). *For every compact subset K of the n-dimensional Euclidean space X we have* (cf. 9.12.4.5)

$$\mathcal{L}(K) \leq 2^{-n}\beta(n)(\operatorname{diam}(K))^n.$$

Proof. If $\mathcal{L}(K) = 0$, there is nothing to prove; otherwise set

$$\mathcal{F} = \{\, G \in \mathcal{K} \mid \mathcal{L}(G) \geq \mathcal{L}(K) \text{ and } \operatorname{diam}(G) \leq \operatorname{diam}(K)\,\}.$$

By 9.11.5, 9.13.4 and 9.13.6 (where a is arbitrary), either $\{a\} \in \mathcal{F}$ or $\mathcal{F}$ contains a ball $B(a, r)$ with $r > 0$. The first case is impossible since $\mathcal{L}(K) > 0$; thus $\mathcal{L}(B(a, r)) = \beta(n)r^n \geq \mathcal{L}(K)$, and

$$\operatorname{diam}(B(a, r)) = 2r \leq \operatorname{diam}(K),$$

achieving the demonstration since $r > 0$.   □

9.13.9. REMARKS. Observe that, in general, a compact set K is not contained in any ball of radius $\operatorname{diam}(K)/2$ (consider an equilateral triangle, for instance).

The isodiametric inequality clearly becomes an equality when K is a ball. The converse is true but much harder to prove; see [EN, 106–107].

For applications and generalizations of the Steiner symmetrization, see [P–S], as well as 12.11.

9.14. Exercises

9.14.1. Generalize 9.2.6.5 to the case of two arbitrary subspaces.

9.14.2. Write equations of the form $f(x, y, z) = 0$ (in an appropriately chosen system of coordinates) for the sets of points in three-dimensional space defined by the equation $d(x, A) = d(x, B)$, where A and B are, respectively: two lines, a line and a point, a line and a plane, a point and a plane, and two planes. Classify the sets according to the terminology of 15.3 and 15.6. What happens if we consider the equation $d(x, A) = kd(x, B)$ instead?

* **9.14.3.** BISECTORS. Let D, D' be two non-parallel lines in a Euclidean plane X; show that $\{\, x \in X \mid d(x, D) = d(x, D')\,\}$ is formed by the two bisectors of D, D'. What is this set when X is higher-dimensional? (Observe that D and D' may not intersect.)

9.14.4. Let X be a Euclidean affine space. Determine all maps $f \in \mathrm{GA}(X)$ such that $f^2 = f$ and f does not increase distance (i.e., $f(x)f(y) \leq xy$ for all $x, y \in X$).

9.14.5. Let X be three-dimensional and oriented. Define the oriented angle of a rotation so as to characterize rotations unambiguously by their oriented axis and oriented angle.

9.14.6. Let X be three-dimensional and take points (a_i) and (a_i') $(i = 1, 2, 3)$ in X such that $a_i a_j = a_i' a_j'$ for all $i, j = 1, 2, 3$. Assuming the a_i are affine independent, show that there exists a unique $f \in \mathrm{Is}^+(X)$ (resp. $\mathrm{Is}^-(X)$) such that $f(a_i) = a_i'$ for all i. Find a geometric procedure to obtain the characteristic elements of f (cf. 9.3.5).

9.14.7. CONTINUOUS SCREW MOTIONS. A *continuous screw motion* of a three-dimensional Euclidean affine space X is a one-parameter family of isometries $f : \mathbf{R} \to \mathrm{Is}(X)$ that can be expressed in an appropriate orthonormal frame as

$$f(t) : (x, y, z) \mapsto (\cos kt \cdot x + \sin kt \cdot y, -\sin kt \cdot x + \cos kt \cdot y, z + ht),$$

where k, h are scalars. Show that, for any map $g : \mathbf{R} \to \mathrm{Is}(X)$ of class C^1 (consider $\mathrm{Is}(X)$ as a subset of $\mathrm{GA}(X)$ to define differentiability), there exists a continuous screw motion f such that $(g - f)'(0) = 0$. Determine f when g is the map defined by the tangent, normal and binormal vectors of a curve. See also [SY, 282].

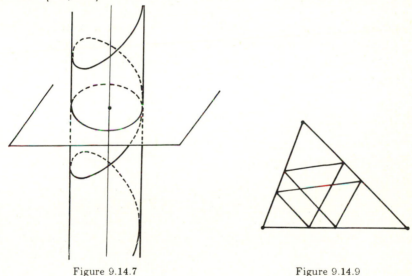

Figure 9.14.7   Figure 9.14.9

9.14.8. Show that the isometry f in 9.4.2.3 is never the identity. While at it, make a critical study of [GR, 179 ff.].

9.14.9. Study billiard trajectories that close after going around k times, according to the value of k and the number of sides of the polygon. See a beautiful picture in [GR, 176].

9.14.10. Make a complete study of least-perimeter polygons inscribed in a quadrilateral whose vertices lie on a circle.

9.14.11. Show that rectangular billiards are never strongly ergodic, but always weakly ergodic. Show that an equilateral triangle is not strongly ergodic.

9.14.12. Study the structure of similarities in three dimensions.

9.14.13. In dimension two, take four points a, a', b, b' such that $a \neq b$ and $a' \neq b'$. Show that there exists a unique $f \in \mathrm{Sim}^-(X)$ such that $f(a) = a'$ and $f(b) = b'$. Construct its center and axis geometrically.

9.14.14. Use 9.7.6.5 to find a geometric method different from 9.6.2 for the construction of the map $f \in \mathrm{Sim}^+(X)$ (resp. $\mathrm{Sim}^-(X)$) such that $f(a) = a'$ and $f(b) = b'$.

9.14.15. HARMONIC QUADRILATERAL. Let x, y, z, t be points in a Euclidean plane such that $[x, y, z, t] = -1$ (cf. 9.6.5.2). Let a (resp. b) be the midpoint of x and y (resp. z and t). Show that x, y, z, t are cocyclic, that the pole of $\langle x, y \rangle$ lies in $\langle z, t \rangle$, that $\langle x, y \rangle$ is the bisector of $\langle a, z \rangle$ and $\langle a, t \rangle$, that $az \cdot at = ax^2 = ay^2$, that $xy \cdot zt = xz \cdot yt + xt \cdot yz$ and that $az + at = bx + by$. Show that, if u is the intersection of $\langle x, y \rangle$ and $\langle z, t \rangle$, we have

$$\frac{ux}{uv} = \left(\frac{zx}{zy}\right)^2 \left(\frac{tx}{ty}\right)^2 \qquad \text{and} \qquad \frac{uz}{ut} = \left(\frac{xz}{xt}\right)^2 \left(\frac{yz}{yt}\right)^2.$$

State and prove converses for these statements. Show that inversions (cf. 10.8) transform harmonic quadrilaterals into harmonic quadrilaterals.

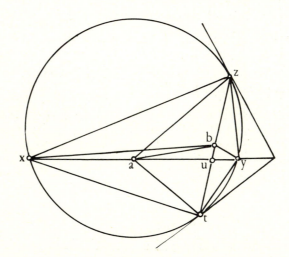

Figure 9.14.15

* **9.14.16.** Inscribe a square inside a given triangle (figure 9.14.16).

9.14.17. Discuss the form and relative position of the Pascal limaçons in figure 9.6.8.0.4, according to the relative position of the two circles.

* **9.14.18.** Study the convexity characteristics of limaçons (figure 9.6.8.0.3) as the position of the point a varies with respect to the circle. (Hint: use the formula $\rho^2 + 2\rho'^2 - \rho\rho''$, which gives the concavity of the curve $\rho(\theta)$ in polar coordinates.)

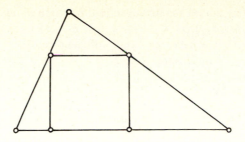

Figure 9.14.16

9.14.19. Let C, C' be two subsets of the Euclidean plane X such that, for any α, there exists a similarity f of angle α taking C into C'. What can be said about C and C'?

9.14.20. Are there any differentiable group homomorphisms $\mathbf{R} \to \mathrm{Sim}^+_\omega(X)$ other than the ones in 9.6.9.1?

9.14.21. LOGARITHMIC SPIRALS. What are the plane curves whose tangent makes a constant angle with the line joining the foot of the tangent to a constant point? What are the plane curves whose radius of curvature is proportional to arclength? Can a logarithmic spiral coincide with its evolute?

9.14.22. PASCAL LIMAÇONS, DESCARTES OVALS AND STIGMATIC DIOP-TERS. Let u and v be points and S a surface of revolution around the axis $\langle u, v \rangle$. Show that a necessary condition for S, considered as a diopter satisfying Descartes's law $\dfrac{\sin i}{\sin r} = $ constant, to be perfectly stigmatic for the two points u, v is that the points x of S satisfy a relation $a \cdot xu + b \cdot xv = c$, for some $a, b, c \in \mathbf{R}$. A *Descartes oval* is a plane curve of the form $\{ x \in X \mid a \cdot xu + b \cdot xv = c \}$, where $u, v \in X$ and $a, b, c \in \mathbf{R}$ are given. Show that Pascal limaçons are Descartes ovals. Study the shape of Descartes ovals. For more details, see [BP].

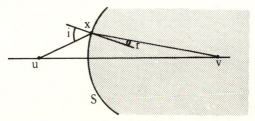

Figure 9.14.22

9.14.23. CAYLEY–MENGER DETERMINANTS. Show that Γ satisfies the following relation, where M_{ik} is the cofactor of the entry d^2_{ik} in Γ:

$$\Gamma(x_0, \ldots, x_{k-1}) \Gamma(x_0, \ldots, \hat{x}_i, \ldots, x_k)$$
$$= M^2_{ik} + \Gamma(x_0, \ldots, \hat{x}_i, \ldots, x_{k-1}) \Gamma(x_0, \ldots, x_k).$$

Deduce that, if $\Gamma(x_0, \ldots, x_k)$ is non-zero and has sign $(-1)^{k+1}$, all the subdeterminants $\Gamma(x_{i_1}, \ldots, x_{i_h})$ $(i_1 < \cdots < i_h, h = 2, \ldots, k)$ are non-zero and have sign $(-1)^h$. Show that a necessary and sufficient condition for the existence of points $(x_i)_{i=0,1,\ldots,k}$ in X such that $x_i x_j = d_{ij}$ for $i, j = 0, 1, \ldots, k$ is that all the determinants $\Gamma(x_{i_1}, \ldots, x_{i_h})$ $(i_1 < \cdots < i_h, h = 2, \ldots, k)$ either be zero or have sign $(-1)^h$.

9.14.24. Give a simple criterion to guarantee that the case $\{a\} \in \mathcal{F}$ does not occur in 9.13.6.

* **9.14.25.** SYLVESTER'S THEOREM. Prove that if a set of n points $(x_i)_{i=1,\ldots,n}$ on an affine plane has the property that every straight line containing two of the x_i also contains a third, then all the points are on the same line. (Hint: introduce an arbitrary Euclidean structure and consider a triple of non-collinear points such that the distance from one to the line passing through the other two is minimal in the set of all such distances.)

9.14.26. Find $f \in \mathrm{Is}(X)$ and a bounded subset A of X such that $f(A) \subset A$ but $f(A) \neq A$. Show that if A is compact $f(A) \subset A$ implies $f(A) = A$.

9.14.27. Show that $\mathrm{Is}_A(X) \subset \mathrm{Is}_{\overline{A}}(X)$. Find an example where the two groups are different.

9.14.28. Find examples of unbounded (resp. infinite) sets A such that $\mathrm{Is}_A(X)$ is compact (resp. finite).

9.14.29. PLANE LATTICES. A *lattice* of $\mathbf{R}^2$ is a subset of the form $\mathbf{Z}x + \mathbf{Z}y$, where x, y are linearly independent (cf. 1.7.5.2). Show that every lattice A of $\mathbf{R}^2$ is similar to a unique lattice of the form $\mathbf{Z}u + \mathbf{Z}v$, where $u = (1, 0)$ and v belongs to the region

$$\mathcal{D} = \left\{ (a, b) \mid 0 \leq a \leq \tfrac{1}{2},\ a^2 + b^2 \geq 1,\ b > 0 \right\}.$$

Study $\mathrm{Is}_A(\mathbf{R}^2)$ according to the position of v in $\mathcal{D}$. Show that two lattices A and A' of $\mathbf{R}^2$ satisfying

$$\#\{\, x \in A \mid \|x\| = r \,\} = \#\{\, x \in A' \mid \|x\| = r \,\}$$

for all $r \in \mathbf{R}_+$ are necessarily isometric. This is not true for lattices in higher dimension; see [SE2, 177, lines 4–8].

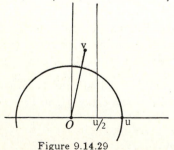

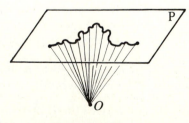

Figure 9.14.29   Figure 9.14.30

9.14.30. INTRINSIC METRICS. Prove 9.9.7.1. Study the metric $\bar{d}$, where (M, d) is the following metric space: d is the distance induced from X on the cone M with vertex O and containing the curve given in 9.9.3.3 (drawn on a plane P not passing through O). Find an example to show that if (M, d) is such that $\bar{d}$ is still a distance, the topologies of $(M, \bar{d})$ and (M, d) may differ.

9.14.31. MENGER CURVATURE. Let f be a regular curve of class C^2 in a Euclidean space. Let x, y, z be distinct points of f, and set

$$K(x, y, z) = \frac{\sqrt{(xy + yz + zx)(xy + yz - zx)(xy - yz + zx)(xy - yz + zx)}}{xy \cdot yz \cdot zx}$$

(cf. 10.3.4). Show that, as y and z approach x on f, $K(x, y, z)$ tends towards the curvature of f at x (see [B–G, 8.4], for example). Find examples of regular curves of class C^1 for which $K(x, y, z)$ does not have a limit, or becomes infinite, as y and z approach x.

9.14.32. (This exercise is optional, as the author cannot take upon himself the costs involved.) Visit the Basel cathedral and decide whether the curve engraved on Bernoulli's tomb is really a logarithmic spiral.

* **9.14.33.** Let C be a plane curve in a Euclidean plane, of class C^1 and strictly convex. Show that for any integer $n \geq 3$ there is at least one n-sided light polygon inscribed in C.

9.14.34. HYPOCYCLOIDS AND EPICYCLOIDS (a.k.a. Spirograph).

9.14.34.1. Definition. A *hypocycloid* (resp. *epicycloid*) is the set C of points of the Euclidean plane occupied by a given point on a circle Γ' that rolls (without sliding) outside (resp. inside) a fixed circle Γ of commensurable radius (figure 9.14.34.1). Study the shape of C according to the ratio between the radii. How many cusps does C have?

9.14.34.2. Equivalent definitions. Let Σ be the unit circle of $\mathbf{R}^2 = \mathbf{C}$, and let $r \neq 0$ be rational. Show that the envelope of the lines $D(\theta)$ joining the points $e^{i\theta}$ and $e^{ir\theta}$, for $\theta \in \mathbf{R}$, is a hypo- or epicycloid. Discuss what kind it is and how many cusps it has, according to the value of r. Are all hypo- and epicycloids obtained in this way?

9.14.34.3. Examples.

A) What does the figure look like when Γ' has half the radius of Γ and rolls inside Γ? (It is called *Lahire's cogwheel*.)

B) Show that, if Γ' and Γ have the same radius and Γ' rolls outside Γ, one obtains a Pascal limaçon, called a *cardioid*. Let a and D be a point and a line in the plane, with $a \notin D$; show that the envelope of the family of lines Θ defined by $m \in \Theta$ and $\widehat{D\Theta} = 3\widehat{ma, D}$, for $m \in D$, is a cardioid, whose *center* is, by definition, the point a.

C) Show that the caustic of a plane spheric mirror (that is, the envelope of the light rays reflected from a beam parallel to the axis) is a piece of a two-cusped epicycloid (called a *nephroid*).

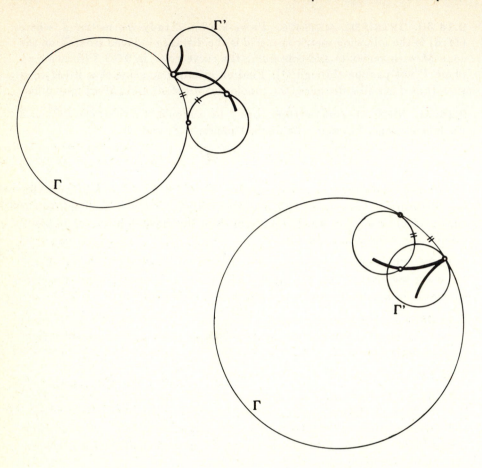

Figure 9.14.34.1

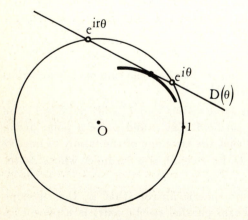

Figure 9.14.34.2

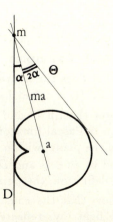

Figure 9.14.34.3.B

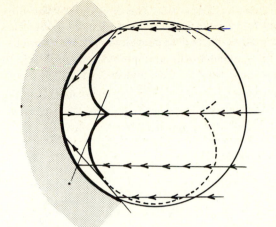

Figure 9.14.34.3.C

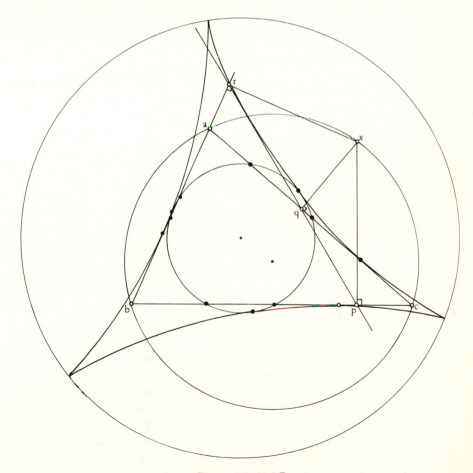

Figure 9.14.34.3.D

D) Let T be a triangle. Show that the envelope of the Simson lines of T (cf. 10.4.5.5, 10.9.7.1) drawn from a point that describes the circle circumscribed to T is a three-cusped hypocycloid. Show that the circle defining this hypocycloid is tangent to the nine-point circle of T (cf. 10.11.3) at three points; determine these three contact points.

E) Show that horizontal projections of *spherical helices* with a vertical axis (that is, curves on the sphere whose tangent makes a constant angle α with the axis) are epicycloids for appropriate values of α.

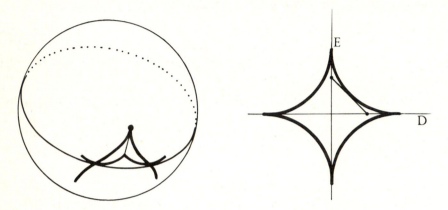

Figure 9.14.34.3.E   Figure 9.14.34.3.F

F) Show that the envelope of a segment of fixed length whose endpoints describe two orthogonal lines is a four-cusped hypocycloid (called an *astroid*).

9.14.34.4. Properties. Show that the evolute of a hypo- or epicycloid is similar to the original curve. Show that the arclength s (computed from an appropriated starting point) and the curvature K of a hypo- or epicycloid satisfy a relation of the form

$$as^2 + bK^{-2} = c,$$

for a, b, c constant (this is called the intrinsic equation, cf. [B–G, 8.5.7]). Conversely, what are the curves satisfying such a relation? Find the total length of a hypo- or epicycloid.

9.14.34.5. Cardioids and Morley's theorem. Let T be a triangle. Show that the centers of all cardioids tangent to the three sides of T form 27 lines, lying in three directions making an angle of $2\pi/3$ (use 9.14.33.3.B and 10.13.18). Show that the sides of the Morley triangle of T (cf. 10.3.10) lie on three of those lines (notice that each vertex of the Morley triangle is the center of a cardioid lying inside T and tangent to all its sides, one of them twice.) See also 10.13.23.

9.14.34.6. For more information on hypo- and epicycloids, see [LM1]. See also [ZR], a very agreeable text on plane curves an their links with mechanics, optics and electricity; in particular, see chapter XXI, where the connection between epicycloids and cogwheel design is discussed. Finally, [LF–AR, 413–435] gives an analytic presentation of the various cycloids, including, in the last three pages, a derivation of the shape of the carter of the Wankel engine (cf. 12.10.5).

9.14.35. Let a, b, c, d be points in a Euclidean plane, and assume b, c, d to be collinear. Prove the *Stewart relation*

$$(ab)^2 \, \overline{cd} + (ac)^2 \, \overline{db} + (ad)^2 \, \overline{bc} + \overline{bc} \cdot \overline{cd} \cdot \overline{db} = 0,$$

where the bars stand for signed distances on the line containing b, c, d.

9.14.36. Let X be a Euclidean affine space and G a finite subgroup of $\mathrm{Is}(X)$. Show that there exist subsets A of X such that $G = \mathrm{Is}_A(X)$. Deduce that, for any finite group G, there exists a Euclidean affine space X and a subset B of X such that G is isomorphic to $\mathrm{Is}_B(X)$.

9.14.37. Let ϕ be a map from a Euclidean plane E into itself that "preserves the distance 1", that is, such that $d(\phi(x), \phi(y)) = 1$ whenever $d(x, y) = 1$, for any $x, y \in E$. Show that E is an isometry. (Hint: Show first that ϕ preserves the distance $\sqrt{3}$, and use the fact that $\mathbf{Z} + \mathbf{Z}\sqrt{3}$ is dense in $\mathbf{R}$). See [M–P, 152].

9.14.38. Show that a necessary and sufficient condition for three lines D, E, F of a three-dimensional Euclidean space to have a common perpendicular (cf. 9.2.6.5) is that the composition $\sigma_D \circ \sigma_E \circ \sigma_F$ of the reflections through these lines also be a reflection through a line. Deduce the following result, due to Petersen and Morley: If X, Y, Z are three lines in Euclidean three-space, X' (resp. Y', Z') is the common perpendicular to Y and Z (resp. Z and X, X and Y) and X'' (resp. Y'', Z'') is the common perpendicular to X and X' (resp. Y and Y', Z and Z'), the three lines X'', Y'' and Z'' have a common perpendicular. See other proofs in [LF–AR, 681] and [FL, 339].

9.14.39. Given six points a_i, a_i' $(i = 1, 2, 3)$ of $\mathbf{R}^3$ satisfying $d(a_i, a_j) = d(a_i', a_j')$, find a geometric construction for the isometries f taking a_i into a_i' for all i.

∗ **9.14.40.** Given four points a, b, a', b' in $\mathbf{R}^2$, construct the centers of the similarities taking a to a' and b to b'.

9.14.41. Given six points a_i, a_i' $(i = 1, 2, 3)$ of $\mathbf{R}^3$, give necessary and sufficient conditions for the existence of a similarity f taking a_i to a_i' for all i. Assuming those conditions are satisfied, find a geometric construction for the center of the similarities f. Extend to arbitrary dimension.

9.14.42. Given three circles in a Euclidean plane, study the figure formed by the centers of the homotheties taking one of the circles into another. Same study for four spheres in three-space. Same question for similarities.

9.14.43. Take a plane triangle, and consider the three equilateral triangles built on its sides and outside it. Show that the centers of these three triangles form an equilateral triangle.

9.14.44. (NAPOLEON'S THEOREM). Take a parallelogram and consider the four squares built on its sides and outside it. Show that the centers of the squares form a square.

Chapter 10
Triangles, spheres and circles

This chapter deals with the domestic animals of geometry: triangles, polygons, tetrahedra, circles, spheres. We have collected here basic definitions and results, but also more intricate results that have a simple statement—or illustration. A presentation for this chapter seems unnecessary; the reader who leafs through it will glean from it whatever strikes his fancy, and will turn back when necessary for enlightenment on notation or facts used in the proofs. In any case, we do not tarry overlong on the proof of classical results, as they are likely to be familiar to the reader, and the purpose of the chapter is just to provide a number of illustrations, exercises and problems on the material covered in chapter 9. Section 10.12 serves as introduction and motivation to material to be covered later.

See [COO] for more on these domestic animals.

> All spaces considered are Euclidean affine. The dimension is two in sections 10.1 to 10.5 and 10.9 to 10.11, and three in section 10.12.

10.1. Triangles: definitions and notation

10.1.1. According to 2.4.7, a *triangle* consists of three affine independent points x, y, z. By restricting to the affine plane spanned by the three points, we can assume that we're working in a Euclidean plane X; we do so in sections 10.1 through 10.4.

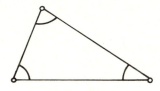

Figure 10.1.1

10.1.2. Let $\mathcal{T} = \{x, y, z\}$ be a fixed triangle. We shall make consistent use of the following notation:

$$a = yz, \quad b = zx, \quad c = xy \qquad \text{(cf. 9.1.1)},$$

$$A = \overline{\overrightarrow{xy}, \overrightarrow{xz}}, \quad B = \overline{\overrightarrow{yx}, \overrightarrow{yz}}, \quad C = \overline{\overrightarrow{zy}, \overrightarrow{zx}} \qquad \text{(cf. 8.6)}.$$

The points x, y, z are called *vertices*. The *sides* of $\mathcal{T}$ are either the segments joining each pair or vertices (cf. 2.4.7) or their lengths a, b, c (or yet, by a harmless abuse of language, the lines supporting them). The *angles* of $\mathcal{T}$ are A, B, C, real numbers in $]0, \pi[$.

10.1.3. A triangle is called *isosceles* if two of its sides are equal, *equilateral* if all three are. It is called a *right* triangle if one of its angles is $\pi/2$, an *acute* triangle if all its angles are in $]0, \pi/2[$, and an *obtuse* triangle otherwise.

10.1.4. The *altitudes* of $\mathcal{T}$ are the lines containing one vertex and perpendicular to the opposite side; the length of the corresponding segments is *denoted* by h_a, h_b, h_c (figure 10.1.4.1). The *medians* of $\mathcal{T}$ are the lines $\left\langle x, \dfrac{y+z}{2} \right\rangle$, $\left\langle y, \dfrac{z+x}{2} \right\rangle$, $\left\langle z, \dfrac{x+y}{2} \right\rangle$, and the lengths of the corresponding segments are *denoted* by m_a, m_b, m_c (figure 10.1.4.2). The *interior bisectors* of $\mathcal{T}$ are the lines containing the bisectors of the oriented lines $\langle \overrightarrow{xy}, \overrightarrow{xz} \rangle$, $\langle \overrightarrow{yz}, \overrightarrow{yx} \rangle$, $\langle \overrightarrow{zx}, \overrightarrow{zy} \rangle$; the *exterior bisectors* are the lines perpendicular to the interior bisectors at the corresponding vertices. The length of the segments determined by interior and exterior bisectors is *denoted* by i_a, i_b, i_c and e_a, e_b, e_c (figure 10.1.4.3).

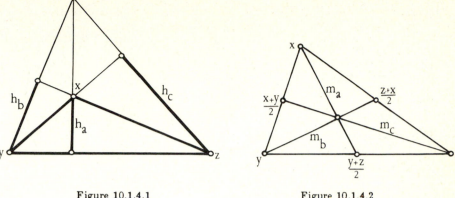

Figure 10.1.4.1   Figure 10.1.4.2

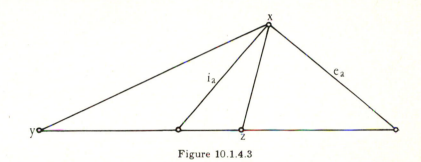

Figure 10.1.4.3

10.1.5. By 9.7.5, there exists a unique circle circumscribed around $\mathcal{T}$, whose radius we *denote* by R (figure 10.1.5.1). There exist four circles tangent to the three sides of $\mathcal{T}$; one inside the triangle, called the *inscribed* circle, and three outside. Their radii are *denoted* by r, r_a, r_b, r_c (figure 10.1.5.2). Finally, S stands for the area of $\mathcal{T}$ (cf. 9.12.4).

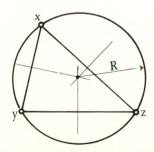

Figure 10.1.5.1

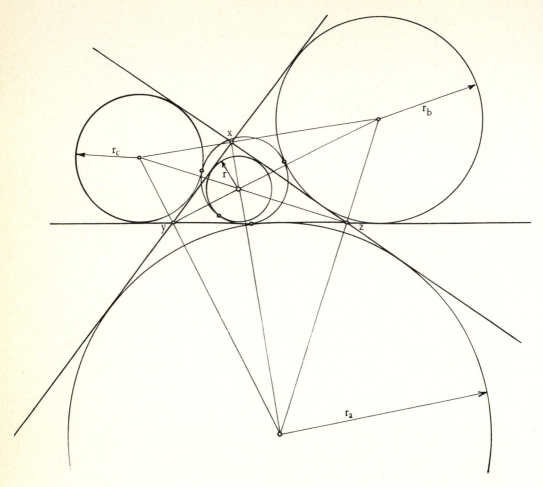

Figure 10.1.5.2

10.2. Classical results

10.2.1. EXISTENCE. For there to be a triangle $\{x, y, z\}$ with sides $xy = c$, $yz = a$ and $zx = b$, it is necessary and sufficient that x, y, and z satisfy the strict triangle inequalities

$$a < b + c, \quad b < c + a, \quad c < a + b,$$

which can be summarized as

$$|b - c| < a < b + c.$$

In fact, formula 10.3.1 shows that it is enough to find $A \in \left]0, \pi\right[$ such that

$$\cos A = \frac{b^2 + c^2 - a^2}{2bc},$$

and this can be done if and only if this fraction is in $]-1,1[$. For another proof, use 9.7.3.4 and 9.7.3.8.

10.2.2. ISOSCELES TRIANGLES. For the triangle $\mathcal{T}$ to be isosceles, it is necessary and sufficient that two of its angles be equal. This follows from formula 10.3.1; necessity is trivial, and for sufficiency, observe that the condition

$$\frac{b^2 + c^2 - a^2}{2bc} = \frac{c^2 + a^2 - b^2}{2ca}$$

can be written as $(a - b)((a + b)^2 - c^2) = 0$, and 10.2.1 forces $a = b$. It follows that a triangle is equilateral if and only if its three angles are equal; they must have the value $\pi/3$ by 10.2.4 or 10.3.1.

10.2.3. PYTHAGOREAN THEOREM. A necessary and sufficient condition for $\mathcal{T}$ to have a right angle at x is that $a^2 = b^2 + c^2$ (cf. 9.2.3 or 10.3.1).

10.2.4. SUM OF THE ANGLES. One always has $A + B + C = \pi$. In fact, we can apply 8.7.5.3; $\overrightarrow{xz}$ is between $\overrightarrow{xy}$ and $\overrightarrow{yz}$, and similarly $\overrightarrow{yz}$ is between $\overrightarrow{xz}$ and $\overrightarrow{yx}$, so we have

$$\overline{\overrightarrow{xy}, \overrightarrow{xz}} + \overline{\overrightarrow{xz}, \overrightarrow{yz}} + \overline{\overrightarrow{yz}, \overrightarrow{yx}} = \overline{\overrightarrow{xy}, \overrightarrow{yx}} = \pi.$$

We shall encounter geometries where the sum of the angles of a triangle is always greater than π (cf. 18.3.8.4) or less than π (cf. 19.5.4).

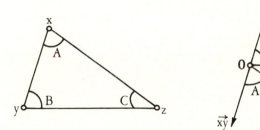

Figure 10.2.4

10.2.5. CONCURRENCE. The reader may have remarked in figures 10.1.4 and 10.1.5 that the altitudes are concurrent, and so are the medians, the interior bisectors (or one interior and two exterior bisectors), and the perpendicular bisectors of the sides (9.7.5). This follows from 3.4.10 for the medians, from 9.14.3 for the bisectors and from 9.7.5 for the perpendicular bisectors. Only the altitudes require more thought; one solution is to consider the triangle $\mathcal{T}'$ whose sides are the parallels to the sides of $\mathcal{T}$ passing through the opposite vertex: the altitudes of $\mathcal{T}$ become the perpendicular bisectors of the sides of $\mathcal{T}'$. See also 10.13.1 and 17.5.4. The intersection point of the altitudes of $\mathcal{T}$ is called the *orthocenter* of $\mathcal{T}$.

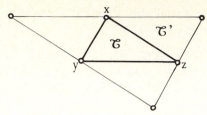

Figure 10.2.5

10.2.6. CONGRUENCE. Let T, T' be triangles whose elements (cf. 10.1.2) we denote by unprimed and primed letters, respectively. The following conditions are equivalent:

i) there exists $f \in \mathrm{Is}(X)$ such that $f(x) = x'$, $f(y) = y'$, $f(z) = z'$;
ii) $a = a'$, $b = b'$, $c = c'$;
iii) $A = A'$, $b = b'$, $c = c'$;
iv) $A = A'$, $B = B'$, $c = c'$.

This follows from 9.7.1 and the formulary below. Watch out, though, when using the formulary: the sine of an angle is not enough to uniquely determine the angle in the interval $]0, \pi[$, though equality between cosines can be used to show equality between angles.

Triangles that satisfy the conditions above are called *congruent*.

10.2.7. SIMILARITY. With the same notation, the following conditions are equivalent:

i) there exists $f \in \mathrm{Sim}(X)$ such that $f(x) = x'$, $f(y) = y'$, $f(z) = z'$;
ii) $\dfrac{a'}{a} = \dfrac{b'}{b} = \dfrac{c'}{c}$;
iii) $A = A'$, $B = B'$, $C = C'$.

Use 9.5.3.1 and the formulary to see that (i) $\Rightarrow$ (iii) $\Rightarrow$ (ii). To see that (ii) $\Rightarrow$ (i), use a similarity of ratio a'/a to reduce the problem to congruent triangles.

10.3. Formulary

In the formulary below we use the notation introduced in 10.1, plus the symbol p for the *half-perimeter* of T, defined as $p = (a + b + c)/2$. The triangle $\{x, y, z\}$ is arbitrary.

10.3.1

$$\begin{cases} \cos A = \dfrac{b^2 + c^2 - a^2}{2bc} \\[2ex] \sin A = \dfrac{2}{bc}\sqrt{p(p-a)(p-b)(p-c)} \\[2ex] \sin \dfrac{A}{2} = \sqrt{\dfrac{(p-b)(p-c)}{bc}} \end{cases}$$

10.3.2
$$\frac{a}{\sin A} = \frac{b}{\sin B} = \frac{c}{\sin c} = 2R$$

10.3.3. (HERON'S FORMULA)

$$S = \frac{1}{2}ah_a = \frac{1}{2}bc\sin A = pr = \sqrt{p(p-a)(p-b)(p-c)}$$

10.3.4
$$R = \frac{abc}{\sqrt{(a+b+c)(a+b-c)(a-b+c)(-a+b+c)}}$$

10.3.5
$$r = \frac{S}{p}, \qquad r_a = \frac{S}{p-a}$$

10.3.6
$$i_a = \frac{2}{b+c}\sqrt{p(p-a)bc}, \qquad e_a = \frac{2}{|b-c|}\sqrt{(p-b)(p-c)bc}$$

10.3.7
$$m_a^2 = \frac{2(b^2+c^2) - a^2}{4}.$$

The first formula in 10.3.1 follows from 8.1.2.4 by applying 8.6.3 and 9.1.1. The second comes from $\sin A = \sqrt{1 - \cos^2 A}$ and the third from

$$\sin^2 \frac{A}{2} = \frac{1 - \cos A}{2}.$$

For 10.3.2, let ω be the center of the circle circumscribed around $\mathcal{T}$ (figure 10.3.2). By 10.9.3, the triangle ωyz has angle $2A$ at ω; by definition of sine, $a = \frac{1}{2}R\sin A$. If you don't want to resort to 10.9.3, use 9.7.3.7, which also gives 10.3.4 and the last equality in 10.3.3.

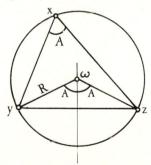

Figure 10.3.2

The first equality in 10.3.3 is just 9.12.4.4, and the second comes from $h_b = c\sin A$. The third is obtained by dividing $\mathcal{T}$ into three triangles:

$$\mathcal{T} = \{x, y, \alpha\} \cup \{y, z, \alpha\} \cup \{z, x, \alpha\},$$

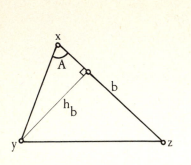

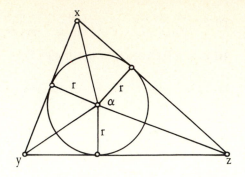

Figure 10.3.3.1   Figure 10.3.3.2

where α is the center of the triangle inscribed in T (figure 10.3.3.2). Since each of these has altitude r and the opposite sides are a, b, c, the equality follows from 9.12.4.4. The last equality is a consequence of 10.3.1.

Equality 10.3.4 could be obtained from 9.7.3.7, as mentioned; but it also follows directly from 10.3.1 and 10.3.2.

For 10.3.5 we use the same trick that shows $S = pr$, choosing the center α' of the exinscribed circle (figure 10.3.5). Here $T \cup \{y, z, \alpha'\} = \{x, y, \alpha'\} \cup \{x, z, \alpha'\}$, so

$$\text{area}(T) = \text{area}(\{x, y, \alpha'\}) + \text{area}(\{x, z, \alpha'\}) - \text{area}(\{y, z, \alpha'\}).$$

The three triangles have altitude r_a and sides b, c, a, respectively.

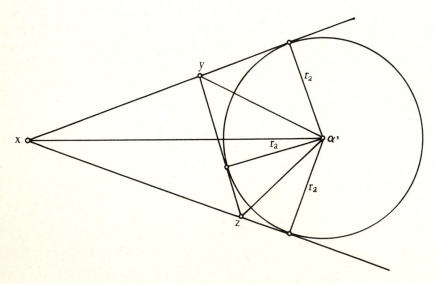

Figure 10.3.5

Formula 10.3.7 is just 9.7.6.5. As for 10.3.6, we leave e_a to the reader, and find i_a. Let s be the point where the bisector from x intersects the side

$[y, z]$. Applying 10.3.2 and observing that the angles at s of the triangles $\{s, y, x\}$ and $\{s, z, x\}$ have the same sine, we get

10.3.8
$$\frac{sy}{b} = \frac{sz}{c} = \frac{sy + sz}{b + c} = \frac{a}{b + c}.$$

We now apply 10.3.2 to the triangle $\{x, s, z\}$, for example, and 10.3.1 twice.

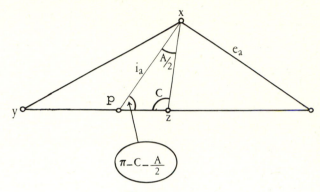

Figure 10.3.8

10.3.9. NOTE. It is a good exercise to find out what these formulas become for right triangles.

10.3.10. APPLICATION: MORLEY'S THEOREM. This is an apparently simple result, but its proof is not obvious; the reader is encouraged to attempt a geometric proof, or even a trigonometric one, before reading what follows, so as to appreciate its difficulty. Exercises 10.13.4 and 9.14.34.5 give geometric proofs; 10.13.23 is a related exercise. See [LB1, 173–194] for the figure formed by all the trisectors of a triangle, and [CAL] for a generalization to polygons.

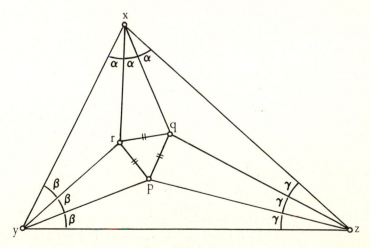

Figure 10.3.10

Take a triangle $\{x, y, z\}$ and trisect its angles; let p (resp. q, r) be the point where the trisectors adjacent to side a (resp. b, c) intersect. Morley's theorem says that the triangle $\{p, q, r\}$ is equilateral.

Set $\alpha = A/3$, $\beta = B/3$, $\gamma = C/3$, and apply 10.3.2 to $\{x, y, z\}$ and $\{x, y, r\}$. We obtain

$$rx = 2R \frac{\sin \beta \sin(3\alpha + 3\beta)}{\sin(\alpha + \beta)},$$

using 10.2.4 also. A trigonometric calculation (cf. 8.7.8 and 8.12.8) using $\alpha + \beta + \gamma = \pi/3$ leads to

$$rx = 8R \sin \beta \sin \gamma \sin(\pi/3 + \gamma).$$

Similarly,

$$qx = 8R \sin \beta \sin \gamma \sin(\pi/3 + \beta),$$

so that

$$\frac{rx}{\sin(\pi/3 + \gamma)} = \frac{qx}{\sin(\pi/3 + \beta)} = 8R \sin \beta \sin \gamma;$$

but since $(\pi/3 + \gamma) + (\pi/3 + \beta) + \alpha = \pi$, formula 10.3.2, applied to a triangle of angles $(\pi/3 + \gamma)$, $(\pi/3 + \beta)$ and α and one side equal to rx, shows that such a triangle is congruent to $\{x, r, q\}$. Then 10.3.2 implies $rq = 8R \sin \alpha \sin \beta \sin \gamma$, and this must also be the value of qp and pr.

10.3.11. THE GEOMETRY OF THE TRIANGLE. There is an immense number of results on triangles; see the wonderful book [COO], and also [R–C, volume I, note III]. Some examples are given in 10.11.

10.4. Inequalities and problems of minimum

10.4.1. THE ISOPERIMETRIC INEQUALITY. *For every triangle we have $S \le p^2/3\sqrt{3}$, and equality holds if and only if the triangle is equilateral.*

Proof. It is enough to apply 10.3.3 and 11.8.11.6 to obtain

$$(p - a)(p - b)(p - c) \le \left(\frac{(p - a)(p - b)(p - c)}{3} \right)^3 = \left(\frac{p}{3} \right)^3;$$

equality is satisfied if and only if $p - a = p - b = p - c$. □

10.4.2. NOTE. We shall encounter two generalizations of 10.4.1, one in 10.5 and one in 12.11.1. As for an "isodiametric" inequality for triangles (cf. 9.13.8), it follows immediately from 10.3.3, since the diameter of T is $\sup \{a, b, c\}$.

10.4.3. FERMAT'S PROBLEM. From 9.7.6.3 we deduce that

$$tx^2 + ty^2 + tz^2 \ge gx^2 + gy^2 + gz^2,$$

where $g = (x + y + z)/3$ is the center of mass of T and x is arbitrary. Equality holds only when $t = g$. The value of the minimum follows from 10.3.7 and 3.4.10.

Finding the minimum of the function $t \mapsto tx + ty + tz$ is, surprisingly enough, a much more complicated problem (the reason being that the distance function is not bilinear, whereas its square is); this is the so-called Fermat problem. The solution is indicated by 9.10.6; we are looking for points t of X from which the three sides of T are seen under an angle $2\pi/3$. More precisely, either some angle of T, say A, is $\geq 2\pi/3$, and then the minimum is achieved for $t = x$ only; or all the angles of T are $< 2\pi/3$, and then there exists a unique point $\omega \in X$ from which the three sides of T subtend an angle $2\pi/3$, and the minimum is attained for $t = \omega$ only.

To begin with, the point t where the minimum is reached must be in the closed triangle, for otherwise the foot u of the perpendicular dropped from t onto a side $\langle y, z \rangle$ of T such that t and x are on different half-planes would satisfy $tx + ty + tz > ux + uy + uz$ (apply 9.2.2).

If $A \geq 2\pi/3$, for every $t \neq x$ in the interior of T, the angle at t of $\{t, y, z\}$ is $> 2\pi/3$, and t cannot minimize the sum. Since there is a minimum by compactness, it must be at x.

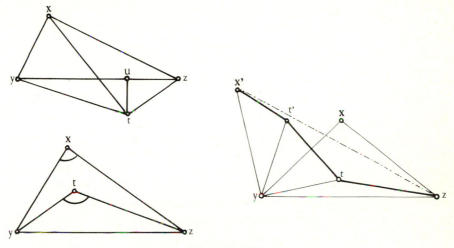

Figure 10.4.3.1

Now assume that all three angles of T are strictly less than $2\pi/3$. Assume X oriented, and let f be the rotation with center y and angle $\pi/3$. Set $x' = f(x)$, $t' = f(t)$ for $t \in X$ arbitrary. Then

$$tx + ty + tz = x't' + t't + tz \geq x'z,$$

and equality holds if and only if x', t', t, z are aligned and in the right order. Next we construct $z'' = f^{-1}(z)$; since the angles if T are $< 2\pi/3$, the lines $\langle x, z'' \rangle$ and $\langle z, x' \rangle$ intersect at a single point ω, such that $x', \omega' = f(\omega), \omega, z$ are aligned and in the right order. Thus ω satisfies $\omega x + \omega y + \omega z = x'z$, concluding the proof.

This incidentally shows that the three lines $\langle x', z \rangle$, $\langle x, z'' \rangle$ and $\langle y, x'' \rangle$ in figure 10.4.3.2 intersect in the same point, forming three angles $\pi/3$.

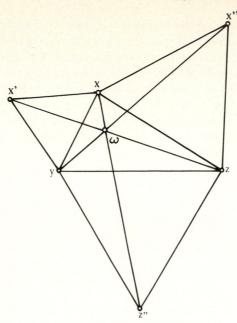

Figure 10.4.3.2

10.4.4. NOTE. We have studied in section 9.4 the problem of least-perimeter triangles inscribed in T.

10.4.5. AREA OF PEDAL TRIANGLES. If $T = \{x, y, z\}$ is a triangle and $t \in X$ a point, we define the *pedal triangle* of t as the triangle $\{p, q, r\}$ formed by the orthogonal projections of t onto the sides of T (this is an abuse of language, for p, q, r can be collinear). We are interested in the area of pedal triangles, for example, for what position of t inside T is the area maximized? To do this we need to introduce oriented triangles and their area.

10.4.5.1. Definition. *An oriented triangle in an oriented plane X is a triple (x, y, z) of points of X. The (signed) area of (x, y, z) is $\mathcal{A}(x, y, z) = \frac{1}{2}\lambda_X(\overrightarrow{xy}, \overrightarrow{xz})$ (cf. 8.11.3).*

It is immediately seen that $\mathcal{A}$ is invariant under cyclic permutations and that

10.4.5.2 $\mathcal{A}(x, y, z) = \mathcal{A}(t, x, y) + \mathcal{A}(t, y, z) + \mathcal{A}(t, z, x).$

Further, if $\sigma \in \mathrm{Is}^-(X)$, we have $\mathcal{A}(\sigma(x), \sigma(y), \sigma(z)) = -\mathcal{A}(x, y, z)$.

Now back to our triangle T and our point t in the plane X, assumed oriented. In addition to p, q, r, we introduce the reflections p', q', r' of t through the sides of T. Then, if T has positive area,

10.4.5.3 $\mathcal{A}(p, q, r) = \dfrac{1}{2}\mathcal{A}(x, y, z) - \dfrac{1}{8}\left(\sin 2A \cdot tx^2 + \sin 2B \cdot ty^2 + \sin 2C \cdot tz^2\right).$

Proof. Refer to figure 10.4.5.3, where one possible relative position for the points is given; the others are treated analogously. The idea is to decompose

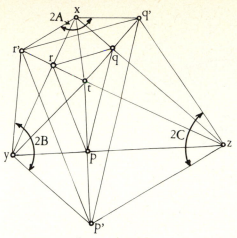

Figure 10.4.5.3

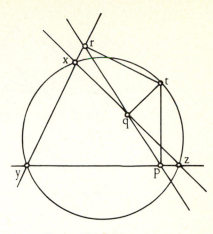

Figure 10.4.5.4.1

the hexagon $\{x, r', y, p', z, q'\}$ into triangles in two different ways: $\{x, y, t\}$, $\{x, y, r'\}$, $\{y, z, t\}$, $\{y, z, p'\}$, $\{z, x, t\}$, $\{z, x, q'\}$ on the one hand, $\{p', q', r'\}$, $\{x, r', q'\}$, $\{y, p', r'\}$, $\{z, q', p'\}$ on the other. In algebraic terms we have

$$
\begin{aligned}
2A(x, y, z) &= 2A(t, x, y) + 2A(t, y, z) + 2A(t, z, x) \\
&= A(t, x, y) - A(r', x, y) + A(t, y, z) \\
&\quad - A(p', y, z) + A(t, z, x) - A(q', z, x) \\
&= A(p', q', r') + A(x, r', q') + A(y, p', r') + A(z, q', p').
\end{aligned}
$$

But $A(p', q', r') = 4A(p, q, r)$ because $\{p', q', r'\}$ can be derived from $\{p, q, r\}$ by a homothety of center t and ratio 2, and

$$
A(x, r', q') = \frac{1}{2} \sin 2A \cdot xr' \cdot xq' = \frac{1}{2} \sin 2A \cdot tx^2
$$

from 10.3.3 and 8.7.7.8; this implies 10.4.5.3.   □

By 10.4.5.3 and 9.7.6 the least-area problem will be solved if we know the barycenter of the punctual masses $(\sin 2A, x)$, $(\sin 2B, y)$, $(\sin 2C, z)$. But this is none other than the center ω of the circle circumscribed around T; it is enough to check that

$$
\sin 2A \cdot \overrightarrow{\omega x} + \sin 2B \cdot \overrightarrow{\omega y} + \sin 2C \cdot \overrightarrow{\omega z} = 0
$$

(cf. 3.4.6.5), and this follows from the proof of 10.3.2.

10.4.5.4. Without delving too deep into the problem, we can still extract some conclusions from 9.7.6.2: for certain values of k, the locus of the points of X for which the pedal triangle with respect to T has (unsigned) area k is made up of two circles of center ω (figure 10.4.5.4.2); for other values of k, of one circle only. The following unexpected fact comes up as a bonus (figure 10.4.5.4.1): the locus of the points of X such that p, q, r are collinear is the circle circumscribed around T (the "Simson line", see 10.9.7.1). In fact, p, q, r collinear means $A(p, q, r) = 0$, so the locus is a circle of center ω; but this circle

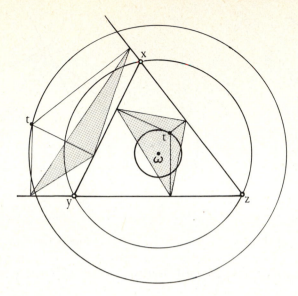

Figure 10.4.5.4.2

must go through x, y, z, which satisfy the zero area condition; this proves the assertion. We also get the formula $4S = R^2(\sin 2A + \sin 2B + \sin 2C)$.

10.4.5.5. We mention that, as x describes the circle circumscribed around $\mathcal{T}$, the Simpson line envelopes a three-cusped hypocycloid (9.14.34.3.D), thus revealing a symmetry of order three in any triangle.

10.4.6. THE THEOREM OF ERDÖS-MORDELL. The next theorem has an interesting history: it was conjectured in 1935 by Erdös and proved in 1937 by Mordell, but the methods used were not at all elementary, in spite of the simplicity of the statement. It was only in 1945 that D. K. Kazarinoff gave an elementary demonstration. Here we present the theorem's statement only; the reader can amuse himself by tackling the demonstration, or read the whole story in [KF, 78] or [FT1, 12].

10.4.7. THEOREM. *Let $\mathcal{T}$ be a triangle and t a point in the interior of $\mathcal{T}$. For p, q, r as above, we have*

$$tx + ty + tz \geq 2(tp + tq + tr),$$

and equality holds if and only if $\mathcal{T}$ is equilateral.

10.4.8. NOTES. For other inequalities concerning triangles, see 10.13.5 and [GS].

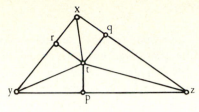

Figure 10.4.7

10.5. Polygons

We just give a few references on this subject. In 9.4 we have already studied convex polygonal billiards. For a polygonal analogue of 10.4.3, see 10.13.8.

10.5.1. ISOPERIMETRIC INEQUALITY FOR POLYGONS. The generalization of 10.4.1 is the following: if P is a convex n-sided polygon with area S and perimeter P, we have

$$S \leq \frac{1}{4n \tan(\pi/n)} P^2,$$

and equality holds if and only if P is regular. For a proof and refinements, see 12.12.15, [FT1, 8–11] or [GR, 192].

10.5.2. If P is an n-sided convex polygon, the sum of the angles of P is equal to $(n-2)\pi$ (cf. 12.1.12).

To see this, use the fact that the polygon is convex to divide it into $n-2$ triangles, starting from an arbitrary vertex. Applying 10.2.4 to each of the small triangles and adding up, we find that the sum of all the angles of P is $(n-2)\pi$.

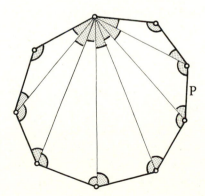

Figure 10.5.2

10.6. Tetrahedra

We won't talk a lot about tetrahedra, just enough to remark that there are things about them that have no counterpart for triangles. We discuss in detail one phenomenon peculiar to tetrahedra, arising from the study of spheres tangent to their four faces; we also mention two other particularities not shared with triangles. This is all done from 10.6.6 on, while 10.6.1 through 10.6.5 present generalizations of results already studied for triangles.

For references on the geometry of tetrahedra, the reader can consult [COO], [C–B] and [R–C, volume II, note IV]. (Incidentally, as can be seen from the preface of [C–B], already in 1935 there were teachers complaining that the geometric proficiency of students was declining. I'm offering a prize to anyone who can find an Egyptian tablet containing the same plaint...)

10.6.1. Let $\mathcal{T} = \{x, y, z, t\}$ be a tetrahedron (cf. 2.4.7). In this whole section, we shall assume that $\mathcal{T}$ is in three-space. An *edge* of $\mathcal{T}$ is both a line like $\langle x, y \rangle$ and a distance like xy; similarly, a *face* means either a plane like $\langle x, y, z \rangle$, a triangle like $\{x, y, z\}$ considered in this plane, or the area of this triangle in this plane (cf. 10.1.5). The volume of $\mathcal{T}$ is *denoted* by V (cf. 9.12.4), and the radius of the sphere circumscribed around $\mathcal{T}$ by R (cf. 9.7.5).

10.6.2. By 9.7.1, a tetrahedron is defined, up to isometry, by its edges. For the existence of a triangle with given edges, it is sufficient that the strict triangle inequality (10.2.1) be satisfied for each face, and that $\Gamma(\cdot, \cdot, \cdot, \cdot)$ be strictly positive (9.7.3.4). In fact (cf. 9.14.23), this latter condition implies the strict triangle inequalities.

10.6.3. MEDIANS. From 3.4.10 we deduce that the four medians of $\mathcal{T}$, that is, the segments connecting a vertex to the center of mass of the opposite face, intersect. The length of the medians can be found by applying 9.7.6.

10.6.4. Discussing angles at the vertices of a tetrahedron is more difficult and awkward, involving ideas we won't discuss until chapter 18. It is easier to introduce the dihedral angles between faces, but we don't do this here; see 10.13.10.

10.6.5. From 9.7.3.3, 9.7.3.7 and 9.12.4.3 we deduce V and R as a function of the edges:

$$V^2 = \frac{1}{288}\Gamma(x, y, z, t), \qquad R^2 = -\frac{1}{2}\frac{\Gamma(x, y, z, t)}{\Delta(x, y, z, t)}.$$

10.6.6. ALTITUDES. The altitudes of a tetrahedron, that is, the lines going through one vertex and orthogonal to the opposite face, do not intersect in general. See [C–B, 121] and [R–C, volume II, p. 643].

10.6.7. MÖBIUS TETRAHEDRA. There exist pairs of tetrahedra such that each vertex of one belongs to a face of the other. This phenomenon can be elegantly explained in the context of projective geometry, cf. 4.9.12 and 14.8.12.4.

10.6.8. ROOFS AND THEIR SPHERES. We consider a tetrahedron $T = \{x_1, x_2, x_3, x_4\}$ and search for spheres S tangent to the four faces of T, that is, such that their center ω is equidistant from the faces of T. A tempting analogy with figure 10.1.5 would lead us to believe that there exist five such spheres, one interior to T and four inside the frusta determined by T (see figure 10.6.8).

| the interior | a frustum | a trihedron | a pair of opposite roofs |

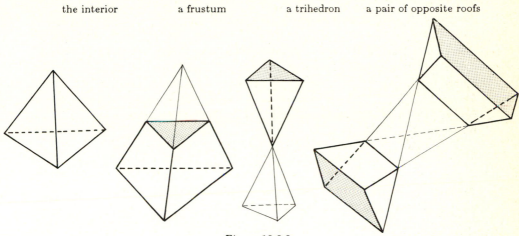

Figure 10.6.8

A more accurate analysis shows these are not the only possibilities. We can start by using formula 9.2.6.4; let f_i $(i = 1, 2, 3, 4)$ be affine forms on X whose kernels are each a face of T. We can assume that $\|\vec{f_i}\| = 1$ for all i. Then ω must satisfy

$$|f_1(\omega)| = |f_2(\omega)| = |f_3(\omega)| = |f_4(\omega)|,$$

which is equivalent to eight equalities between affine forms. By 2.4.8.5, this gives us eight solutions in the "generic" case, that is, when the f_i are linearly independent. Checking independence is non-trivial; we tackle the problem by introducing barycentric coordinates with respect to T. In such coordinates, the following lemma holds in any dimension:

10.6.8.1. Lemma. *Let $(x_i)_{i=1,\dots,n+1}$ be a simplex of X, and set*

$$H_i = \langle x_1, \dots, \hat{x}_1, \dots, x_{n+1} \rangle$$

(the i-th face of the simplex). Let V be the volume of $(x_i)_{i=1,\dots,n+1}$ and a_i the volume of the simplex $(x_j)_{j \neq i}$ in H_i. Then the barycentric coordinates (λ_i) of a point $x \in X$ with respect to this simplex are given by

$$|\lambda_i| = \frac{\alpha_i}{nV} d(x, H_i)$$

and we have $\lambda_i \geq 0$ (resp. ≤ 0) if x and x_i lie in the same half-space (resp. in opposite half-spaces) with respect to H_i (cf. 2.7.3).

Proof. Let f_i be an affine form such that $\|\vec{f_i}\| = 1$, with $H_i = f_i^{-1}(0)$ and $f_i(x_i) > 0$. By 9.2.6.4, we have

$$d(x, H_i) = \left| f_i(x) \right| = \left| f_i\left(\sum_j \lambda_j x_j \right) \right| = |\lambda_i| |f_i(x_i)|.$$

In particular, taking $x = x_i$, we have $d(x_i, H_i) = f_i(x_i)$, which is equal to nV/a_i by 9.12.4.4.   □

To proceed with our investigation, we divide the space into various pieces determined by the tetrahedron T: its *interior*, the set of x all of whose barycentric coordinates are ≥ 0; four *frusta*, sets where three λ_i are ≥ 0 and one is ≤ 0; four *trihedra*, where three λ_i are ≤ 0 and one is ≥ 0; and six *roofs*, where two are ≥ 0 and two are ≤ 0. Frustra and trihedra are associated with a vertex each; roofs are associated with edges, and matched in pairs. If r is the radius and ω the center of a sphere tangent to the four faces of T, the barycentric coordinates λ_i of ω satisfy

$$\frac{|\lambda_1|}{a_1} = \frac{|\lambda_2|}{a_2} = \frac{|\lambda_3|}{a_3} = \frac{|\lambda_4|}{a_4} = \frac{r}{3V}, \qquad \lambda_1 = \lambda_2 = \lambda_3 = \lambda_4 = 1.$$

Thus we see there is exactly one sphere in the interior, of radius $3V/(a_1 + a_2 + a_3 + a_4)$; exactly one in each frustum, with radius $3V/(a_1 + a_2 + a_3 - a_4)$ for the frustum adjacent to the face $\langle x_1, x_2, x_3 \rangle$, for example (this is because a_4 must be less than $a_1 + a_2 + a_3$, since a_4 is the sum of the projections of the other faces onto H_4, and such a projection decreases area, cf. 9.12.4.9). The case of the roofs is different, because nothing can be said about $a_1 + a_2 - a_3 - a_4$, for example; all we know is that if there is a sphere in one roof, there is none in the opposite one. To deal with this, assume $a_1 \geq a_2 \geq a_3 \geq a_4$. If all four faces are equal (i.e., have the same area), there are no spheres in the roof; the total number of tangent spheres is five. If $a_1 = a_2 > a_3 = a_4$, there are six spheres in all, one being inside the roof $\lambda_1 > 0$, $\lambda_2 > 0$, $\lambda_3 < 0$, $\lambda_4 < 0$. If $a_1 \geq a_2 \geq a_3 = a_4$ and $a_1 + a_4 = a_2 + a_3$, there are seven spheres, two in roofs; and, finally, there are eight spheres in all other cases.

   For more details on these spheres, see [COO] or [R–C, volume II, p. 653]. Here we just remark that $a_1 = a_2 = a_3 = a_4$ for the regular tetrahedron (cf. 12.5.4.1), but it is easy to construct non-regular tetrahedra all of whose faces have the same area.

10.7. Spheres

   In the next sections, we study circles and spheres in their environment, and their relationships among themselves. The sphere for its own sake is studied in chapter 18. For more, see [COO]. We will implicitly assume $\dim X \geq 2$, since in dimension one a sphere is just two points and not very interesting.

10.7.1. Definition. *The sphere of center a and radius r is the subset* $S(a, r) = \{ x \in X \mid ax = r \}$ *(also denoted by $S_X(a, r)$, is there is danger of confusion). If $n = \dim X = 2$, we talk about a circle.*

10.7.2. INTERSECTION WITH A SUBSPACE. Let $S = S(a, r) \subset X$, and let Y be an affine subspace of X. Denote by x the projection of a on Y (cf. 9.2.4). Then, by 9.2.3, we have (figure 10.7.2):

$$S \cap Y = \begin{cases} \emptyset & \text{if } d(a, Y) = ax > r; \\ \{x\} & \text{if } d(a, Y) = ax = r; \\ S_Y\left(x, \sqrt{r^2 - az^2}\right) & \text{if } d(a, Y) = ax < r. \end{cases}$$

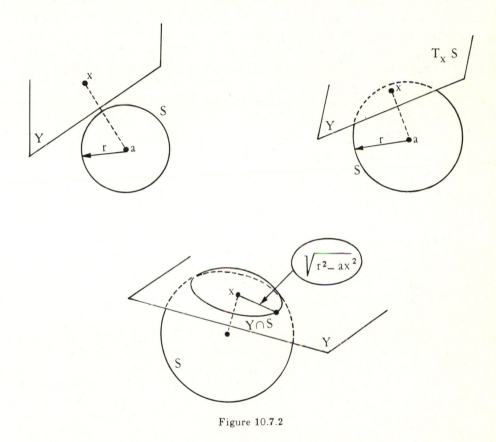

Figure 10.7.2

10.7.3. NOTE. Depending on the nature of the problem, it can be useful to exclude spheres of zero radius. For the sake of notational simplicity, we leave it to the reader to specify which case is being considered, the task not being an arduous one.

10.7.4. TANGENT HYPERPLANES. The subspace Y is said to be *tangent* to $S = S(a, r)$ (at the point $Y \cap S$) if $d(a, y) = r$. For each $x \in S$ there exists a unique tangent hyperplane to S at x, denoted by $T_x S$; it is characterized by the fact that it contains x and is orthogonal to $\overrightarrow{xa}$. (It can also be considered as the vector subspace $x^\perp$, cf. 18.1.2.4.)

This definition coincides with the notions of tangent hyperplanes stemming from other theories; for example, if S is considered as a quadric, its
tangent hyperplane in the sense of a non-degenerate quadric is the same as
the one just defined (cf. 14.3.8). Again, if S is considered as a differentiable
submanifold of X, the tangent space to S at $x \in S$, defined as the set of
vectors tangent at x to C^1 curves whose image is in X, coincides with our
tangent hyperplane (cf. 18.3.3).

10.7.5. INTERSECTION OF TWO SPHERES. Consider in X two spheres
$S = S(a, r)$ and $S' = S'(a', r')$ with $a \neq a'$ (that is, S and S' are *eccentric*).
It follows immediately from 10.2.1 that
 — if $aa' < |r - r'|$ or $aa' > r + r'$ we have $S \cap S' = \emptyset$;
 — if $aa' = r + r'$ or $aa' = |r - r'|$ there is a point x such that $S \cap S' = \{x\}$,
   and $T_x S = T_x S'$;
 — if $|r - r'| < aa' < r + r'$, there exists a hyperplane H, orthogonal to aa',
   such that $S \cap S' = S \cap H = S' \cap H$ is a sphere in H, of center $\langle a, a' \rangle \cap H$.

If $aa' = r + r'$, we say that S and S' are *externally tangent*; if $aa' = |r - r'|$,
they are *internally tangent*. In the third case we say that S, S' are *secant*.

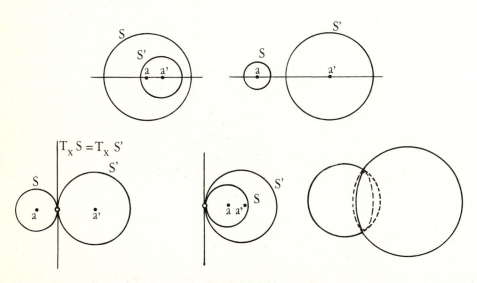

Figure 10.7.5

10.7.6. ANALYTIC GEOMETRY. The analytic geometry of spheres is simple:
if we vectorialize X, the equation of $S = S(a, r)$ is $\|x - a\|^2 = r^2$, or

$$\|x\|^2 - 2(a \mid x) + \|a\|^2 - r^2 = 0.$$

Conversely, the equation $k\|x\|^2 - (\alpha \mid x) + h = 0$, where $k \neq 0$ and h are scalars
and α is a vector of X (which we have vectorialized at an arbitrary point)
represents the empty set, if $\|\alpha\|^2 - 4kh < 0$; the point $\{-\alpha/2k\}$, if $\|\alpha\|^2 -$

$4kh = 0$; and the sphere of center $\{-\alpha/2k\}$ and radius $\left(\dfrac{\|\alpha\|^2 - 4kh}{4k^2}\right)^{\frac{1}{2}}$ otherwise.

The intersection $S \cap S'$ of two spheres (cf. 10.7.5) must be contained in the set of equation

$$\|x\|^2 - 2(a \mid x) + \|a\|^2 - r^2 = \|x\|^2 - 2(a' \mid x) + \|a'\|^2 - r'^2,$$

which turns out to be the hyperplane $2(a' - a \mid x) + \|a\|^2 - \|a'\|^2 + r^2 - r'^2$.

For applications of analytic geometry to spheres, see 9.7.3.7, the whole of chapter 20 and [COO].

10.7.7. ANGLE BETWEEN TWO SPHERES. If $S = S(a, r)$ and $S' = S(a', r')$ intersect, the angle $\phi = \overrightarrow{xa}, \overrightarrow{xa'}$ does not depend on x for $x \in S \cap S'$. We call this the *angle between S and S'*; it is given by

$$\cos \phi = \frac{r^2 + r'^2 - aa'^2}{2rr'}.$$

If $\phi = \pi$ (resp. $\phi = 0$), S and S' are *externally* (resp. *internally*) tangent; if $\phi = \pi/2$, we say that S and S' are *orthogonal*, and we write $S \perp S'$.

10.7.8. PROPOSITION. *Given $n + 1$ spheres S_i $(i = 1, \ldots, n + 1)$ in an n-dimensional space X and $n + 1$ angles $\phi_i \in [0, \pi]$, there exist at most two spheres that intersect each S_i at an angle ϕ_i $(i = 1, \ldots, n + 1)$.*

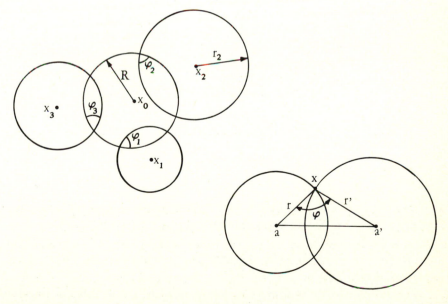

Figure 10.7.8

Proof. Assume $S_i = S(x_i, r_i)$ and let $S = (x_0, R)$ be the desired sphere; with the notations and results of 9.7.3, we have $\Gamma(x_0, x_1, \ldots, x_{n+1}) = 0$ and $d_{0i}^2 = R^2 + r_i^2 - 2Rr_i \cos \phi_i$. Plugging the values of d_{0i} $(i = 1, \ldots, n+1)$ into $\Gamma(x_0, x_1, \ldots, x_{i+1})$ and carrying out some row and column subtractions, we find a second-degree equation for R, say $\alpha R^2 + \beta R + \gamma$, so there are only two possible values for the radius. But knowing R determines $d_{0i} = x_0 x_i$, so by 9.7.1 there are at most two possibilities for x_0. $\square$

10.7.9. EXAMPLES. The case of orthogonal spheres is special and has at most one solution, for if $\cos \phi_i$ is zero for all i, the equation in R we obtain is of the form $\alpha R^2 + \gamma = 0$. We shall return to this point in 10.7.10.2.

A search for spheres tangent to $n + 1$ given spheres would yield 2^{n+1} solutions at most, since there are 2^{n+2} choices of $\phi_i = 0$ or π; but if we replace ϕ_i by $\pi - \phi_i$ everywhere, the equation $\alpha R^2 + \beta R + \gamma$ is not changed, so there are at most 2^{n+1} solutions. This upper bound actually occurs, and each combination of $\phi_i = 0$ or π admits two or zero solutions—see figures 10.7.9.1 and 10.7.9.2 for the two-dimensional case. But the lower bound zero for the number of solutions also occurs!

Figure 10.7.9.1

10.7.10. POWER OF A POINT WITH RESPECT TO A SPHERE. Consider $S = S(a, r)$ and $x \in X$. For every line D containing x and intersecting S at t and t' (where we may have $t = t'$), the inner product $(\overrightarrow{xt} \mid \overrightarrow{xt'})$ is a constant, equal to $xa^2 - r^2$. This number is called the *power* of x with respect to S, and *denoted* by $P_x S$.

The absolute value of $P_x S$ is $xt \cdot xt'$, and its sign is positive or negative depending on whether x is inside or outside the segment $[t, t']$.

Figure 10.7.9.2

The proof is very simple: call h the midpoint of $[t, t']$, and write

$$\left(\overrightarrow{xt} \mid \overrightarrow{xt'}\right) = xh^2 - ht^2 = xa^2 - ah^2 - ht^2 = xa^2 - r^2.$$

In practice, $P_x S$ is obtained by plugging in x into the equation of S (assuming the coefficient of $\| \cdot \|^2$ is 1).

10.7.10.1. Let $S = S(a, r)$ and $S' = S(a', r')$ be eccentric spheres $(a \neq a')$. The set $\{ x \in X \mid P_x S = P_x S' \}$ is a hyperplane (use 9.7.6.5 or analytic geometry), orthogonal to $\overrightarrow{aa'}$, and called the *radical hyperplane* (or *radical axis* if $n = 2$) of S and S'. If the two spheres are secant, the radical hyperplane is merely the hyperplane containing their intersection; otherwise, it can be constructed geometrically as shown in figure 10.7.10.1, which uses the defining property of the power of a point.

10.7.10.2. Orthogonality between spheres (cf. 10.7.7) can be expressed in various ways, and is, for this reason, a useful notion. Two spheres $S = S(a, r)$ and $S' = S(a', r')$ are orthogonal if and only if $aa'^2 = r^2 + r'^2$, or $P_a S' = r^2$, or $P_{a'} S = r'^2$, or there exists a line D going through a and intersecting S' (resp. S) at points t, t' (resp. x, x') such that $[x, x', t, t'] = -1$ (harmonic division). And, sure enough, if and only if $T_z S \perp T_z S'$ for some (or any) $z \in S \cap S'$.

Figure 10.7.10.0

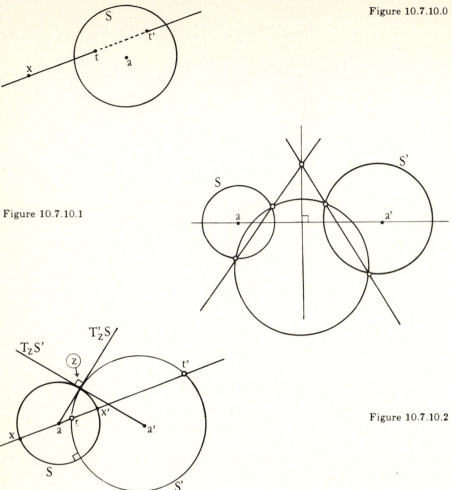

Figure 10.7.10.1

Figure 10.7.10.2

We deduce from this that the center of a sphere orthogonal to two others lies in the radical hyperplane of the two; a sphere orthogonal to $n + 1$ others (in an n-dimensional space X) has its center in the intersection of all the radical planes determined by the others, which is a point, called the *radical center* of the $n + 1$ spheres. This gives a simple proof of the example discussed in 10.7.9.

10.7.10.3. The notion of power enables us to solve a number of construction problems. For example, finding a circle in the plane containing two points and tangent to a given circle (figure 10.7.10.3.1).

Another example: the theorem of the "sixth circle" demonstrated in figure 10.7.10.3.2 (not to be confused with the theorem of six circles of Miguel, whose proof is much more involved). For a recent reference on such theorems, see [DI, 256].

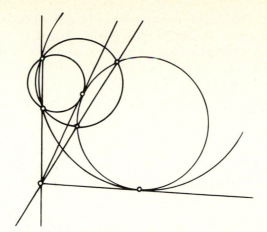

Figure 10.7.10.3.1

Figure 10.7.10.3.2

10.7.11. POLARITY. Polarity with respect to a sphere is a particular case of polarity with respect to a non-degenerate quadric (cf. 14.5); it presents no special interest, and in fact has the drawback (in the affine case) of being subject to exceptions, caused by the points at infinity. We just mention two particularities of this case of polarity: first, the polar hyperplane H of x with respect to $S(a, r)$ is orthogonal to $\overrightarrow{ax}$, and determined by

$$d(a, H) = \frac{r^2}{ax};$$

second, polarity with respect to a sphere can be defined in an elementary way as follows: two points x, y are conjugate with respect to S if the sphere of diameter $[x, y]$ is orthogonal to S (this is a consequence of 10.7.10.2).

See an application in 10.13.14.

10.8. Inversion

Our treatment of inversions here will be elementary. In chapter 20 it will be expanded in order to avoid exceptional cases and allow the use of quadratic forms, a powerful tool. Thus this section can be read for its own sake or as motivation for chapter 20.

10.8.1. DEFINITION. *The inversion of pole c and power $\alpha \in \mathbf{R}^*$ is the map* $i = i_{c.\alpha} : X \setminus c \to X \setminus c$ *defined by*

$$i(x) = \frac{\alpha}{\|x\|^2} \cdot x,$$

where we have vectorialized X at c.

10.8.1.1. We always have $i^2 = \mathrm{Id}_{X \setminus c}$. If $\alpha < 0$, there are no fixed points; if $\alpha > 0$, the set of fixed points is the so-called *inversion sphere* $S(c, \sqrt{\alpha}) = \{ x \in X \mid i(x) = x \}$, and we say that $i_{c.\alpha}$ is the *inversion through S*. Every sphere S such that $P_c S = \alpha$ is globally invariant under $i_{c.\alpha}$, as can be seen from 10.7.10. Every set of the form $Y \setminus c$, for Y a subspace of X containing c, is also invariant, and $i|_{Y \setminus c}$ is the inversion of pole c and power α in Y (heredity).

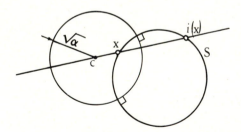

Figure 10.8.1.1

10.8.1.2. We have

$$i_{c.\alpha} \circ i_{c.\beta} = H_{c.\alpha\beta^{-1}}|_{X \setminus c}.$$

10.8.1.3. For $x, y \in X \setminus c$, we have

$$i(x)i(y) = |\alpha| \frac{xy}{cx \cdot cy}.$$

This follows from a trivial calculation in the vectorialization of X at c.

10.8.2. TRANSFORMS OF SPHERES AND HYPERPLANES. Let $i = i_{c,\alpha}$ be an inversion in X. The image $f(H)$ of a hyperplane H not containing c is a punctured sphere $S \setminus c$, where S is a sphere of X going through c. Conversely, the image of $S \setminus c$, where S is a sphere containing c, is a hyperplane not containing c. The image of a sphere not containing c is a sphere not containing c.

To see this, Use an appropriate homothety and 10.8.1.2 to reduce to the case $\alpha = ch^2$, where h is the orthogonal projection of c onto H. Putting $i(x) = x'$ for $x \in H$ and calculating in X_c, we get

$$x \in H \Longleftrightarrow (x \mid h) = \|h\|^2;$$

but $x = \dfrac{\|h\|^2 x'}{\|x'\|^2}$, so that

$$x \in H \Longleftrightarrow (x' - h \mid x') = 0,$$

which means that x' belongs to the sphere of diameter $[c, h]$. If S is a sphere not containing c, we get $i(S) = S$ if $P_c S = \alpha$, by 10.8.1.1; the general case can be reduced to this one by applying an appropriate homothety and 10.8.1.2.

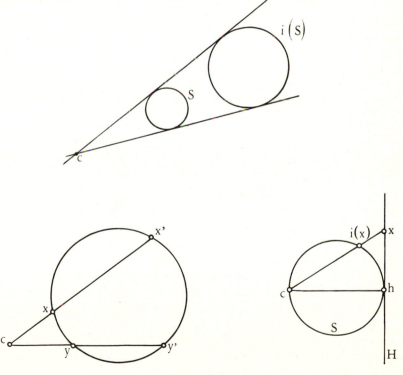

Figure 10.8.2

10.8.2.1. **Notes.** For the study of sphere-preserving maps, see 9.5.3.2, 9.5.3.6 and 18.10.4.

10.8.3. MECHANICAL LINKAGES. There exist mechanical linkages that realize inversions; their use lies in transforming a circular movement into a rectilinear one, since a circular movement is mechanically very easy to achieve (think of a compass or a crank), whereas straight lines are not (in order to obtain a straight ruler, one has to rub one ruler against another). The reader can check that the mechanisms in figures 10.8.3, where the point c is fixed and the others move around, act as inverters. For other mechanical linkages, see [LB1, 64–88]. See also 9.14.34.A. It is interesting to notice that the problem of finding a linkage yielding straight lines stayed open during the nineteenth century for many years.

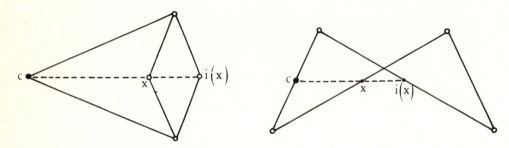

Figure 10.8.3

10.8.4. CONJUGATION BY INVERSIONS. Let g be an inversion, and f a hyperplane reflection σ_S or an inversion through the sphere S (cf. 10.8.1.1). The map gfg^{-1} is a reflection $\sigma_{g(S)}$ in the first case, and an inversion through $g(S)$ in the second.

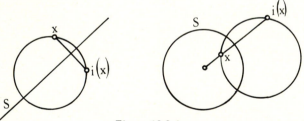

Figure 10.8.4

10.8.4.1. *Proof.* We use the following criterion, a consequence of 10.7.10.2: two points x, x' are inverse to each other under the inversion through the sphere S, or symmetric to one another with respect to the hyperplane S, if and only if every sphere containing x and x' is orthogonal to S (a sphere and a hyperplane are orthogonal if the hyperplane contains the center of the sphere). This criterion implies 10.8.4 once we know that inversions map orthogonal spheres and hyperplanes into orthogonal spheres and hyperplanes; this will be proved in 10.8.5.3.   □

10.8.4.2. Note. The reader will have noticed how cumbersome it is to always have to say "sphere or hyperplane", and also that 10.8.4 is incorrect, in that

gfg^{-1} is not defined everywhere (one must exclude the poles of f and g). For these two reasons, we begin to feel the need for a structure unifying spheres and hyperplanes, and for being able to compose inversions and hyperplane reflections without restriction; this will be one of the aims of chapter 20.

10.8.5. INVERSIONS AND DIFFERENTIAL GEOMETRY. Let $i = i_{c.\alpha}$, and vectorialize X at c. The map i is a C^∞ diffeomorphism, and its derivative is given by

10.8.5.1
$$i'(x)(y) = \frac{\alpha}{\|x\|^2}\left(y - \frac{2(x\mid y)}{\|x\|^2}x\right).$$

This follows from the definition of i and differential calculus (cf. [CH1]).

10.8.5.2. Corollary (inversions preserve angles). *Let $i = i_{c,\alpha}$ be an inversion in an n-dimensional space X. The derivative $i'(x)$ at any point $x \in X \setminus c$ is a similarity, preserving orientation if $\alpha^n < 0$, and reversing it if $\alpha^n > 0$. In particular, the derivative preserves angles between lines and oriented lines, and switches the sign of oriented angles when X is a plane.*

Proof. From 10.8.5.1, $i'(x)$ is the composition of the reflection through the vector hyperplane $x^\perp$ and the homothety of ratio $\alpha/\|x\|^2$; the corollary follows from 8.8.5.   □

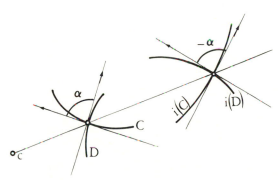

Figure 10.8.5.2

10.8.5.3. In particular, inversions preserve orthogonality relations between spheres and hyperplanes. This could be shown directly, of course, without resorting to differential calculus.

10.8.5.4. Note. In 9.5.4.6 (Liouville's theorem) we classified diffeomorphisms whose derivative is everywhere a similarity.

10.8.5.5. Inversions and osculating circles. Let X be a plane, i an inversion of pole c, and f a C^2 curve in X. The image $i(C)$ of the osculating circle C to f at $f(t)$ is the osculating circle to $i \circ f$ at $i(f(t))$ (at least if $c \notin C$). In fact, i preserves circles, and the osculating circle is the limit of the circle passing through three points of the curve as they get ever closer.

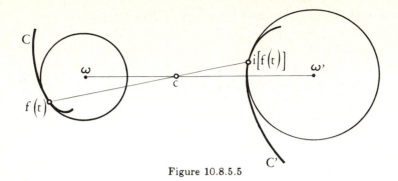

Figure 10.8.5.5

In any case, the result above gives a construction for the center of curvature of the inverse C' of a curve C, given the center of curvature of C (figure 10.8.5.5).

An interesting corollary of 10.8.5.5 is the following: the torsion of a simple closed curve on the sphere vanishes at four points at least (see [B–G, 9.7.9]).

10.8.5.6. See the interesting article [BA–WH].

10.9. Circles in the plane

> In this section and the next X is a plane.

This, after lines in the plane, is the simplest situation in geometry, and perhaps the simplest of all, since, as observed in 10.8.3, a compass is more natural than a ruler. There are myriads of results, of which we mention only a few (see [COO] for more). The essential tools are the power of a point, inversions and oriented angles (cf. 8.7). In fact, the principal application of oriented angles is exactly to such problems, mostly via 10.9.5.

10.9.1. Notation and convention. Until the end of this chapter we denote the line passing through two distinct points x, y by $\overline{xy} = \langle x, y \rangle$. If x belongs to a circle C determined by the context, we denote by $\overline{xx}$ the tangent $T_x C$ to C at x.

Our underlying theme will be the cocyclicity of four points x, y, z, t of X. The first criterion we give is purely metrical:

10.9.2. Proposition (Ptolemy's theorem). If (x_i) $(i = 1, 2, 3, 4)$ are points in the plane, and their distances are denoted by $d_{ij} = x_i x_j$, the following inequality always obtains:

$$d_{12} d_{34} \leq d_{13} d_{42} + d_{14} d_{23}.$$

If equality holds, the points are on the same circle or line. For four points in the plane to be cocyclic or collinear it is necessary and sufficient that one of *the three equalities*

$$\pm d_{12}d_{34} \pm d_{13}d_{42} \pm d_{14}d_{23} = 0$$

be satisfied.

Proof. We have already shown this in 9.7.3.8, but a more elementary demonstration is welcome. Set $x_1 = x$, $x_2 = y$, $x_3 = z$, $x_4 = c$, and let i be the inversion of pole c and power 1. Then the points $x' = i(x)$, $y' = i(y)$, $z' = i(z)$ satisfy

$$x'y' = \frac{xy}{cx \cdot cy}, \qquad y'z' = \frac{yz}{cy \cdot cz}, \qquad z'x' = \frac{zx}{cz \cdot cx}.$$

Thus

$$cz \cdot xy + cx \cdot yz - cy \cdot zx = (cx \cdot cy \cdot cz)(x'y' + y'z' - x'z'),$$

and the conclusion follows form 9.1.1.1 and 10.8.2 if we remark that, among three collinear points, one is always between the other two. □

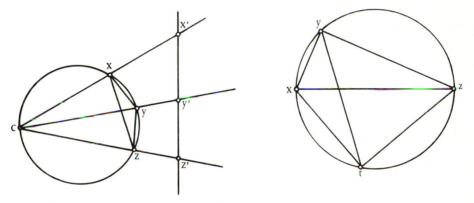

Figure 10.9.2

10.9.2.1. Historical note. Ptolemy used this relation to produce a "table of chords": if $[x, z]$ is a diameter of the circle, the formula $xz \cdot yt = xy \cdot tz + xt \cdot yz$ can be rephrased as the trigonometric formula $\sin(a + b) = \sin a \cos b + \sin b \cos a$ (cf. 8.7.8.1 and 10.3.2). See also 9.7.3.8.

10.9.2.2. The most useful condition for cocyclicity involves *oriented* angles. We state it under the conventions established in 8.7 and 10.9.1; thus the statements below are still valid even if some of the points are not distinct. We give the proofs in the general case; the reader can take care of the particular cases, using the fact that a tangent is orthogonal to the radius.

10.9.3. PROPOSITION. *Let C be a circle of center w, and a, b two points on C. For every $x \in C$ we have* (cf. 8.7.7.7)

$$\overrightarrow{wa}, \overrightarrow{wb} = 2\overset{\frown}{xa}, \overset{\frown}{xb}.$$

Proof. Let y be the point diametrically opposed to x; by 8.7.2.4 (iv), it is enough to verify that

$$\widehat{\overrightarrow{\omega a}, \overrightarrow{\omega y}} = 2\widehat{\overrightarrow{xa}, \overrightarrow{xy}}.$$

Let D be the perpendicular bisector of $[x, a]$, and consider the reflection σ_D. By 8.7.2.4 (v),

$$\widehat{\overrightarrow{\omega a}, \overrightarrow{\omega y}} = \widehat{\overrightarrow{\omega a}, \overrightarrow{xa}} + \widehat{\overrightarrow{xa}, \overrightarrow{\omega y}} = \widehat{\overrightarrow{xa}, \overrightarrow{x\omega}} + \widehat{\overrightarrow{xa}, \overrightarrow{x\omega}} = 2\widehat{\overrightarrow{xa}, \overrightarrow{x\omega}}. \qquad \square$$

10.9.4. COROLLARY. *Let $a, b \in X$ be distinct points. For any $\alpha \in \mathcal{A}(\vec{X}) \setminus 0$, the set*

$$\{\, x \in X \mid \widehat{\overrightarrow{xa}, \overrightarrow{x\omega}} = \alpha \,\}$$

is a circle containing a and b.

Proof. Given a and b, one easily constructs a point ω such that $\omega a = \omega b$ and $\widehat{\overrightarrow{\omega a}, \overrightarrow{\omega b}} = 2\alpha$; for example, ω is the intersection of the perpendicular bisector of $[a, b]$ and the line D containing a such that $\widehat{\overrightarrow{ab}, D} = \delta - \alpha$ (cf. 8.7.7.4). The circle C of center ω and passing through a and b is such that $\widehat{\overrightarrow{xa}, \overrightarrow{xb}} = \alpha$; conversely, if the two angles are equal, the center ω' of the circle containing a, b and x satisfies $\widehat{\overrightarrow{\omega' a}, \overrightarrow{\omega' b}} = 2\alpha$, and can be no other than ω. $\qquad \square$

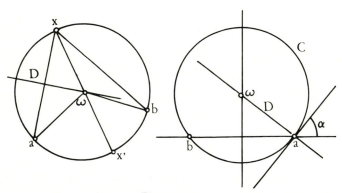

Figure 10.9.4

10.9.5. COROLLARY. *Four distinct points a, b, c, d in X are cocyclic if and only if $\widehat{\overrightarrow{ca}, \overrightarrow{cb}} = \widehat{\overrightarrow{da}, \overrightarrow{db}}$.*

Proof. This is obvious if a, b, c are collinear (see 8.7.7.3, for example). Otherwise $\widehat{\overrightarrow{ca}, \overrightarrow{cb}} \neq 0$ and we apply the two previous results. $\qquad \square$

10.9.6. REMARKS. The condition in corollary 10.9.5 is false for non-oriented angles. What is true is that cocyclicity implies $\overline{\overline{ca}, \overline{cb}} = \overline{\overline{da}, \overline{db}}$ or $\overline{\overline{ca}, \overline{cb}} + \overline{\overline{da}, \overline{db}} = \pi$, but the converse does not hold. Thus 10.9.5 provides a significant

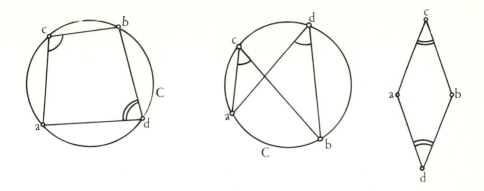

Figure 10.9.5   Figure 10.9.6

advantage, especially in the case of complicated figures, where the number of cases to consider would be large. See also 10.13.15.

Notice the special case $\alpha = \delta$, which yields the circle with diameter $[a, b]$.

We remark that 10.9.3 and 10.9.5 will be obtained as particular cases of a general theorem on conic sections in 17.4.2.

10.9.7. EXAMPLES. Corollary 10.9.5 has numerous applications. The reader can tackle exercises 9.14.3 and 10.13.18 and leaf through the exercises in [D–C1], [D–C2], [I–R], [R–C], [HD]. Here we discuss only two applications, starting with the Simson line, already mentioned in 10.4.5.4 (see also 9.14.34.3.D, 10.4.5.5, 10.9.7.10, 10.11.3, 10.13.27, 17.4.3.5, 17.8.3.2).

10.9.7.1. Let $\{a, b, c\}$ be a triangle. In order that the projections p, q, r of a point x on the sides be collinear it is necessary and sufficient that x belong to the circle circumscribed around $\{a, b, c\}$.

For this kind of proof, one can often just look at the figure and read out chains of equal angles. Here we work as follows: since $\widehat{pc, px} = \delta = \widehat{qc, qx}$ and $\widehat{qa, qx} = \delta = \widehat{ra, rx}$, we get, applying 10.9.5 four times, $\widehat{cb, cx} = \widehat{cp, cx} = \widehat{qp, qx}$ and $\widehat{ab, ax} = \widehat{ar, ax} = \widehat{qr, qx}$. But the condition "$p, q, r$ are collinear" is equivalent to $\widehat{qr, qx} = \widehat{qp, qx}$, so also to $\widehat{cb, cx} = \widehat{ab, ax}$, and this is equivalent to "a, b, c, x are cocyclic" by 10.9.5.

This theorem was attributed to Simson for a long time, but it is actually due to Wallace.

10.9.7.2. Miguel's six circle theorem. Let C_i ($i = 1, 2, 3, 4$) be circles such that $C_1 \cap C_2 = \{a \cap a'\}$, $C_2 \cap C_3 = \{b \cap b'\}$, $C_3 \cap C_4 = \{c \cap c'\}$ and $C_4 \cap C_1 = \{d \cap d'\}$. Then a, b, c, d are cocyclic if and only if a', b', c', d' are.

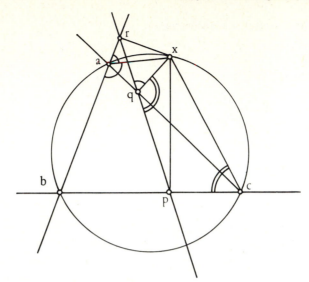

Figure 10.9.7.1

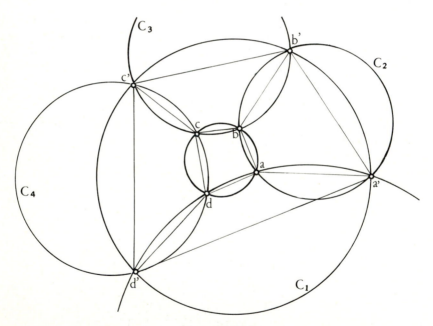

Figure 10.9.7.2

Proof. Using Chasles's relation and 10.9.5 (four times), we get

$$\widehat{\overline{ba}, \overline{bc}} = \widehat{\overline{ba}, \overline{bb'}} + \widehat{\overline{bb'}, \overline{bc}} = \widehat{\overline{a'a}, \overline{a'b'}} + \widehat{\overline{c'b'}, \overline{c'c}},$$

$$\widehat{\overline{da}, \overline{dc}} = \widehat{\overline{da}, \overline{dd'}} + \widehat{\overline{dd'}, \overline{dc}} = \widehat{\overline{a'a}, \overline{a'd'}} + \widehat{\overline{c'd'}, \overline{c'c}}.$$

Subtracting and applying Chasles again,

$$\overset{\frown}{ba,bc} - \overset{\frown}{da,dc} = \overline{a'd',a'b'} - \overline{c'd',c'b'},$$

and the conclusion follows form 10.9.5.

10.9.8. NOTE. Miguel's theorem is the first in a "chain of theorems", cf. 10.13.19. On chains of theorems see [PE, 431] and [CR1, 262, 258]. See also 10.11.7.

10.10. Pencils of circles

The following facts are simple consequences of the discussion in 10.7.10.

10.10.1. DEFINITION AND PROPOSITION. Let C, C' be non-concentric circles. Any two elements of the set $\mathcal{F} = \{\text{circles } \Gamma \mid \Gamma \perp C \text{ and } \Gamma \perp C'\}$ have the same radical axis, the line $\langle \omega, \omega' \rangle$ determined by the centers ω, ω' of C, C'. The set $\mathcal{F}$ is called a *pencil of circles*. If Γ, Γ' belong to a pencil $\mathcal{F}$, the pencil formed by circles C such that $C \perp \Gamma$ and $C \perp \Gamma'$ is *denoted* by $\mathcal{F}^\perp$ and called the pencil *orthogonal* to $\mathcal{F}$; in this case $C \perp \Gamma$ for every $C \in \mathcal{F}^\perp$ and $\Gamma \in \mathcal{F}$, and we have $\mathcal{F}^{\perp\perp} = \mathcal{F}$. Let C and C' be two distinct circles in a pencil $\mathcal{F}$. If C and C' are tangent, $\mathcal{F}$ consists of circles tangent to one another, and so does $\mathcal{F}'$ (figure 10.10.1.1). If $C \cap C' = \emptyset$, there exist points x and y (on $\langle \omega, \omega' \rangle$) such that $\mathcal{F}^\perp = \{\text{circles } C \mid x, y \in C\}$ (figure 10.10.1.2); we call x and y the *limit points* of $\mathcal{F}$. If $C \cap C'$ contains more than one point, the elements of $\mathcal{F}^\perp$ are pairwise disjoint (figure 10.10.1.3).   □

More need not be said, if only because we will discuss pencils of circles at length in chapter 20.

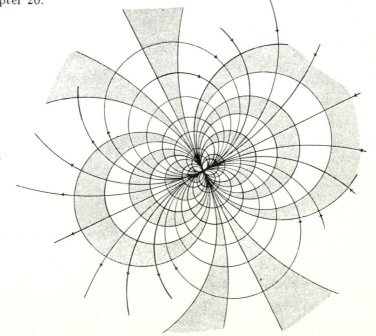

Figure 10.10.1.1

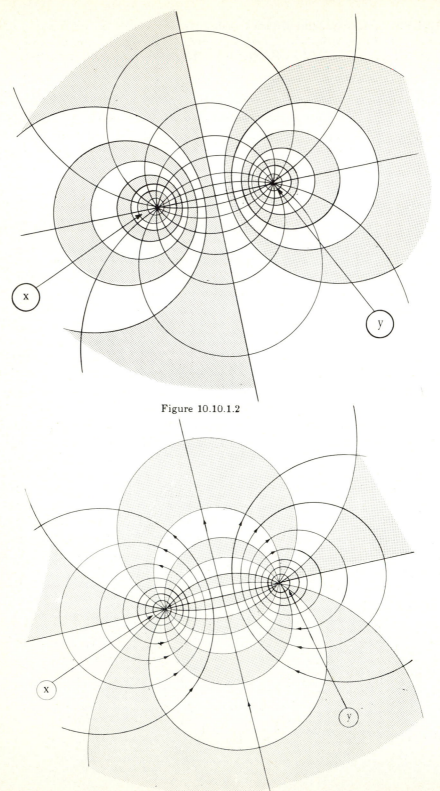

Figure 10.10.1.2

Figure 10.10.1.3

10.10.2. REDUCTION OF TWO CIRCLES. One use for 10.10.1 and the inversion of circles is afforded by the following fact: two circles C, C' can always be transformed by an inversion into either two concentric circles or two lines.

If $C \cap C' \neq \emptyset$, make one of the intersection points the pole of the inversion; if $C \cap C' = \emptyset$, choose instead a limit point, as defined above. Since inversions preserve orthogonality (cf. 10.8.5.3), the image of the second limit point under the inversion will be such that all lines going through it are orthogonal to the image circles; this can only happen if that point is the center of both image circles.

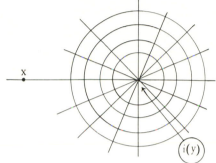

Figure 10.10.2

Observation 10.10.2 solves a number of problems about circles, for example, the study of circles intersecting two given circles at the same angle, or the construction of a circle intersecting three given circles at given angles. All this is possible because the conditions in question are preserved under inversion. Another, deeper way of phrasing this remark is by saying that the set of two concentric circles has a big stabilizer G, containing all isometries that leave invariant the common center of the circles; similarly, two intersecting lines are left invariant by any homothety centered at their intersection point. Conjugated by an inversion, the group G becomes a group $\hat{G}$ that stabilizes the two circles originally considered, and $\hat{G}$ is made up of maps which preserve circles, lines and angles. (They are not really maps from E into itself, but from $\hat{E}$ into itself, $\hat{E}$ being introduced in chapter 20 exactly to make precise the kind of argument we have been presenting.)

Now an example of an application of 10.10.2.

10.10.3. STEINER'S ALTERNATIVE. Let C and C' be two circles, with C inside C', and let Γ_1 be a circle externally tangent to C and internally tangent to C'. Construct a chain of circles Γ_i $(i = 1, 2, \ldots)$ as follows: for each i, let Γ_{i+1} be tangent to Γ_i, C and C', and distinct from Γ_{i-1}. If we have $\Gamma_i \neq \Gamma$ for every $i > 1$, too bad; but if $\Gamma_n = \Gamma_1$ for some n, then for any other initial circle Γ_n we'll have $\Gamma'_n = \Gamma'_1$.

The proof is trivial if we transform C, C' into two concentric circles, for a pair of concentric circles is invariant under the group of rotations around their common center.

Figure 10.10.3

The *alternative* is that either no chain closes up, or they all do. Here is another example of an alternative:

10.10.4. THE GREAT PONCELET THEOREM FOR CIRCLES. As in 10.10.3, we start from two circles C, C' with C inside, and we construct a sequence of points x_i $(i \geq 1)$, beginning with a given $x_1 \in C'$, by the following condition: $\langle x_i, x_{i+1} \rangle$ is tangent to C for every i, and $\langle x_i, x_{i+1} \rangle \neq \langle x_i, x_{i-1} \rangle$. Then it can be shown that either there exists n such that $x_n = x_1$ for all $x_1 \in C'$, or $x_i \neq x_1$ for all $x_1 \in C'$ and $i > 1$.

This result is much harder to prove than Steiner's alternative, in spite of the simplicity of its statement (notice that straight lines are not preserved under inversions, so the previous reasoning does not apply). We shall see a proof of it in 16.6. See also 10.13.3, 17.6.5, and [B–H–H].

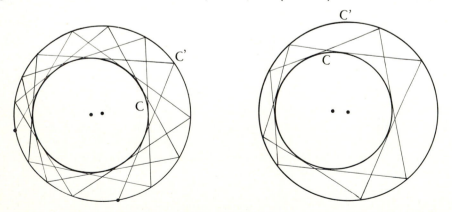

Figure 10.10.4

10.11. Classical problems

We collect here some classical results about circles. The reader can try his hand at proving them or consult the references, especially [COO], an excellent and systematic book.

10.11.1. APPOLONIUS' PROBLEM. The question is to find a circle Γ tangent to three given circles C, C', C''. By 10.7.9 we know there are at most eight solutions.

An elementary method consists in adding or subtracting from the radii of the three circles the radius of the smallest one, thus reducing the problem to finding a circle containing a given point and tangent to two given circles (figure 10.11.1.1); this can be done by using 10.10.2 (see 9.6.6 if necessary).

Figure 10.11.1.1

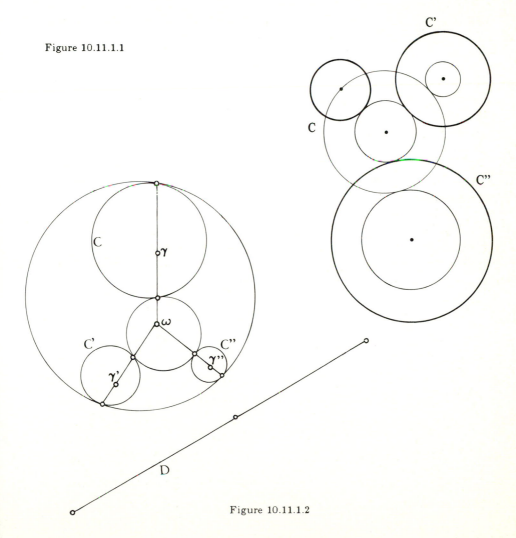

Figure 10.11.1.2

A much more elegant method, due to Gergonne, gives the contact points with C, C', C'' directly. In figure 10.11.1.2, the point ω is the radical center (cf. 10.7.10.2) of C, C', C'', that is, the point having the same power with respect to each of the three circles. This can be easily found using 10.7.10.1. The line D is *one* of the four lines connecting the six centers of the homotheties taking C into C', C' into C'' or C'' into C (the six centers do fall into four lines). Finally, γ, γ' and γ'' are the poles of D with respect to C, C' and C'' (cf. 10.7.11). The desired contact points are given by the intersection of C (resp. C', C'') with the line $\langle \omega, \gamma \rangle$ (resp. $\langle \omega, \gamma' \rangle$, $\langle \omega, \gamma'' \rangle$).

See [CR5] for a recent reference.

10.11.2. PROBLEM OF NAPOLEON-MASCHERONI. Using a compass only (cf. 10.8.3), find the center of a given circle. See [LB1, 25 ff.], more precisely, the historical note at the bottom of page 25 and the statement of the theorem of Mohr–Mascheroni on page 26: "Every construction possible with a straightedge and compass is possible with the compass alone." A good portion of [LB1] reads like a novel.

10.11.3. THE NINE-POINT CIRCLE AND THE FEUERBACH THEOREM. Let $T = \{x, y, z\}$ be a triangle, p, q, r the feet of the altitudes, u, v, w the midpoints of $[x, h]$, $[y, h]$, $[z, h]$, where h is the intersection of the altitudes (cf. 10.2.5), ω the center of the circle circumscribed around T and g its center of mass. Prove that the nine points $u, v, w, l, m, n, p, q, r$ all lie on the same circle Γ whose center O is the midpoint of h and ω. Show also that g is on the line $\langle h, O, \omega \rangle$ and that $\overrightarrow{gh} = -2\overrightarrow{g\omega}$. The circle Γ is tangent to the inscribed and exinscribed circles (cf. 10.1.5 and figure 10.1.5).

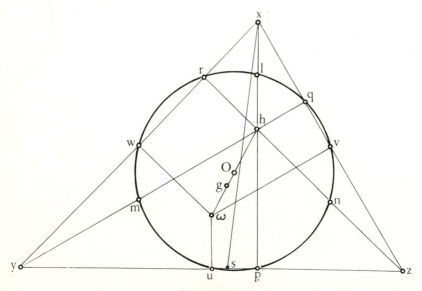

Figure 10.11.3

The proof utilizes the inversion of pole u and power $up \cdot us$, where s is the foot of the interior bisector stemming from x. Notice also that Γ is the circle tritangent to the three-cusped hypocycloid found as the envelope of the Simson lines of $\mathcal{T}$: cf. 10.4.5.5, 10.9.7.1 and 9.14.34.

10.11.4. CASTILLON'S PROBLEM. Given a circle C and points $(x_i)_{i=1.....n}$ not on C, find a polygon $(z_i)_{i=1.....n}$ such that $x_i \in \langle z_i, z_{i+1} \rangle$ for all $i = 1, \ldots, n$. One rather complicated solution consists in taking the product of the inversions f_i of pole x_i that leave C globally invariant; then $(f_n \circ f_{n-1} \circ \cdots \circ f_2 \circ f_1)(z_1) = z_1$, and we just have to find the fixed points of $f_n \circ f_{n-1} \circ \cdots \circ f_2 \circ f_1$. A better method, which holds for all conics, will be given in 16.3.10.3. See also [DO, 144].

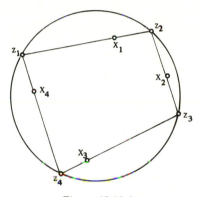

Figure 10.11.4

10.11.5. MALFATTI'S PROBLEM. Given a triangle $\mathcal{T}$, find three circles C, C', C'' that are each tangent to two sides of $\mathcal{T}$ and to the other two circles. For the solution, see [HD, volume 1, 310], [R–C, volume I, p. 311–314], [DO, 147] or [COO].

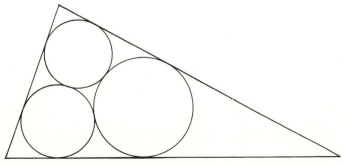

Figure 10.11.5

10.11.6. LAGUERRE'S CIRCLES. This deals with oriented circles and lines, tangent among themselves with the right orientation; the result would be

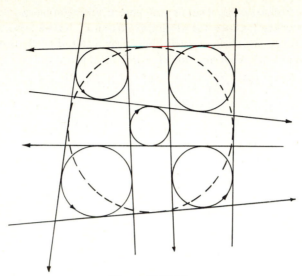

Figure 10.11.6.1

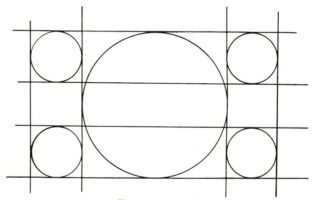

Figure 10.11.6.2

false if matching orientation were not required. A complete exposition can be found in [PE, 426]. For an algebraic formulation of this theory, see [BZ, 251 ff.]. See also [BLA3, chapter 4].

10.11.7. THE SEVEN CIRCLE THEOREM. For this very nice result, see [E–M–T].

10.12. Parataxis: prelude to sections 18.9, 20.5 and 20.7

This section discusses a number of surprising phenomena relating to circles and spheres in ordinary three-space. They will be explained in two ways: geometrically in 18.9, and algebraically in 20.5.4.

10.12.1. Let T be a *torus*, that is, the surface or revolution created when a circle rotates around a line disjoint from and coplanar with it. A torus contains two families of circles, the parallels of latitude and the meridians. The first surprising fact is that T contains other circles as well. These are sometimes known to mathematicians as Villarceau circles, but they were known much before he was around (1848), as attested by the l'Oeuvre Notre-Dame museum in Strasbourg, where the spiral staircase is topped with a torus sculpted in such a way that its live edges are exactly these circles (figure 10.12.1.2). Villarceau circles are found by intersecting T with a bitangent plane (figure 10.12.1.3).

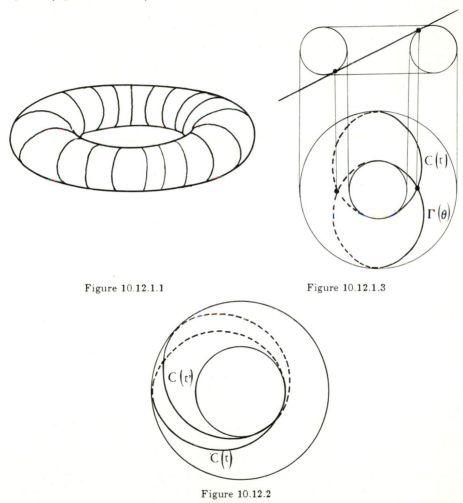

Figure 10.12.1.1   Figure 10.12.1.3

Figure 10.12.2

10.12.2. These exotic circles are divided into two families $\{C(t)\}$ and $\{\Gamma(t)\}$. Two circles $C(t)$ and $\Gamma(\theta)$ from different families intersect in exactly two points. Two circles from the same family not only do not intersect, but

Figure 10.12.1.2 (Source: [HG])

they are always linked (for the definition of linkedness, see [B–G, 7.4.8], for example), as demonstrated in the Strasbourg torus.

10.12.3. Villarceau circles satisfy strong angular properties. First, they are *helices* of the torus, that is, they intersect each meridian at a constant angle. What is more, two different circles $C(t)$ and $C(t')$ (or $\Gamma(t)$ and $\Gamma(t')$) form a *paratactic annulus*, that is, every sphere S containing $C(t)$ intersects $C(t')$ at a constant angle α, and every sphere S' containing $C'(t)$ intersects $C(t)$ at a constant angle α.

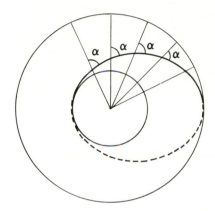
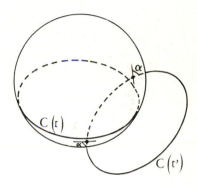

Figure 10.12.3

10.13. Exercises.

* **10.13.1.** THE ALTITUDES OF A TRIANGLE ARE CONCURRENT.

10.13.1.1. Deduce this fact from the following property (show it): for every four points $\{a, b, c, d\}$ in a Euclidean plane, we have

$$\left(\overrightarrow{ab} \mid \overrightarrow{cd}\right) + \left(\overrightarrow{ac} \mid \overrightarrow{db}\right) + \left(\overrightarrow{ad} \mid \overrightarrow{bc}\right) = 0.$$

10.13.1.2. Deduce it now from the Theorem of Ceva (2.8.1).

10.13.1.3. Deduce it from 16.5.4 by using cyclic points.

10.13.2. MORE FORMULAS. Show that the following relations hold for a triangle:

$$S^2 = r r_a r_b r_c, \quad 4R = r_a + r_b + r_c - r, \quad \frac{1}{2Rr} = \frac{1}{ab} + \frac{1}{bc} + \frac{1}{ca},$$

$$\frac{1}{r} = \frac{1}{r_a} + \frac{1}{r_b} + \frac{1}{r_c} = \frac{1}{h_a} + \frac{1}{h_b} + \frac{1}{h_c}, \quad p^2 = r_a r_b + r_b r_c + r_c r_a,$$

$$2R = \frac{a + b + c}{\sin A + \sin B + \sin C}.$$

* **10.13.3.** THE GREAT PONCELET THEOREM FOR CIRCLES $(n = 3, 4)$.
Given a triangle $\mathcal{T}$, show that the radius R of the circumscribed circle C, the
radius r of the inscribed circle Γ and the distance d between the two centers
satisfy the relation $R^2 - 2Rr = d^2$. Show that if, conversely, two circles C
and Γ of radii R and r whose centers are d units apart satisfy the condition
$R^2 - 2Rr = d^2$, then for any $x \in C$ there is a triangle $\mathcal{T}$ inscribed in C and
circumscribed around Γ, and having x as a vertex. Deduce from this fact the
theorem of Poncelet for the case of triangles (16.10.4). Make a similar study
for a quadrilateral inscribed in C and circumscribed around Γ (see [COO]).

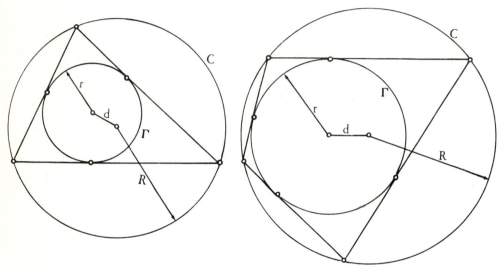

Figure 10.13.3

10.13.4. MORLEY'S THEOREM (cf. 10.3.10). First assume the theorem to be
true, and show that, in the figure, the triangles $\{u, r, q\}$, $\{v, r, p\}$ and $\{w, p, q\}$
are isosceles; find their angles as a function of the angles of $\{x, y, z\}$. Deduce
from this a proof of Morley's theorem, starting from an equilateral triangle
$\{r, p, q\}$ and constructing u, v, w and finally x, y, z with the desired angles,
thus obtaining a triangle similar to the original one.

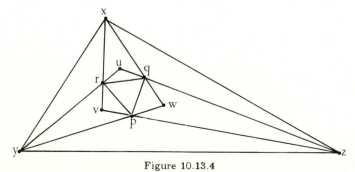

Figure 10.13.4

10.13.5. Show that, for every t in the plane and every triangle $T = \{x, y, z\}$, we have $tx + ty + tz \geq 2\sqrt{\sqrt{3}S}$, and that equality holds if and only if T is equilateral. Deduce that $tx + ty + tz \geq 6r$.

10.13.6. Fill in the details of the discussion in 10.4.5.4.

10.13.7. BRAHMAGUPTA'S FORMULA. For a quadrilateral inscribed in a circle, calculate the diagonals as a function of the sides a, b, c, d. Show that the area of the quadrilateral is

$$\sqrt{(p-a)(p-b)(p-c)(p-d)},$$

where $p = (a + b + c + d)/2$. Show that a formula like

$$\text{area} = p^2 \prod_{i=1}^{n} \left(1 - \frac{a_i}{p}\right)^{\frac{1}{2}},$$

generalizing Heron's formula (10.3.3) and Brahmagupta's formula to the case of arbitrary polygons incribed in a circle, is false when they have more than four sides. See [LEV].

* **10.13.8.** Given a convex quadrilateral, find the minimum of the function "sum of the distances from a variable point to the vertices of the quadrilateral". Do the same for a convex hexagon circumscribed around an ellipse (cf. 16.2.13).

10.13.9. Calculate the radius of a sphere circumscribed around a regular tetrahedron.

10.13.10. Calculate the volume of a tetrahedron as a function of:
 i) two faces, the common edge and the dihedral angle between the two;
 ii) the four faces, two opposite dihedral angles and the corresponding edges;
 iii) one face and the adjacent dihedral angles.

10.13.11. Show that if the four faces of a tetrahedron have the same area, they are congruent.

10.13.12. In a tetrahedron, call α, β, γ the distances between pairs of opposite edges (cf. 9.2.5), and h_i ($i = 1, 2, 3, 4$) the distances between a vertex and the opposite face. Then

$$\frac{1}{\alpha^2} + \frac{1}{\beta^2} + \frac{1}{\gamma^2} = \sum_i \frac{1}{h_i^2}.$$

10.13.13. In a tetrahedron, the product of the sines of two opposite dihedral angles is proportional to the product of the corresponding edges.

10.13.14. SALMON'S PRINCIPLE. Consider in a Euclidean plane a circle C centered at a, two points x, y and their polars D_x, D_y with respect to C. Show that

$$\frac{ax}{ay} = \frac{d(x, D_y)}{d(y, D_x)}.$$

Show that if P is a $2n$-sided polygon inscribed in C and z is a point on C, the product of the distances from z to the even-numbered sides of P is equal to the product of the distances to the odd-numbered sides. If P is circumscribed around C, show that the product of the distances from even-numbered vertices to a fixed tangent to C is proportional to the product of distances from odd-numbered vertices, the ratio not depending on the tangent.

10.13.15. Given two distinct points a and b in a Euclidean plane and an angle α, oriented or not, between lines or half-lines, study the sets $\{\,x\,|\,\overrightarrow{xa},\,\overrightarrow{xb} = \alpha\,\}$, $\{\,x\,|\,\overrightarrow{xa},\,\overrightarrow{xb} = \alpha\,\}$, $\{\,x\,|\,\widehat{xa,\,xb} = \alpha\,\}$, $\{\,x\,|\,\overline{xa,\,xb} = \alpha\,\}$.

10.13.16. LINE OF THE IMAGES. Let T be a triangle and D a line. Show that the reflections of D through each side have a point in common if and only if D goes through the intersection of the altitudes of T. What happens when D rotates around this intersection point? (See also 17.6.2.2.)

10.13.17. Give an algebraic proof of 10.9.7.2 using cross-ratios (cf. 9.6.5.2).

10.13.18. THE PIVOT. Consider a fixed triangle $\{x, y, z\}$ in the plane. Show that, given three points p, q, r on the sides of this triangle and a point π in the plane, $\{z, q, \pi, p\}$ are cocyclic if $\{x, r, \pi, q\}$ and $\{y, p, \pi, r\}$ are. Now assume π, called the *pivot*, is fixed, and that p, q, r slide along the sides so that the cocyclicity conditions are satisfied. Show that the various positions of $\{p, q, r\}$ are all similar (under orientation-preserving similarities). Show that any point linked to $\{p, q, r\}$ by an orientation-preserving similarity (for example, the intersection of the altitudes, or of the bisectors, etc.) describes a line, and every line linked to $\{p, q, r\}$ by an orientation-preserving similarity has as its envelope a parabola of center π (cf. 9.6.7).

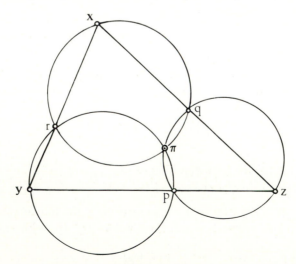

Figure 10.13.18

* **10.13.19.** A CHAIN OF THEOREMS. In this problem the objects are supposed to be "in general position". Let $(D_i)_i = 1, 2, 3, 4$ be four lines in a Euclidean plane, and let C_i be the circle circumscribed around the triangle formed by the three lines D_j $(j \neq i)$; show that the four circles C_i have a point in common. (See another explanation in 17.4.3.5.) Draw an illustrative picture.

Let $(D_i)_{i=1,\ldots,5}$ be five lines in a Euclidean plane, and let p_i be the point associated in the preceding theorem with the four lines D_j $(j \neq i)$; show that the five points p_i belong to the same circle. Draw a picture.

State and prove a chain of theorems whose two first elements are the results above. On the subject of theorem chains, see [PE, 431, exercise 94.7] and [CR1, 258].

10.13.20. ISOGONAL CIRCLES. A circle Γ is said to be *isogonal* with respect to two circles C, C' if it makes the same angle with the two. Show that the circles isogonal with respect to two fixed secant circles are all orthogonal to the same circle or line. Analyze the situation when the two circles are not secant. Replace isogonality by the condition that the angles of Γ with C, C' have sum π. Study circles isogonal with respect to three and then four fixed circles. Can your results be used in problem 10.11.1?

10.13.21. Let C, C' be two circles as in 10.3.10, with radii R, R' and centers a, a' such that $aa' = d$. Show that if C and C' admit a chain of n circles as in 10.10.2, we have $(R - r)^2 - d^2 = 4Rr\tan^2(\pi/n)$.

10.13.22. Prove that the bitangent plane in figure 10.12.1.3 intersects the torus along two circles, and that these circles are helices for the torus.

10.13.23. Using the technique of 10.3.10, show how the intersection points of the internal and external trisectors of a triangle fall into 27 lines. See 9.14.34.5 and [LB1, 173–208].

10.13.24. Given a plane triangle $\mathcal{T}$, study the circles with respect to which $\mathcal{T}$ is self-polar, that is, its vertices are pairwise conjugate with respect to the circle. Same question in dimension three.

10.13.25. Consider three pairwise tangent circles, with radii a, b, c, and set $\alpha = a^{-1}$, $\beta = b^{-1}$, $\gamma = c^{-1}$. Show that the radii of the two circles tangent to these three are $(\alpha + \beta + \gamma + 2\sqrt{\beta\gamma + \gamma\alpha + \alpha\beta})^{-1}$ and $|\alpha + \beta + \gamma - 2\sqrt{\beta\gamma + \gamma\alpha + \alpha\beta}|^{-1}$. What does the sign of $\alpha + \beta + \gamma - 2\sqrt{\beta\gamma + \gamma\alpha + \alpha\beta}$ signify?

10.13.26. Let a, b, c be the lengths of the sides of a triangle $\mathcal{T}$ in a Euclidean plane. For every integer n, consider the point $X(n)$ whose barycentric coordinates with respect to $\mathcal{T}$ are

$$\left(\frac{a^n}{a^n + b^n + c^n}, \frac{b^n}{a^n + b^n + c^n}, \frac{c^n}{a^n + b^n + c^n} \right).$$

Find a geometric characterization for $X(0)$, $X(1)$, $X(2)$. What does $X(n)$ become when n approaches $+\infty$ (resp. $-\infty$)?

10.13.27. THE SIMSON LINE STRIKES AGAIN. Let D, D' be distinct lines in a Euclidean plane intersecting at a, and take $x \notin D \cup D'$. Show that the map $D \ni m \mapsto m' \in D'$, obtained by intersecting D and D' at m and m' by a circle of variable radius containing a and x, is the restriction to D of an orientation-preserving similarity of center x. What line does the orthogonal projection h of x onto mm' describe? Deduce theorem 10.9.7.1.

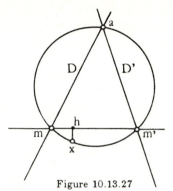

Figure 10.13.27

* **10.13.28.** Given four points a, b, c, d in a Euclidean plane, construct a square $ABCD$ such that $a \in AB$, $b \in BC$, $c \in CD$ and $d \in DA$.

More generally, given two quadrilaterals Q and Q' in the plane, construct a quadrilateral Q'' similar to Q and inscribed in Q' (resp. circumscribed around Q').

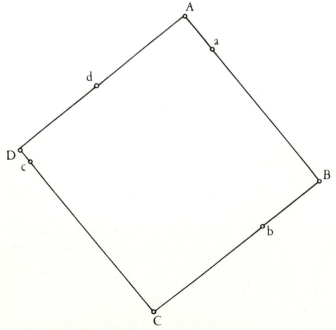

Figure 10.13.28

* **10.13.29.** FORD CIRCLES. Consider three pairwise tangent circles, and also tangent to a line D, in the situation shown in figure 10.13.29.1. If r and s are the radii of γ and δ, and ρ is the radius of the small circle, find ρ, $\overline{ax}$ and $\overline{bx}$ as functions of $\overline{ab}$, r, and s, where a, b, and x are the tangency points of γ, δ and the small circle with D.

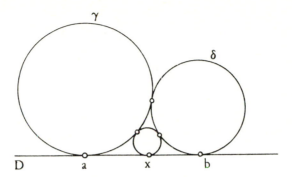

Figure 10.13.29.1

Starting now from two circles whose equations, in an orthonormal frame, are $x^2+y^2-y=0$ and $x^2+y^2-2x-y+1=0$, use induction to construct circles as shown in figure 10.13.29.2. Show that the tangency points of these circles with the x-axis always have rational abscissas. Are all rational numbers in $[0,1]$ obtained in this way?

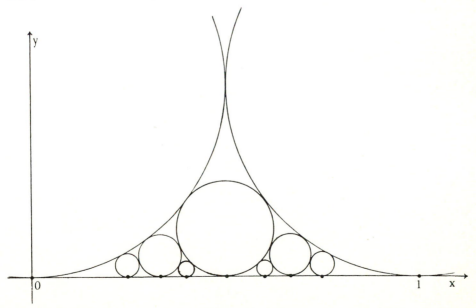

Figure 10.13.29.2

Ford circles have found a nice application in arithmetic—one performs complex integrations along them ([RA, 267]).

10.13.30. NAGEL'S POINT. Prove, in at least two different ways, that the three lines joining the vertices of a triangle with the point of contact of the inscribed circle with the opposite sides are concurrent.

Chapter 11
Convex sets

Convex sets arise naturally in geometry, but they also play a fundamental role in analysis, and even in arithmetic and differential geometry. Here we discuss only the geometric aspect of the notion of convexity, apart from a few words about convex functions. Convex polyhedra, a particular class of convex sets, will be discussed in detail in the next chapter.

Section 11.1 presents some non-trivial constructions for convex sets, to be used later: the Minkowski sum, the polar convex body of a convex set with respect to a sphere, and the Stein symmetrization of a convex set. In section 11.3 we classify convex sets and their boundaries up to homeomorphisms: the algebraic topology of convex sets is very simple. (When it's so easy to classify one's objects of study, one may as well make the best of it...)

Section 11.4 is classical and includes separation theorems derived from the Hahn–Banach theorem. Such theorems are used, among other things, to show that polarity between convex sets is a good duality relation, and, in section 11.6, to study the boundary points of a convex set. Boundary points and duality play an essential role in chapter 12.

In section 11.7 we offer Helly's theorem, a spectacular result with a very simple statement. There follows Krasnosel'skii's theorem, a pretty application of Helly's theorem. Finally, we define convex functions and give some of their properties in section 11.8. Two of their applications are used later in the book: the theorem of Brunn–Minkowski in the proof of the isoperimetric inequality, and the theorem of Loewner–Behrend in a characterization of affine quadrics.

This chapter only deals with real affine spaces of finite dimension d.

11.1. Definition and examples

11.1.1. DEFINITION. *A subset S of an affine space X is called convex if, for any $x, y \in S$, we have $[x, y] \subset S$, where $[x, y] = \{ \lambda x + (1 - \lambda)y \mid \lambda \in [0, 1] \}$* (cf. 3.4.3).

11.1.2. EXAMPLES

11.1.2.1. Refer to the figures below: figure 11.1.2.1.1 represents a strip in the plane, 11.1.2.1.3 an infinite cylinder in space, and figures 11.1.2.1.4, 11.1.2.1.5 and 11.1.2.1.6 represent the regions determined by a hyperbola, an ellipse and a parabola, respectively (cf. 17.1.4). Furthermore, in figures 11.1.2.1.7 and 11.1.2.1.8, the thick parts of the frontier belong to the set, and the thin parts don't. Among these sets, the only ones that are not convex are 11.1.2.1.2 and 11.1.2.1.7.

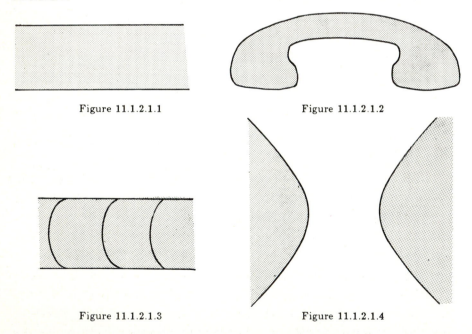

Figure 11.1.2.1.1   Figure 11.1.2.1.2

Figure 11.1.2.1.3   Figure 11.1.2.1.4

11.1.2.2. The following are convex sets: X itself, any affine subspace of X (cf. 2.4), and in particular points, lines and hyperplanes. The empty set is convex, but, as one surmises, it gives rise to all kinds of false statements; we thus enact the following bailing-out maxim (cf. [VE, 198]): "If a theorem is false when the set A is empty, we tacitly assume A is non-empty. We sincerely hope never to have to resort to this S.O.S."

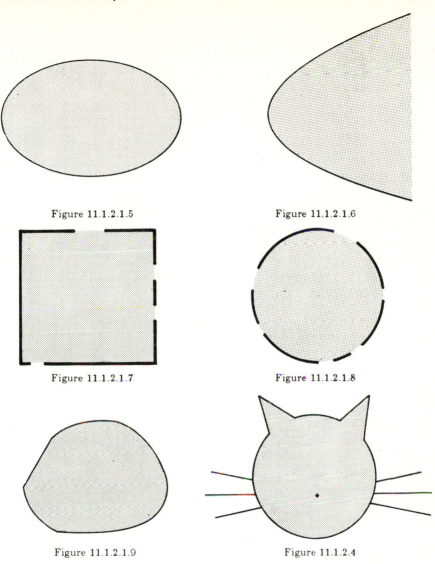

Figure 11.1.2.1.5   Figure 11.1.2.1.6

Figure 11.1.2.1.7   Figure 11.1.2.1.8

Figure 11.1.2.1.9   Figure 11.1.2.4

11.1.2.3. The convex sets of **R** (or of X, when $d = 1$) are the intervals (of all sorts).

11.1.2.4. A subset E of X is called *star-shaped* (at $x \in E$) if $[x, y] \subset E$ for every $y \in E$. Thus a convex set is star-shaped at any of its points. See 11.7.7 for a very nice characterization of star-shaped sets. Star-shaped, and consequently convex sets, are connected and path-connected.

11.1.2.5. In a Euclidean affine space, all open and closed balls $B(a, r)$ and $U(a, r)$ are convex. In fact (cf. figure 11.1.2.1.8), for any subset A of the sphere $S(a, r)$, the difference set $B(a, r) \setminus A$ is convex. This is false for cubes (cf. figure 11.1.2.1.7).

11.1.2.6. Let X, Y be affine spaces, $S \subset X$ and $T \subset Y$ convex sets, and $f : X \to Y$ an affine map. Then $f(S) \subset Y$ and $f^{-1}(T) \subset X$ are again convex (apply 3.5.1). In particular, if $H \subset X$ is a hyperplane, the (open and closed) half-spaces determined by it are convex: just take f to be an affine form on X such that $H = f^{-1}(0)$ (cf. 2.7.3), so the half-spaces are inverse images of intervals (like $[0, \infty[$, for instance), and intervals are convex in $\mathbf{R}$ (cf. 11.1.2.3).

11.1.2.7. Every (finite or infinite) intersection of convex sets is convex. Every intersection of half-spaces (cf. 11.1.2.6) is convex. Every convex polyhedron (a finite intersection of half-spaces) is convex. Convex polyhedra will be studied in chapter 12.

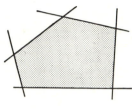

Figure 11.1.2.7

An increasing union of convex sets is convex.

11.1.2.8. Positive definite quadratic forms over a vector space E form a convex subset of $\mathcal{P}_2^\bullet(E)$. This example will be met again in 11.8.9.

We now give three somewhat more intricate examples of convex sets, obtained through different procedures:

11.1.3. THE MINKOWSKI SUM. The following operation is due to Minkowski: *If S and T are convex sets in a vector space X, the set*

11.1.3.1 $$\lambda S + \mu T = \{ \lambda s + \mu t \mid s \in S,\, t \in T \}$$

is convex, for any two real numbers λ, μ.

The reader can prove this directly, or use 11.1.2.6 with $f : X \times X \ni (x, y) \mapsto \lambda x + \mu y \in X$.

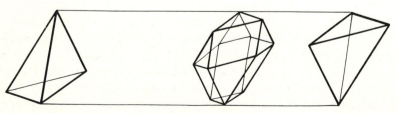

Figure 11.1.3.1

If X is an affine space, this definition only makes sense for $\lambda + \mu = 1$ (cf. 3.4.1). Otherwise we can only define $\lambda S + \mu T$ up to translations, or consider it as a "shape" (cf. 2.1.8). The reader is encouraged to draw the set $\lambda S + \mu T$, and especially $S - T$, for sets S and T to his taste. A particular case of the Minkowski sum will be essential in the proof of the isoperimetric inequality (cf. 12.10.10).

11.1.3.2. Proposition. *Let X be a Euclidean vector space, A a subset of X and $\epsilon > 0$. The sets $U(A, \epsilon)$ and $B(A, \epsilon)$ (cf. 0.3) satisfy $U(A, \epsilon) = A + U(0, \epsilon)$ for A arbitrary and $B(A, \epsilon) = A + B(0, \epsilon)$ for A compact.*

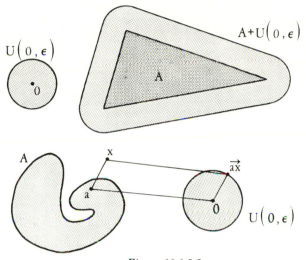

Figure 11.1.3.2

Proof. If $d(x, A) < \epsilon$ there exists a point $a \in A$ such that $d(x, a) < \epsilon$; thus $x = a + \overrightarrow{ax} \in A + U(0, \epsilon)$. Conversely, if $x \in A + U(0, \epsilon)$, we have $x = a + \overrightarrow{ax}$, where $\|\overrightarrow{ax}\| < \epsilon$, so that $d(x, A) \leq d(x, a) < \epsilon$. In the compact case, we know that, for any $x \in X$, there exists $y \in A$ such that $d(x, y) = d(x, A)$. $\square$

11.1.3.3. Corollary. *If A is convex, so is $U(A, \epsilon)$ for every $\epsilon > 0$. If A is compact and convex, so is $B(A, \epsilon)$.* $\square$

11.1.4. STEIN SYMMETRIZATION OF A COMPACT SET. With the definitions and notations from section 9.13, we have:

11.1.4.1. Proposition. *If S is convex, so is $\mathrm{st}_H(S)$.*

Proof. Let $x, x' \in \mathrm{st}_H(S)$, let D and D' be the perpendiculars to H through x and x', respectively, and let $[u, v] = \mathrm{st}_H(S) \cap D$, $[u', v'] = \mathrm{st}_H(S) \cap D'$. By construction and the convexity of S, the segments $[u, v]$ and $[u', v']$ derive from segments $[a, b]$ and $[a', b']$ of S. Consider, in the affine plane determined by D and D', the affine map f such that $f(a) = u$, $f(b) = v$, $f(a') = u'$ and $f(b') = v'$. This map preserves the length of segments perpendicular to H; thus, if T denotes the trapezoid with vertices a, b, a', b', the trapezoid with vertices u, v, u', v' will be $f(T)$. But $T \subset S$ because S is convex, so $f(T) \subset \mathrm{st}_H(S)$, and, in particular, $[x, y] \subset \mathrm{st}_H(S)$.   □

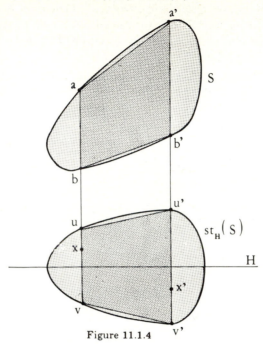

Figure 11.1.4

11.1.5. POLAR BODY OF A CONVEX SET. DUALITY

11.1.5.1. Definition. *Let A be an arbitrary subset of a Euclidean vector space X. The polar body of A is the set*

$$A^* = \big\{\, y \in X \mid (x \mid y) \le 1 \text{ for any } x \in A \,\big\}.$$

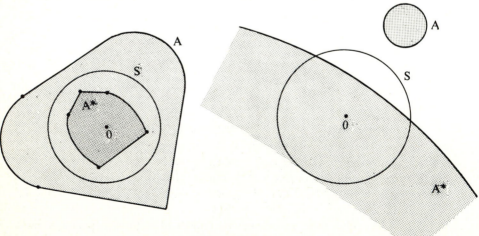

Figure 11.1.5

By 11.1.2.7, A^* is convex for every A. Definition 11.1.5.1 is intimately connected with the polarity transformation with respect to the unit sphere $S = S(0, 1)$ of X; in fact, the polar hyperplane of $x \in X$ is exactly the set $\{ y \in X \mid (x \mid y) = 1 \}$, cf. 10.7.11 and 15.5. The reader is encouraged to draw various sets A and find the corresponding polar bodies A^*, as in figure 11.1.5. We shall see in 11.4.8 that, if A ranges over convex compact sets whose interior contains the origin, the map $A \mapsto A^*$ is a duality with excellent properties, which will be abundantly used in the study of polyhedra (chapter 12).

11.1.6. CONVEX CONES. This is an important notion: a convex set C is called a *convex cone with vertex* x if C is invariant under all homotheties $H_{x,\lambda}$ of center x and ratio $\lambda_{\mathbf{R}_+}^*$. We won't have the time to study this notion, but see, for example, [BI3, 46] or [VE].

11.1.7. CONVEXITY CRITERION. We first mention a simple but useful property of convex sets:

11.1.7.1. Proposition. *Let S be a convex set in a Euclidean affine space X, and take $x \in X$. There exists at most one point $y \in S$ such that $d(x, y) = d(x, S)$ (it is obvious that there exists at least one if S is closed and non-empty).*

Proof. This is a consequence of 9.2.2 and the following lemma:

11.1.7.2. Lemma. *Let S be a convex set in X, and let $x \in X$ and $y \in S$ be such that $x \neq y$ and $d(x, S) = d(x, y)$. If we denote by H the hyperplane containing y and orthogonal to $\overrightarrow{xy}$, the closed half-space determined by H and not containing x contains the whole set S. This is still true if S is merely star-shaped at y.*

Proof. By contradiction: let $z \in X$ be a point not belonging to that half-space. The angle $\overrightarrow{yx}, \overrightarrow{yz}$ is acute, and there exist on $[y, z]$ points t such that $d(x, t) < d(x, y)$. $\square$

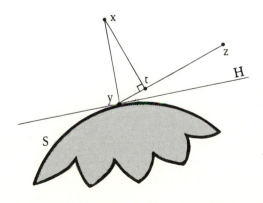

Figure 11.1.7

The so-called Motzkin theorem states that this property characterizes convex sets:

11.1.7.3. Theorem. *Let S be a non-empty closed set in a Euclidean affine space X. Assume that, for any $x \in X$, there exists a unique $y \in S$ such that $d(x, y) = d(x, S)$. Then S is convex.*

For a proof, see [VE, 94]. In the same book there is a discussion of the points of S whose distance to x is maximal (p. 98 ff.).

11.1.7.4. A second characterization of convex sets shall be given in 11.5.4, using the notion of supporting hyperplane.

Again in [VE] there is a characterization of convex sets by various "local convexity" conditions (p. 48 ff.); this is an example of passing from local to global properties. For more examples of this philosophy, see 9.5.4.5, 12.8, 16.4, 18.3.8.6.

11.1.8. CONVEX HULLS

11.1.8.1. From 11.1.2.2 and 11.1.2.7 it is straightforward to see that, for every subset A of an affine space X, there exists a smallest convex set containing A. This smallest convex set is called the *convex hull* of A, and *denoted by* $\mathcal{E}(A)$.

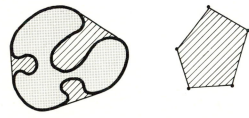

Figure 11.1.8

11.1.8.2. We shall see in 12.1.9 (and using 11.6.8) that the compact polyhedra of X are exactly the convex hulls of finite subsets of X.

11.1.8.3. There are at least two important ways of finding $\mathcal{E}(A)$. One (taking the intersection of all half-spaces containing A) will be studied in 11.5.5; the other employs barycenters:

11.1.8.4. Proposition. *For any subset A of X, the convex hull $\mathcal{E}(A)$ is the set of barycenters of families of points of A with positive masses:*

$$\mathcal{E}(A) = \left\{ \sum_{i \in I} \lambda_i x_i \;\middle|\; x_i \in A, \; \lambda_i \geq 0, \; \sum_i \lambda_i = 1, \; I \text{ arbitrary} \right\}$$

(the notation $\sum_i \lambda_i$ assumes that $\lambda_i = 0$ for all but a finite number of indices).

Proof. Trivial, given definition 11.1.1 and using an induction based on 3.4.9. $\square$

11.1.8.5. It is natural to try to refine 11.1.8.4 in two directions: first, can we limit ourselves to families with a finite number of elements, and, if the answer is yes, can we find an upper bound for that number? Secondly, is it necessary to take the barycenters of all points of A? If not, of which ones? The answer to the second question is afforded by the theorem of Milman–Krein, cf. 11.6.8 (see also 11.2.9). As to the first: any of the figures above will convince the reader that it is necessary to take families I with $\#I = d + 1$. This is also sufficient:

11.1.8.6. Theorem (Carathéodory). *If X is an affine space of dimension d, the convex hull of an arbitrary subset A of X is given by*

$$\mathcal{E}(A) = \left\{ \sum_{i=1}^{d+1} \lambda_i x_i \;\middle|\; x_i \in A,\ \lambda_i \geq 0,\ \sum_{i=1}^{d+1} \lambda_i = 1 \right\}.$$

Proof. Let $x = \sum_{i=1}^{k} \lambda_i x_i$, with $k > d+1$. Take an arbitrary vectorialization of X (cf. 2.1.9). Since $\dim X = d$ and $k > d + 1$, there exist α_i $(i = 1, \ldots, k)$ such that $\sum_{i=1}^{k} \alpha_i x_i = 0$, $\sum_{i=1}^{k} \alpha_i = 0$ and not all the α_i are zero. Set

$$\Theta = \{\, \tau \in \mathbf{R} \mid \tau \alpha_i + \lambda_i \geq 0 \text{ for all } i = 1, \ldots, k \,\}.$$

This is a closed subset of $\mathbf{R}$, non-empty since $0 \in \Theta$, and distinct from $\mathbf{R}$ since the α_i don't all vanish. Let τ be a boundary point of Θ and j an index such that $\tau \alpha_j + \lambda_j = 0$. Then

$$x = \sum_{i=1}^{k} \lambda_i x_i + \tau \sum_{i=1}^{k} \alpha_i x_i = \sum_{i=1}^{k} (\lambda_i + \tau \alpha_i) x_i = \sum_{i \neq j} (\lambda_i + \tau \alpha_i) x_i,$$

and we have managed to express x as the barycenter (with positive masses) of $k - 1$ vectors, since $\sum_{i \neq j} (\lambda_i + \tau \alpha_i) = 1$ by construction. $\square$

11.1.8.7. Corollary. *If A is compact, so is $\mathcal{E}(A)$.*

Proof. The set

$$K = \left\{ (\lambda_1, \ldots, \lambda_{d+1}) \in \mathbf{R}^{d+1} \;\middle|\; \lambda_i \geq 0 \text{ for all } i,\ \sum_{i=1}^{d+1} \lambda_i = 1 \right\}$$

is compact, and 11.1.8.6 shows that $\mathcal{E}(A)$ is the image of $K \times A^{d+1} \subset \mathbf{R}^{d+1} \times X^{d+1}$ under the continuous map

$$(\lambda_1, \ldots, \lambda_{d+1}, x_1, \ldots, x_{d+1}) \mapsto \sum_{i=1}^{d+1} \lambda_i x_i. \qquad \square$$

See 11.9.3 for a refinement of 11.1.8.6; see also 11.2.2. For a very nice application of 11.1.8.6, due to Hilbert, see [EL, especially section 5].

11.1.8.8. *If A is bounded, $\mathrm{diam}\big(\mathcal{E}(A)\big) = \mathrm{diam}(A)$; in particular, $\mathcal{E}(A)$ is still bounded.*

Proof. This is a consequence of 11.8.7.6. $\square$

11.2. The dimension of a convex set

We recall that finite-dimensional affine spaces have a canonical topology (cf. 2.7.1.4).

11.2.1. PROPOSITION. *If S is convex, so is its closure $\overline{S}$.*

Proof. This can be seen in two ways. One is to use the fact that the map $(x, y) \mapsto \lambda x + (1 - \lambda)y$ is continuous, for all λ. Another is to endow the affine space S where X sits with a Euclidean structure; then $\overline{S} = \bigcap_{\epsilon > 0} U(S, \epsilon)$, and we conclude with 11.1.3.3.

11.2.2. On the other hand, just the fact that A is closed does not imply that $\mathcal{E}(A)$ is also closed: see figure 11.2.2, where A is made up of the interior of one branch plus the center of a hyperbola. This justifies the next definition (the proposition is a trivial consequence of 11.2.1):

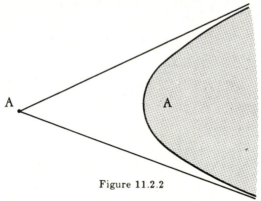

Figure 11.2.2

11.2.3. DEFINITION AND PROPOSITION. *The closed convex hull of a subset A is the intersection of all closed convex sets containing A. The closed convex hull is identical with $\overline{\mathcal{E}(A)}$.* □

11.2.4. LEMMA. *Let S be convex, $x \in \overset{\circ}{S}$ and $y \in \overline{S}$ points. Then $]x, y[$, the open segment with endpoints x and y, lies in $\overset{\circ}{S}$.*

Proof. Take $z \in]x, y[$ and let U be an open set of S containing x (figure 11.2.4). Choose a point y' in the segment $]y, z[$ (possible since $y \in \overline{S}$). The image of U under the homothety of center y' and taking x to z is open and lies entirely in S. □

11.2.5. COROLLARY. *If S is convex, so is $\overset{\circ}{S}$, and one has $\overset{\circ}{S} = \overset{\circ}{\overline{S}}$.* □

It is, of course, false that $\overset{\circ}{S} = \overline{\overset{\circ}{S}}$ for arbitrary sets S (see figure 11.1.2.4, for example). As to the equality $\overline{S} = \overline{\overset{\circ}{S}}$, it may not hold even if S is convex: consider a hyperplane of X, for example.

We now give a simple and nice criterion for the convex set S to have non-empty interior:

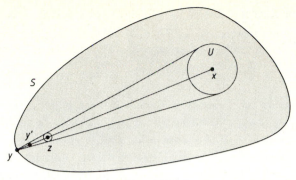

Figure 11.2.4

11.2.6. DEFINITION. *The dimension of a non-empty convex set S, denoted by* dim *S, is the dimension of the affine subspace* $\langle S \rangle$ *spanned by S* (see 2.4.2.5): dim $S = \dim \langle S \rangle$.

11.2.7. PROPOSITION. *If S is a non-empty convex set,* dim $S = \dim X$ *if and only if S has non-empty interior.*

Proof. If $\overset{\circ}{S} \neq \emptyset$ we have dim $S = \dim X$ by 3.5.2, for example. Conversely, let $(x_i)_{i=1,\dots,d+1}$ be a family of affine independent points in S; then

$$\frac{x_1 + \cdots + x_{d+1}}{d+1} \in \overset{\circ}{S}. \qquad \Box$$

11.2.8. We can thus talk about the *relative interior* of a convex set S, the interior of S in the affine subspace $\langle S \rangle$.

11.2.9. PROPOSITION. *If S is a compact convex set,* $S = \mathcal{E}(\operatorname{Fr} S)$.

Proof. Take $x \in S$ and let D be an arbitrary line through x; then $D \cap S$ is a segment $[u, v]$ containing x (cf. 11.1.2.3). Since $u, v \in \operatorname{Fr} S$, we have $x \in \mathcal{E}(\operatorname{Fr} S)$. $\qquad \Box$

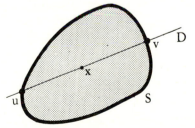

Figure 11.2.9

11.3. Topology of convex sets

In this section we deal with the classification of convex sets up to homeomorphism (what this means will become clear as we go along). We also classify their frontiers.

11.3.1. PROPOSITION. *Let X be a d-dimensional affine space and A a d-dimensional convex subset of X. Then $\overset{\circ}{A}$ is homeomorphic to $\mathbf{R}^d$. In particular, all open, non-empty convex sets in a d-dimensional space are homeomorphic to $\mathbf{R}^d$.*

Proof.

11.3.1.1. We start by giving X a Euclidean vector space structure, with origin $O \in \overset{\circ}{A}$ (this is possible by 11.2.7). Let S be the unit sphere in A. For $y \in S$, let $R(y)$ be the half-line with origin O and containing y; by 11.1.2.3 and 11.1.2.7, $R(y) \cap A$ is an interval, one of whose endpoints is O. If $R(y)$ goes out of A, the other endpoint $f(y)$ of $R(y) \cap A$ is the unique point of $R(y) \cap \mathrm{Fr}(A)$, and we set $\delta(y) = \|f(y)\|$. Otherwise we set $\delta(y) = \infty$, and $f(y)$ is not defined.

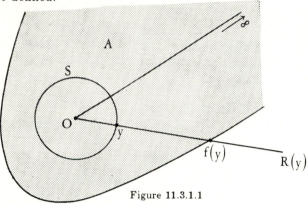

Figure 11.3.1.1

11.3.1.2. *The map $\delta : S \to [0, \infty]$ is continuous.*

By contradiction: assume first that $\delta(y)$ is finite, and let (y_n) be a sequence in S such that $\lim_{n \to \infty} y_n = y$. If it is not true that $\lim_{n \to \infty} \delta(y_n) = \delta(y)$, there exists a subsequence of (y_n), still denoted by (y_n), such that $\delta(y_n) \leq \delta(y) - \eta$ with $\eta > 0$ or $\delta(y_n) \geq \delta(y) + \eta$ with $\eta > 0$. Since $O \in \overset{\circ}{A}$, there exists an open ball $U(0, \alpha)$, for some $\alpha > 0$, contained in A, so A contains the shaded region in figure 11.3.1.2.1, and we get a contradiction if $\delta(y_n) \leq \delta(y) - \eta$. If $\delta(y_n) \geq \delta(y) + \eta$, the contradiction comes from the fact that, by the convexity of A, the shaded region in figure 11.3.1.2.2 is in the interior of A. Finally, if $\delta(y) = \infty$, we use the shaded part of figure 11.3.1.2.3.

11.3.1.3. To conclude the proof, it is enough to recall that the interval $[0, a[\subset \mathbf{R}$, for any $0 < a < \infty$, is homeomorphic to $[0, \infty[$, and that the homeomorphism can be made to vary continuously with a. More precisely, we define a homeomorphism $h : \overset{\circ}{A} \to X$ as follows: if $x = 0$ or if $x \neq 0$ and $\delta(x/\|x\|) = \infty$, put $h(x) = x$; otherwise, put

$$h(x) = \frac{\delta(x/\|x\|) \cdot \|x\|}{\delta(x/\|x\|) - \|x\|} \cdot \frac{x}{\|x\|}. \qquad \square$$

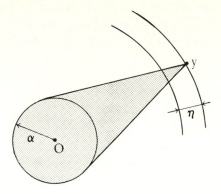

Figure 11.3.1.2.1

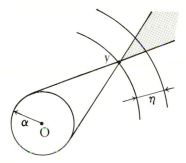

Figure 11.3.1.2.2

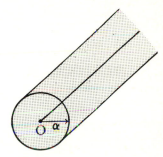

Figure 11.3.1.2.3

11.3.2. NOTE. From this we conclude that the frontier of a convex set has measure zero (cf. 2.7.4.3): on every $R(y)$, the intersection $R(y) \cap \mathrm{Fr}(A)$ is a single point, so it has measure zero; an application of Fubini and 11.3.1.2 proves the result. See another demonstration in 12.9.2.4.

11.3.3. Bearing example 11.1.2.5 in mind, one sees it is useless to try to classify arbitrary convex sets; this explains the restrictions in the statements below. From the proof of 11.3.1 we get:

11.3.4. COROLLARY. *If A is a bounded convex set such that*

$$\dim A = \dim X = d,$$

the frontier $\mathrm{Fr}\,A$ is homeomorphic to the sphere S^{d-1}. If A is also compact, A is homeomorphic to the closed ball of dimension d. In particular, if $d = 2$, $\mathrm{Fr}\,A$ is a simple closed curve.

Proof. The homeomorphism between $\mathrm{Fr}\,A$ and S^{d-1} is the inverse of the map $y \mapsto f(y)$, well-defined because $d(y) < \infty$ for all $y \in S$, and S is always homeomorphic to S^{d-1}.   □

11.3.5. Now assume that A is a convex set of arbitrary dimension, not necessarily the same as that of the ambient space X. Applying 11.3.1 or 11.3.4 to the affine subspace spanned by A, we complete the classification of all open convex sets and all compact convex sets: they are homeomorphic to $\mathbf{R}^{d'}$ and the closed unit ball of $\mathbf{R}^{d'}$, respectively, where d' is the dimension of A.

11.3.6. REMARKS ON STAR-SHAPED SETS.

11.3.6.0. It is tempting to try to make the proof of 11.3.1 work for star-shaped sets (cf. 11.1.2.4): if $E \subset X$ is star-shaped at $x \in \overset{\circ}{E}$, one can give X a Euclidean vector space structure and consider the unit sphere S centered at x and the half-lines $R(y)$, for $y \in S$. Here, however, the endpoint $f(x)$ of the segment $R(y) \cap E$ is not the unique point of $R(y) \cap \mathrm{Fr}\, E$; consider the cat's whiskers in 11.2.2.4. The *cocoon* of the star-shaped set E is the set $\{\, f(y) \mid y \in S, \delta(y) < \infty \,\}$; this set differs from $\mathrm{Fr}\, E$ if E is not convex, as we have just seen. Besides (figure 11.2.2.4 or 11.3.6.0.1) the function δ is no longer necessarily continuous, even if the cocoon is compact (in figure 11.3.6.0.3 the cocoon consists of two points).

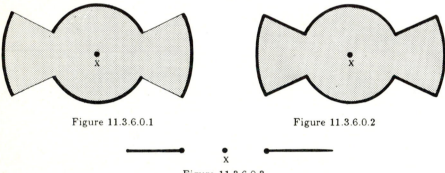

Figure 11.3.6.0.1   Figure 11.3.6.0.2

Figure 11.3.6.0.3

On the other hand, in all the drawings above the star-shaped sets do seem to be homeomorphic to $\mathbf{R}^d$. This is not a coincidence:

11.3.6.1. Theorem. *Every open star-shaped set in X is homeomorphic to X.*

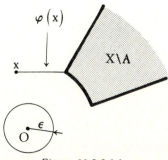

Figure 11.3.6.1.1

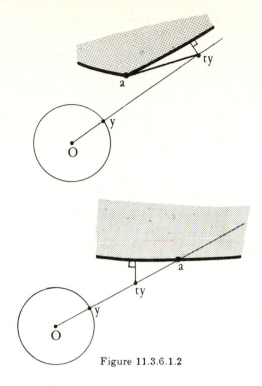

Figure 11.3.6.1.2

Proof. Vectorialize X at the center $O = x$ of the star-shaped set A. If $X = A$ there is nothing to prove. If $\operatorname{Fr} A \neq \emptyset$ define the function

$$\phi : A \ni x \mapsto \phi(x) = d(x, \operatorname{Fr} A) \in \mathbf{R}_+^*.$$

It is obvious that ϕ is continuous. Define a map $F : A \to X$ by setting

11.3.6.2 $$F(O) = O \quad \text{and} \quad F(x) = \left(\int_0^{\|x\|} \frac{dt}{\phi(t(x/\|x\|))} \right) \cdot \frac{x}{\|x\|}.$$

The map F is the desired homeomorphism. It is enough to show that F is continuous, bijective and proper (that is, the inverse image of any compact set is compact).

Continuity: for $x \neq O$ this follows from 11.3.6.2 and the continuity of ϕ. At O, recall that, since A is open, there exist $\epsilon > 0$ and $k > 0$ such that $\phi(x) \geq k$ for all $x \in U(O, \epsilon)$.

Next we show that, for all $y \in S$,

11.3.6.3 $$\int_0^{\delta(y)} \frac{dt}{\phi(ty)} = +\infty,$$

where the notation is the same as in the proof of 11.3.1. We distinguish two cases: if $\delta(y) = +\infty$, let a be an arbitrary point of $\operatorname{Fr} A$. Then $\phi(ty) =$

$d(ty, \mathrm{Fr}\, A) \leq d(ty, a) \leq t + \|a\|$, whence

$$\int_0^\infty \frac{dt}{\phi(ty)} \geq \int_0^\infty \frac{dt}{t + \|a\|} = +\infty.$$

If $\delta(y) = k < +\infty$, we have $\phi(ty) \leq d(ty, ky) = k - t$, and again

$$\int_0^k \frac{dt}{\phi(ty)} \geq \int_0^k \frac{dt}{k - t} = +\infty.$$

To show that F is proper, it is enough to see that the inverse image of a bounded set is bounded, since F is continuous and X is a finite-dimensional vector space. But this follows from the lower bounds given above for $\phi(ty)$: if $\delta(y) = +\infty$ because the bound is uniform in y, and if $\delta(y) < +\infty$ because $\int_0^s \frac{dt}{\phi(ty)}$ can be bounded from below by a uniformly continuous function of y (constructed from $\delta(y)$).

11.3.6.4. Remark. In fact A is even diffeomorphic to X. See also exercise 11.9.6.

11.3.6.5. Corollary. *Let X be a d-dimensional Euclidean affine space, S the unit sphere centered at $O \in X$ and C a closed convex cone with vertex O (cf. 11.1.6). Assume there exists $y \in S \cap C$ with $-y \notin S \cap C$. Then $S \setminus C$ is homeomorphic to $\mathbf{R}^{d-1}$.*

Proof. Set $M = S \cap C$ and $y' = -y$. For any half-great circle γ of S with endpoints y and y' we have $y \in \gamma \cap M$ and $y' \notin \gamma \cap M$; also, $\gamma \cap M$ is convex because C is convex. Thus $\gamma \cap (S \setminus M)$ is a half-open arc $]y', \cdot [$ of γ. In particular, denoting by σ the stereographic projection (cf. 18.1.4.3) with pole y from S onto the hyperplane H tangent to S at y', we conclude that $\sigma(S \setminus M)$ is a subset of H star-shaped at y', and open since C is closed. 11.3.6.5 now follows from 11.3.6.1 because of 18.1.5. $\qquad\square$

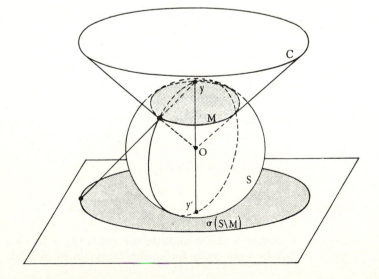

Figure 11.3.6.5

11.3.7. In 11.3.5 we have classified open and compact convex sets. We now must look at unbounded convex sets. Figures 11.2.1.1, 11.2.1.3, 11.2.1.6 and 11.2.1.5 show frontiers homeomorphic to $\mathbf{R} \times \{-1, +1\}$, $\mathbf{R} \times S^1$, $\mathbf{R}$ and S^1, respectively. This phenomenon is general:

11.3.8. PROPOSITION. *Let A be a convex subset of X such that* $\dim A = \dim X = d$ *and* $\operatorname{Fr} A \neq \emptyset$. *Then* $\operatorname{Fr} A$ *is homeomorphic to* $\mathbf{R}^{d-1}$ *or* $S^{d-r-1} \times \mathbf{R}^r$ $(0 \leq r \leq d-1)$.

Proof. Same notation as in 11.3.1.1; if A is bounded, we are done with 11.3.4. Otherwise, set $M = \{\, y \in S \mid \delta(y) = \infty \,\}$. From the convexity of A we know that $C = \bigcup_{y \in M} R(y)$ is a closed convex cone.

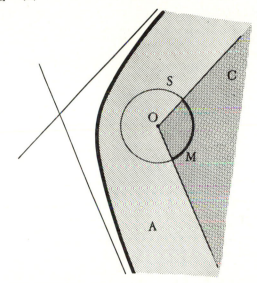

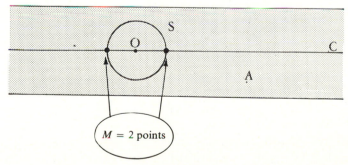

$M = 2$ points

Figure 11.3.8.1

11.3.8.1. First case: $M \cap (-M) = \emptyset$, that is, $y \in M$ implies $-y \notin M$. The set M does not contain antipodal points, or, which is the same, A does not contain entire lines through O. The hypotheses of corollary 11.3.6.5 are satisfied and $\operatorname{Fr} A$ is homeomorphic to $S \setminus M$, hence to $\mathbf{R}^{d-1}$.

11.3.8.2. Second case: there exists $y \in M$ such that $-y \in M$, that is, $A \supset D$, where D is the affine line through O and y. Let C be an affine subspace of maximal dimension contained in A. Its dimension r must be less than d, otherwise $A = X$ and $\mathrm{Fr}\, A = \emptyset$. Let W be a subspace complementary to V; W is an affine subspace of dimension $d - r$ and $W \cap A = B$ is a convex set in W. For $x \in W$, take $y \in V$ and make y go to infinity in V in all possible directions; since $[x, y] \subset A$ and $V \subset A$, an limit argument shows that $\overline{A}$ contains V_x, the affine subspace parallel to V through x. Thus $\overline{A} \supset V \times B$, and the same argument shows that $\overline{A} \subset V \times \overline{B}$. Since $\mathrm{Fr}\, V = \emptyset$, we have

$$\mathrm{Fr}\, A = \mathrm{Fr}\, \overline{A} = \mathrm{Fr}(V \times \overline{B}) = V \times \mathrm{Fr}\, \overline{B} = V \times \mathrm{Fr}\, B.$$

The proposition now follows from 11.3.4 or 11.3.8.1, since the maximality of V implies that no antipodes can occur in B.

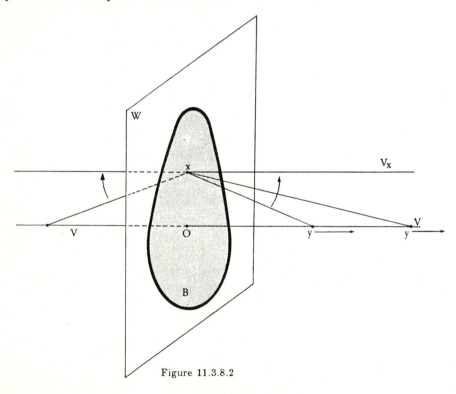

Figure 11.3.8.2

11.3.9. NOTE. The technique used in the demonstration of 11.3.8.2 is quite general, and has recently been utilized in the context of a certain type of Riemannian manifolds ([C–G]). See also [B–B–M].

11.3.10. NOTES

11.3.10.1. A *convex curve* in a Euclidean plane X is, by definition, the boundary of a two-dimensional convex subset of X. Proposition 11.3.8 shows that a convex curve is homeomorphic either to the circle S^1 or to the line $\mathbf{R}$

(compare with the classification of one-dimensional differentiable manifolds, in [B–G, 3.4], for example). The reader can also study, from the present viewpoint, the proof of theorem 9.6.2 in [B–G]; see 11.5.4 below.

11.3.10.2. A *convex surface* in a Euclidean three-dimensional space X is, by definition, a connected set that is the boundary of a convex subset of X. Such a surface is, by 11.3.8, homeomorphic to $\mathbf{R}^2$, $S^1 \times \mathbf{R}$ (the cylinder) or S^2. Convex surfaces have been the object of refined study: see, for example, [PV1], [BU1], [AW1]. For the differentiable case, see [DE4, volume IV, p. 360, problem 3].

11.3.10.3. Again from 11.3.8 it is easy to derive a topological classification of closed convex sets.

11.4. Convex sets and hyperplanes. Separation theorems

This whole section is based on the following result:

11.4.1. THEOREM (HAHN–BANACH). *Let X be an affine space, A an open, non-empty convex subset of X and L an affine subspace of X that does not intersect A. Then there exists a hyperplane of X containing L and not intersecting A.*

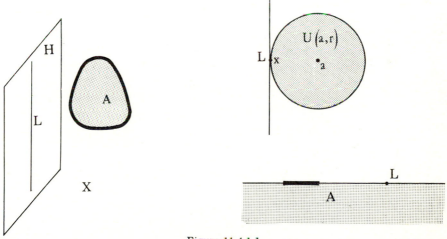

Figure 11.4.1.1

One cannot eliminate the condition that A is open: for instance, take A to be an open half-plane together with a segment in its frontier, and for L a point in the frontier. See also exercise 11.9.9.

A theorem like 11.4.1 is non-trivial; for example, if $A = U(a,r)$ and $L = \{x\}$ is contained in $S(a,r)$, the hyperplane obtained is unique, namely, the hyperplane tangent to $S(a,r)$ at x (cf. 10.7.4).

Proof.

11.4.1.1. The first step is to reduce the problem to the two-dimensional case. Let $M \supset L$ be a subspace of X of maximal dimension satisfying $M \cap A = \emptyset$; we want to show that M is a hyperplane. We first vectorialize X at some point $O \in L$, and let $p : X \to X/M$ be the canonical projection onto the quotient X/M. Then $p(A)$ is open and convex in X/M (11.1.2.6), and $O \notin p(A)$. Let $Z \subset X/M$ be a vector subspace of dimension two intersecting $p(A)$; such a plane exists because $\dim(X/M) \geq 2$, otherwise there is nothing to prove. The intersection $p(A) \cap Z$ is a non-empty, open convex set B.

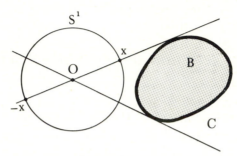

Figure 11.4.1.2.1

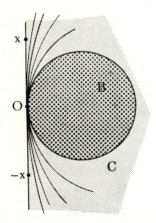

Figure 11.4.1.2.2

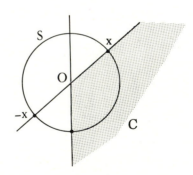

Figure 11.4.1.2.3

11.4.1.2. We now have to show that it is impossible for every line in Z going through O to intersect B (where B is an open convex set in Z not containing O). Heuristically, if we give Z an arbitrary Euclidean structure, the half-lines through O that intersect B determine on the unit circle S^1 of X a subset that cannot contain antipodal points, for a line joining two antipodal points would force $O \in B$. Since B is convex, this subset of S^1 is also connected, so it must be an arc of length less than π, and there is a line that does not intersect it. We now have to make this argument rigorous.

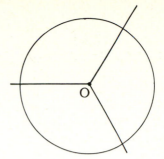

Figure 11.4.1.2.4   Figure 11.4.1.2.5

We introduce the cone $C = \bigcup_{\lambda > 0} \lambda B$; this cone is open, convex (an easy exercise left to the reader), and does not contain O. (In figure 11.4.1.2.2, where B is an open disc whose frontier contains O, C is a half-plane.) There exists at least one frontier point x of C in $Z \setminus O$, otherwise $C \setminus O$ would be a whole connected component of $Z \setminus O$ (this can occur for $\mathbf{R} \setminus O$, but $\dim Z = 2$ forces $Z \setminus O$ connected, cf 8.3.8). Thus $x \notin C$ because C is open; but also $-x \notin C$, otherwise $-x \in \overset{o}{C}$ and $x \in \overline{C}$ would imply $O \in \overset{o}{C}$ by 11.2.4. But now the line D through x and O is entirely outside C, hence outside B, and we've found our line.   □

11.4.2. Hahn–Banach's theorem has fundamental applications in functional analysis, based on the correspondence between hyperplanes and linear forms: see, for example, [LG2, 186], [M–T, 23], [BI3, 65]. In this text, however, we deal only with the "geometric" applications of the Hahn–Banach theorem.

11.4.3. DEFINITION. *Let X be an affine space, A and B subsets of X and H a hyperplane. We say that H separates A and B if A and B lie in different half-spaces determined by H. In addition, if A and B do not intersect H, they are strictly separated by H.*

11.4.4. COROLLARY. *Consider, in an affine space X, two non-empty convex sets A and B, with A open and $A \cap B = \emptyset$. There exists a hyperplane separating A and B.*

Proof. Vectorialize X and define $C = A - B$ as in 11.1.3.1; C is a convex set, open (being the union of open sets) and $0 \notin C$ because $A \cap B = \emptyset$. Now apply 11.4.1 to C and the affine subspace $\{0\}$.   □

Since H is the frontier of each of the subspaces determined by it, we have:

11.4.5. COROLLARY. *If A and B are disjoint, open, non-empty convex sets, there exists a hyperplane strictly separating A and B.*   □

That would be false if the two sets were closed (figure 11.4.5). On the other hand, if one is closed and the other compact, we have:

11.4.6. COROLLARY. *If A, B are disjoint convex sets, A is closed and non-empty and B is compact, there exists a hyperplane strictly separating A and B.*

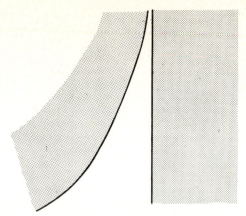

Figure 11.4.5

Proof. From point-set topology, there exists an open ball $U(a, \epsilon)$, with $\epsilon > 0$, such that $\big(A + U(a, \epsilon)\big) \cap \big(B + U(a, \epsilon)\big) = \emptyset$; but these two sets satisfy the hypotheses of 11.4.5. □

In the case of two general closed convex sets, all we can say is this:

11.4.7. COROLLARY. *If A and C are closed convex sets, there is a hyperplane separating A and C.*

Proof. Consider a Euclidean structure on the affine space; take $a \in A$ and consider the closed ball $B(a, n)$, for $n \in \mathbf{N}$. By assumption, the convex sets $A \cap B(a, n)$ and C satisfy the hypotheses of 11.4.6 for any n; let H_n be a hyperplane separating those two sets. It is enough now to show that one can find a subsequence of the H_n converging toward a hyperplane. We first find a subsequence such that the orthogonal directions to the hyperplanes converge; this is possible because the projective space of lines going through a is compact (4.3.3). From this sequence we can now extract a convergent subsequence by considering a convergent subsequence of the sequence of points $H_n \cap [a, c]$, where $c \in C$ is an arbitrary point (remember that $[a, c]$ is compact).

We can now study the polarity operation introduced in 11.1.5:

11.4.8. PROPOSITION. *Let X be a Euclidean vector space with origin 0.*

i) *If A is bounded, $0 \in \overset{\circ}{A^*}$; if $0 \in \overset{\circ}{A}$, the set A^* is bounded.*

ii) *If A is a convex closed set containing 0, we have $A^{**} = A$.*

Proof. Assume $A \subset B(0, r)$ for $r > 0$. Then $A^* \supset \big(B(0, r)\big)^* = B(0, r^{-1})$. Similarly, if there exists $r > 0$ such that $B(0, r) \subset A$, we have $A^* \subset \big(B(0, r)\big)^* = B(0, r^{-1})$.

Let A be a closed convex set containing 0; definition 11.1.5.1 gives the inclusion $A \subset A^{**}$. Take $a \notin A$; by 11.4.6 there exists a hyperplane H that strictly separates A and a, so in particular $0 \notin H$. Let h be the pole of H (cf. 10.7.11), that is, $H = \{ z \in X \mid (z \mid h) = 1 \}$; then $(a \mid h) > 1$ and $(x \mid h) \leq 1$ for every $x \in A$, so $a \notin A^{**}$, and $A^{**} \subset A$. □

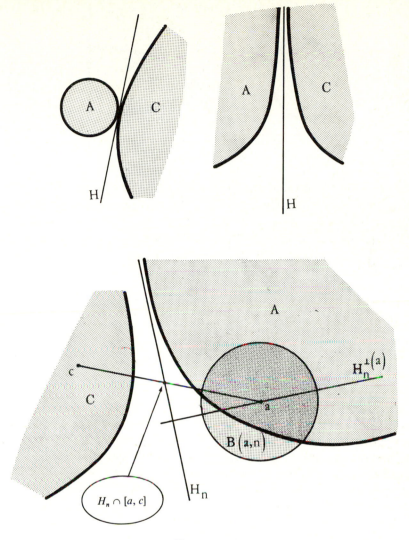

Figure 11.4.7

11.5. Supporting hyperplanes; applications

An important case of separation is when A is convex and B is a point outside A. We're led to the following definition:

11.5.1. DEFINITION. *Let A be an arbitrary subset of an affine space X. A supporting hyperplane for A is any hyperplane H containing a point $x \in A$ and separating $\{x\}$ and A. We say that H is a supporting hyperplane for A at x.*

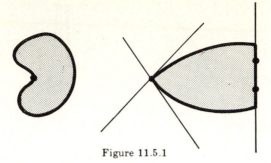

Figure 11.5.1

By the above and the remark preceding 11.4.5, a supporting hyperplane H for a convex set A satisfies $H \cap \overset{\circ}{A} = \emptyset$. On the other hand, if H is a supporting hyperplane at x, we have $x \in \operatorname{Fr} A$.

The figures above show that supporting hyperplanes may not exist at all points of the frontier, that they may not be unique at any given point, and that they may be supporting at more than one point. This will be made precise in 11.6.

11.5.2. PROPOSITION. *Let A be a closed convex set. Then A has a supporting hyperplane at any point of its frontier.*

Proof. A point x in the frontier satisfies $\{x\} \cap \overset{\circ}{A} = \emptyset$; just apply theorem 11.4.1 with $L = \{x\}$.   □

Using the polarity transformation of 11.1.5.1, one can show that there is a duality between points of $\operatorname{Fr} A$ and supporting hyperplanes for A:

11.5.3. PROPOSITION. *Let X be a Euclidean vector space with origin 0, and A a closed convex subset of X containing 0 in its interior. The polar hyperplanes of points of $\operatorname{Fr} A$ make up the set of supporting hyperplanes of A^*. Moreover, if $x \in \operatorname{Fr} A$ has polar hyperplane H, the contact points of H are the poles of the supporting hyperplanes for A at x.*

Figure 11.5.3

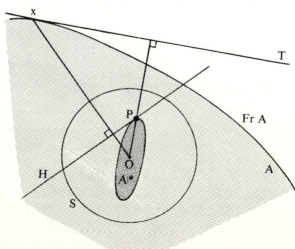

Proof. Take $x \in \operatorname{Fr} A$. By definition 11.1.5.1, A^* is contained in the half-space determined by H, the polar hyperplane of x (cf. 10.7.11). Similarly, if T is a supporting hyperplane for A at x, with pole p, we have $p \in H$ (cf. 10.7.11) and $p \in A^*$ because $A^{**} = A$ and A lies in the half-space determined by H and containing 0. Thus H is indeed a supporting hyperplane for A^* at p. Since $A^{**} = A$, one obtains in this way all the supporting hyperplanes for A^* (cf. 11.4.8). □

We can now fulfil the promise made in 11.1.7.4:

11.5.4. PROPOSITION. *Let A be a closed set with non-empty interior. If A has a supporting hyperplane at every point of its frontier, A is convex.*

Proof. We argue by contradiction. Take $x \in \overset{\circ}{A}$, and assume there exist points $y, z \in A$ and $t \in [y, z]$ such that $t \notin A$. Since $t \notin A$ and $x \in \overset{\circ}{A}$, the segment $[x, t]$ intersects $\operatorname{Fr} A$ in at least one point $u \in \,]x, t[$. Let H be a supporting hyperplane for A at u; its intersection with the affine plane spanned by x, y, z gives a line containing u. Now u is in the interior of the triangle $\{x, y, z\}$, so every line through u strictly separates either x and y or x and z. Contradiction. □

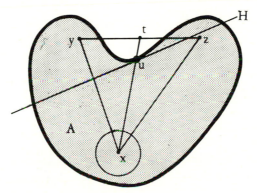

Figure 11.5.4.1

Figure 11.5.4.2

The non-empty interior condition is essential: see figure 11.5.4.2. The reader should compare the demonstration above with the one in [B–G, 9.6].

We can also state a converse:

11.5.5. PROPOSITION: *Let A be a closed convex set. Then A is the intersection of the closed half-spaces containing A.*

Proof. Call that intersection $\mathcal{E}'(A)$. Then $\mathcal{E}'(A)$ is a closed convex set (cf. 11.1.2.7) containing A. Assume there is $x \in \mathcal{E}'(A) \setminus A$: a contradiction is obtained by applying 11.4.6 to A and $B = \{x\}$.   □

11.5.6. REMARKS. There exists a more elementary way, without using Hahn–Banach, to find supporting hyperplanes for A (without specifying the contact point):

11.5.6.1. *Let A be a compact subset of X and V a hyperplane direction of X (cf. 2.4.1). There exists at least one supporting hyperplane for A parallel to V.*

Proof. Let $p : X \to X/V$ be the projection of X onto the quotient (cf. 2.2.4); $p(A)$ is a compact subset of the affine line X/V. If α, β are its extremities, which possibly coincide, $p^{-1}(\alpha)$ and $p^{-1}(\beta)$ are hyperplanes parallel to V, and are supporting at every point of their intersection with A.   □

11.5.6.2. This also shows that there exist at least *two* such hyperplanes if $\overset{\circ}{A} \neq \emptyset$. On the other hand, if A is closed but not compact, such hyperplanes may not exist: figure 11.5.6.2.

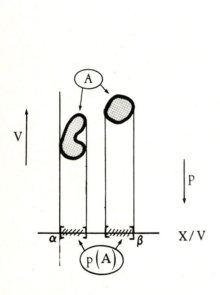

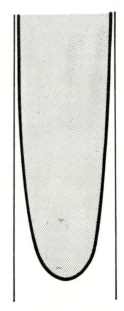

Figure 11.5.6.1   Figure 11.5.6.2

11.5.6.3. If X is also Euclidean, X/V has a natural Euclidean structure, and the length (that is, the diameter) of the compact set $p(A)$ is called the *width* of A in the direction $\xi = V^{\perp}$, and denoted by $\mathrm{wid}_{\xi} A$. Convex sets with same width in all directions have been studied by many people, but there are still many open problems on the subject, even in dimension two. In 12.10.5 we shall calculate the length of a curve of constant width; the reader can find drawings and references there.

11.5.6.4. A notion related to supporting hyperplanes is that of supporting function: see 11.8.12.3.

11.5.7. The notion of convexity allows one to state results about the smallest ball containing a given compact set in a Euclidean space:

11.5.8. THEOREM (JUNG). *Let A be a compact subset of a d-dimensional affine space X. There is a unique ball $B(x, r)$ of minimal radius containing A. Furthermore, we have $x \in \mathcal{E}(A \cap S(x, r))$ and*

$$r \leq \sqrt{\frac{d}{2(d+1)}} \cdot \operatorname{diam}(A),$$

and this is the sharpest possible such inequality.

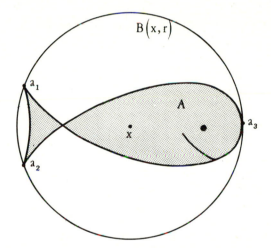

Figure 11.5.8.1

Proof. For every $t \geq 0$, set $Y_t = \{ x \in X \mid B(y, t) \supset A \}$. Since A is bounded, the set $T = \{ t \in \mathbf{R} \mid Y_t \neq \emptyset \}$ is non-empty; observe also that $t \leq t'$ implies $Y_t \subset Y_{t'}$ and that the Y_t are compact. Since a decreasing sequence of compact sets has non-empty intersection (see 0.4), we have $Y_r = \bigcap_{t > r} Y_t \neq \emptyset$.

If we had $x, y \in Y_r$ with $x \neq y$, we would conclude, from figure 11.5.8.2:

$$\sqrt{r^2 - \frac{\overline{xy}^2}{4}} < r \quad \text{and} \quad \sqrt{r^2 - \frac{\overline{xy}^2}{4}} \in T,$$

whence $Y_r = \{x\}$, a contradiction.

Now vectorialize X at x. Let $u \in S(x, 1)$ and $\epsilon > 0$ be arbitrary; by the definition of x and r, there exists $a \in A$ such that

$$d(x, a) \leq r < d(a, \epsilon u).$$

Since A is compact, we can take the limit as $\epsilon \to 0$ to obtain, for every $u \in S(0, 1)$, a point $a \in A$ such that $(a \mid u) \leq 0$ and $d(x, a) = r$. This shows

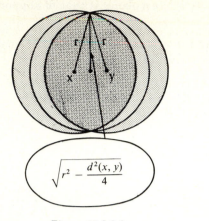

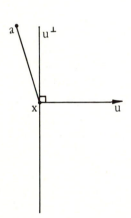

$$\sqrt{r^2 - \frac{d^2(x, y)}{4}}$$

Figure 11.5.8.2

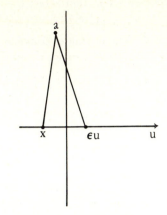

Figure 11.5.8.3

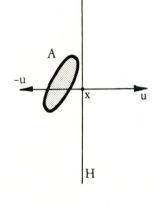

Figure 11.5.8.4

that $x \in \mathcal{E}\big(A \cap S(x, r)\big)$, for otherwise there would be a hyperplane strictly separating x and $A \cap S(x, r)$ (by 11.4.6), contradicting the result above applied to the vectors u and $-u$:

$$\{u\} \cup \{-u\} = S(x, 1) \cap H^{\perp}.$$

To show the inequality in the statement, we use Carathéodory's theorem (11.1.8.6) to write $x = \sum_{i=1}^{d+1} \lambda_i a_i$ with $\lambda_i \geq 0$, $\sum_i \lambda_i = 1$ and each a_i in $S(x, r) \cap A$. In particular, $d(a_i, a_j) \leq \delta = \operatorname{diam} A$ for every $i \neq j$. For a fixed i, we can write $1 - \sum_{j \neq i} \lambda_j = \lambda_i$, whence

$$(1 - \lambda_i)\delta^2 \geq \sum_{j \neq i} \lambda_j d^2(a_i, a_j) = \sum_j \lambda_j d^2(a_i, a_j),$$

since $d(a_i, a_i) = 0$. But

$$d^2(a_i, a_j) = \|a_i\|^2 + \|a_j\|^2 - 2(a_i \mid a_j) \geq 2r^2 - 2(a_i \mid a_j),$$

whence

$$(1 - \lambda_i)\delta^2 \geq \left(\sum_j \lambda_j\right) 2r^2 - 2\left(a_i \mid \sum_j \lambda_j a_j\right) = 2r^2 - 2(a_i \mid x) = 2r^2$$

(remember we vectorialized X at x, so $x = 0$!). Adding over i we get

$$\sum_i (1 - \lambda_i)\delta^2 = d\delta^2 \geq 2(d+1)r^2,$$

as desired. Equality can only obtain if $d(a_i, a_j) = \delta$ for every $i \neq j$, which is the case for the *regular simplex* in the d-dimensional hyperplane $\sum_i \lambda_i = 1$:

$$S_d = \left\{ (\lambda_1, \ldots, \lambda_{d+1}) \in \mathbf{R}^{d+1} \mid \lambda_i \geq 0 \text{ for all } i \text{ and } \sum_i \lambda_i = 1 \right\}$$

(cf. 12.1.2.5).

11.5.9. NOTES.

11.5.9.1. In the bargain, theorem 11.5.8 gives the value of the radius of the sphere circumscribed around the regular simplex S_d (cf. 9.7.3.7). See also 11.9.18.

11.5.9.2. Equality in 11.5.8 can be achieved for sets A other than regular simplices, for example, the Reuleaux triangle (figure 12.10.5.1).

11.5.9.3. Observe that the point x obtained, although canonically associated with A, differs from its centroid (cf. 2.7.5.6).

11.5.9.4. Observe also that one may have $\#(S(x,r) \cap A) < d + 1$, that is, some of the points a_i in the proof may coincide (figure 11.5.9).

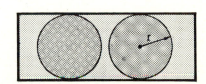

Figure 11.5.9

11.5.9.5. One can also study balls of maximal radius contained in a compact set A; they always exist, but they are not unique. There exists an equality for this maximal radius, involving the minimal width of A: see 11.9.12 or [EN, 112].

11.6. The frontier of a convex set

We shall now examine the different kinds of frontier points a convex set can have.

11.6.1. DEFINITION. *Let A be a closed convex set in a d-dimensional space X, and let x be a frontier point of A. We say that x has order α if the intersection of all supporting hyperplanes to A at x is an affine subspace of dimension α. If x has order 0 we say it is a vertex of A; on the other hand, if α = d − 1, that is, if A has only one supporting hyperplane at x, we say that A is smooth at x.*

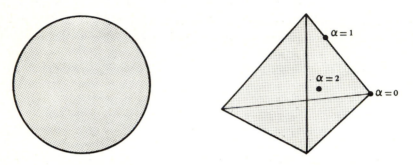

Figure 11.6.1.1   Figure 11.6.1.2

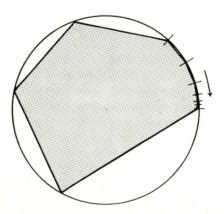

Figure 11.6.1.3

In a circle (figure 11.6.1.1) all frontier points are smooth, and there are no vertices. A d-dimensional simplex has frontier points of every order up to d−1 (figure 11.6.1.2); the same is true of any d-dimensional polytope (12.1.9). Figure 11.6.1.3 shows a convex set with an infinite number of vertices, but no convex set can have more than a countable number of them:

11.6.2. PROPOSITION. *Any convex set has only a countable number of vertices.*

Proof. We start by giving X a Euclidean structure. For any $x \in \mathrm{Fr}\, A$, we define CN_x, the *normal cone* to A at x, as the union of half-lines originating at x, orthogonal to supporting hyperplanes to A at x, and lying on the far side of such hyperplanes with respect to A. In symbols,

$$CN_x = \{\, y \in X \mid (\overrightarrow{xy} \mid \overrightarrow{xz}) \leq 0 \text{ for any } z \in A \,\}.$$

If A is smooth at x, the cone CN_x is just a half-line; on the other hand, saying that x is a vertex is equivalent to saying that CN_x has non-empty interior. If $x, y \in \mathrm{Fr}\, A$ are distinct, the cones CN_x and CN_y cannot intersect, for if they intersected at z, say, the triangle $\{x, y, z\}$ would have two angles $\geq \pi$ (by assumption, $(\overrightarrow{xw} \mid \overrightarrow{xz}) \leq 0$ for any $w \in A$, hence for $w = y \in \overline{A}$; and similarly $(\overrightarrow{yx} \mid \overrightarrow{yz}) \leq 0$). But by a classical argument, involving a countable dense subset, one can only fit a countable number of disjoint sets with non-empty interior in X. $\square$

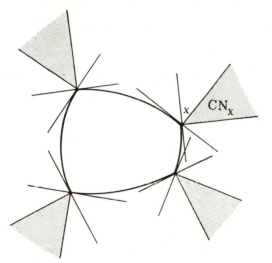 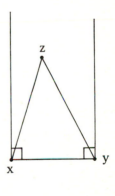

Figure 11.6.2

11.6.3. For a sharper theorem on vertices, see [VE, 136].

11.6.4. DEFINITION. *Let A be convex. A point $x \in \mathrm{Fr}\, A$ is called* exposed *if there exists a supporting hyperplane H at x such that $H \cap A = \{x\}$. A point $x \in A$ is called* extremal *if the equality $x = (y + z)/2$ with $y, z \in A$ implies $y = z$. A convex set is called* strictly convex *if all its frontier points are exposed.*

11.6.5. NOTES.

11.6.5.1. A vertex is always extremal, but the converse is false (see figure 11.6.5.1). But we shall see in 12.1.9 that the converse is true for polyhedra.

11.6.5.2. An exposed point is extremal, but the converse is false (figure 11.6.5.2).

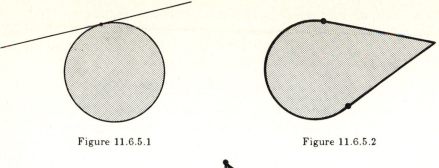

Figure 11.6.5.1   Figure 11.6.5.2

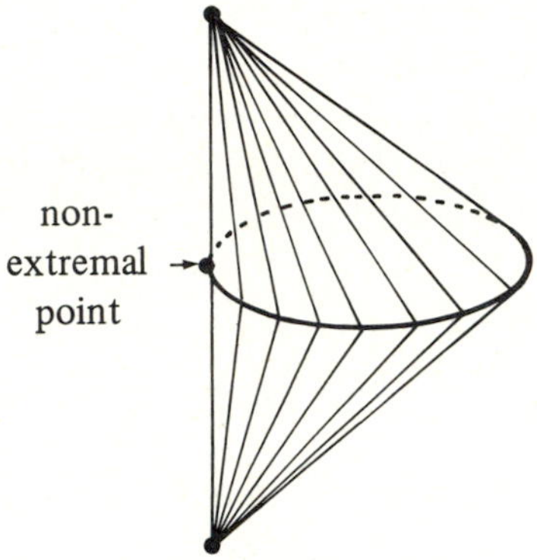

non-
extremal
point

Figure 11.6.5.3

11.6.5.3. The set of extremal points of a compact convex set is not necessarily closed (figure 11.6.5.3), unless $d = 2$ (exercise 11.9.8).

11.6.5.4. If x is extremal, $x \in \operatorname{Fr} A$. Saying that x is extremal is equivalent to saying that $A \setminus x$ is still convex: see figures 11.1.2.7 and 11.1.2.8.

11.6.5.5. One of the reasons for interest in extremal points is proposition 11.8.10.9.

11.6.6. EXAMPLES.

11.6.6.1. The extremal points of a compact interval are its endpoints.

11.6.6.2. Extremal points (in infinite-dimensional spaces) play an important role in measure theory: see [DE4, vol. II, p. 151, problem 8].

11.6.6.3. Extremal points are also encountered in applied math: for example, the set of bistochastic matrices is convex, and its extremal points are the permutation matrices (see 11.9.7, [KE, chapter 21] and [R–V, chapter V]).

11.6.7. We can now answer the question posed in 11.1.8.5:

11.6.8. THEOREM (KREIN–MILMAN). *A compact convex set is the convex hull of its extremal points.*

Proof. For every convex set A, denote by $\operatorname{Extr} A$ the set of extremal points of A. Definition 11.6.4 easily implies that

$$\operatorname{Extr}(A \cap H) = (\operatorname{Extr} A) \cap H$$

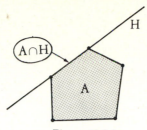

Figure 11.6.8

for every hyperplane H of A; this will be used to prove 11.6.8 by induction on the dimension d of X. The theorem is true for $d = 1$ by 11.6.6.1; assume it is true for $d - 1$. By 11.2.9, the assertion $A = \mathcal{E}(\operatorname{Extr} A)$ will follow from $\operatorname{Fr} A \subset \mathcal{E}(\operatorname{Extr} A)$. Take $x \in \operatorname{Fr} A$ and let H be a supporting hyperplane to A at x (cf. 11.5.2). Since $A \cap H$ is a convex set in H we have, by the induction hypothesis,

$$x \in H \cap A = \mathcal{E}\left(\operatorname{Extr}(A \cap H)\right) = \mathcal{E}\left((\operatorname{Extr} A) \cap H\right)$$
$$= \mathcal{E}(\operatorname{Extr} A) \cap H \subset \mathcal{E}(\operatorname{Extr} A). \qquad \square$$

11.6.9. For other results on frontier points, see [VE, 138–139].

11.7. Helly's theorem and applications

This very geometric theorem has many spectacular applications. It is interesting to observe that it was not proven until 1921, and that Krasnosel'skii's was proved even more recently: 1946.

11.7.1. THEOREM (HELLY). *Let X be a d-dimensional affine space and $\mathcal{F}$ a family of convex subsets of X with more than $d + 1$ elements. If $\mathcal{F}$ satisfies the following two conditions:*
  i) *the intersection of any $d + 1$ sets in $\mathcal{F}$ is non-empty,*
  ii) *$\mathcal{F}$ is finite or all the elements of $\mathcal{F}$ are compact,*
then the intersection of all the sets in $\mathcal{F}$ is non-empty.

11.7.2. REMARKS. The convexity condition is necessary: figure 11.7.2.1, where the elements of $\mathcal{F}$ are arcs of circles. The number $d + 1$ in condition (i) is also best possible: figures 11.7.2.2 and 11.7.2.3, where the sets are faces and edges, respectively. Condition (ii) is necessary: take

$$\mathcal{F} = \{[n, \infty[\mid n \in \mathbf{N}\}$$

(figure 11.7.2.4). Finally, notice that, in some sense, theorem 11.7.1 is a constructive one.

11.7.3. PROOF OF HELLY'S THEOREM

11.7.3.1. **First case:** $\mathcal{F}$ is finite and all its elements are compact. An elementary set-theoretical argument reduces the problem to the case $\#\mathcal{F} =$

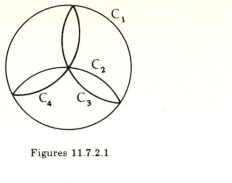

Figures 11.7.2.1

Figure 11.7.2.2

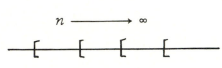

Figures 11.7.2.3

Figure 11.7.2.4

$d + 2$. We then apply induction on $d = \dim X$; the result is trivial if $d = 0$. Let $\mathcal{F} = \{C_1, \ldots, C_{d+2}\}$, and assume that $\bigcap_{i=1}^{d+2} C_i = \emptyset$; by condition (i) we must have

$$A = C_1 \cap \cdots \cap C_{d+1} \neq \emptyset \qquad \text{and} \qquad A \cap C_{d+2} = \emptyset.$$

Let H be a hyperplane strictly separating A and C_{d+2} (cf. 11.4.6), and consider the $d+1$ subsets $H \cap C_1, \ldots, H \cap C_{d+1}$ of H. They satisfy (i) in dimension $d - 1$, because the intersection of any d of the sets $C_1, \ldots, C_{d+1}$ meets both A and C_{d+2} (the latter by the assumptions on $\mathcal{F}$), and so meets H by convexity. By the induction hypothesis we then have $H \cap C_1 \cap \cdots \cap C_{d+1} = \emptyset$, contradicting the choice of H.

11.7.3.2. Second case: $\mathcal{F}$ is not finite but all its elements are compact. By 11.7.3.1 we know that every finite subfamily of $\mathcal{F}$ has non-empty intersection; by a simple result from point-set topology the intersection of $\mathcal{F}$ itself is also non-empty. We recall the proof: suppose that for any $x \in F_1$ there exists $F_x \in \mathcal{F}$ such that $x \notin F_x$. Choose an open neighborhood U_x of x such that $U_x \cap F_x = \emptyset$. By compactness, we can cover F_1 with finitely many open sets $U_{x_1}, \ldots, U_{x_n}$, contradicting the assumption that $F_1 \cap F_{x_1} \cap \cdots \cap F_{x_n} \neq \emptyset$.

11.7.3.3. Third case: The sets in $\mathcal{F}$ are not necessarily compact but $\mathcal{F}$ is finite. As in 11.7.3.1, we can reduce the problem to the case $\#\mathcal{F} = d + 2$. Let $\mathcal{F} = \{C_1, \ldots, C_{d+2}\}$ be such that every subfamily with $d+1$ elements has non-empty intersection. We shall construct compact sets $K_i \subset C_i$ ($i = 1, \ldots, d+2$) such that the intersection of every subfamily with $d + 1$ elements is still non-empty; the result will then follow from 11.7.3.1. The construction of K_1 is as follows: by assumption, there exists $p_i \in \bigcap_{j \neq i} C_j$ ($i = 2, \ldots, d + 2$); we set $K_1 = \mathcal{E}(p_2, \ldots, p_{d+2})$. Clearly $K_1 \subset C_1$ and the family $\{K_1, C_2, \ldots, C_{d+2}\}$

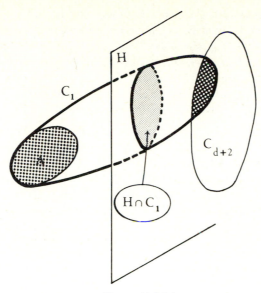

Figure 11.7.3.1

still satisfies (i), since $K_1 \cap \bigcap_{\substack{j \geq 2 \\ j \neq i}} C_i \ni p_i$ for every $i = 2, \ldots, d+2$. The sets K_i $(i = 2, \ldots, d+2)$ are constructed similarly.   □

11.7.4. REMARKS.

11.7.4.1. For a linear algebra proof of the third case, due to Radon, see 11.9.11.

11.7.4.2. There is a close link between Helly's and Carathéodory's theorems: see [EN, 39]. See also a third proof of Helly's theorem in [VE, 72–73].

11.7.5. COROLLARY. *Consider, in an affine plane X, a finite family $\mathcal{F}$ of disjoint parallel segments such that any three of them enjoy a common secant. Then the whole family $\mathcal{F}$ has a common secant.*

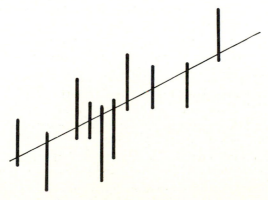

Figure 11.7.5

Proof. We first deal with the case where two segments lie on the same line: then all the segments must lie on that line, and we're done. Otherwise, take coordinates in X so that the y-axis lies in the direction of the lines. For $S \in \mathcal{F}$, set $S' = \{ (\alpha, \beta) \in \mathbf{R}^2 \mid$ line $y = \alpha x + \beta$ intersects $S \}$. It is immediate to verify that S' is convex. But, by assumption, any three sets S' have a common point, so they all have a common point (α, β), and we've found a line $y = \alpha x + \beta$ intersecting all the sets in $\mathcal{F}$. $\square$

The next corollary can be summarized by saying that a convex compact set in a d-dimensional affine space is d-quasi-symmetric. Naturally, if $d = 1$, every segment is symmetric with respect to its center. For other results along these lines, see, for example, exercise 12.12.21 or [VE, 190, proposition 12.5]. See especially the extraordinary result of Dvoretsky [DV] which says that, for appropriately chosen dimensions, a convex sets always has quasi-spherical sections.

11.7.6. COROLLARY. *For every compact convex set A in a d-dimensional affine space X, there exists at least one point $z \in A$ such that every chord $[u, v]$ of A containing z satisfies (cf. 11.2.9 and 2.4.6):*

$$\frac{1}{d} \le \frac{\overrightarrow{zu}}{\overrightarrow{vz}} \le d.$$

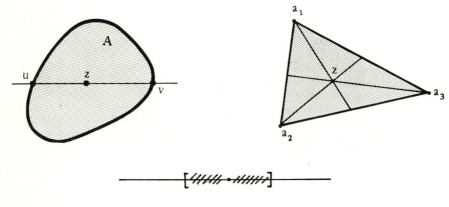

Figure 11.7.6

Observe first that d is the best possible bound, as attested by the example of a simplex, for which, as the reader should check, the only possible z is the centroid.

Proof. For every $x \in A$, introduce the set

$$A_x = \frac{1}{d+1} \cdot x + \frac{d}{d+1} \cdot A$$

(cf. 11.1.3), which is simply the image of A under the homothety $H_{x,d/(d+1)}$ (cf. 2.3.3.8). The sets A_x, for $x \in A$, satisfy the assumptions of Helly's theorem, in the compact case: if $\{x_i\}_{i=1,\ldots,d+1} \subset A$, their center of mass

$$y = \frac{1}{d+1} \sum_i x_i$$

belongs to all the sets A_{x_i}, since A is convex and

$$y = \frac{1}{d+1} x_i + \frac{d}{d+1} \sum_{j \neq i} \frac{1}{d} x_j \in A_{x_i}$$

(cf. 11.1.8.4). Now take $z \in \bigcap_{x \in A} A_x$, and let $[u, v]$ be a chord through z; to say that $z \in A_u$ means that $z \in H_{u.d/(d+1)}([u, v])$, so that

$$\frac{\overrightarrow{uz}}{\overrightarrow{uv}} \leq \frac{d}{d+1}, \qquad \text{whence} \qquad \frac{\overrightarrow{zu}}{\overrightarrow{vz}} \leq d.$$

Switching u and v completes the proof. $\square$

Here's a useful application of the next corollary in two dimensions: if, for any three paintings in a museum, there is a point from which all three can be seen, then there is a point from which all the paintings in the museum can be seen at once (you may need binoculars, but at least you can sit down...) Of course this is the same as saying that the museum is star-shaped.

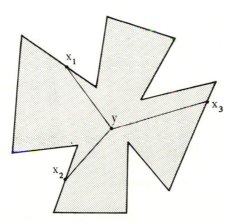

Figure 11.7.7.1

11.7.7. COROLLARY (KRASNOSEL'SKII). *Let A be a compact set of the d-dimensional affine space d. Assume that, for any subset $\{x_i\}_{i=1,\ldots,d+1} \in A$, there exists $y \in X$ such that $[x_i, y] \subset A$ for $i = 1, \ldots, d+1$. Then A is star-shaped.*

Proof. The basic idea is to introduce, for every $x \in X$, the largest star-shaped set V_x centered at x and contained in A, that is,

$$V_x = \{y \in A \mid [x, y] \subset A\}.$$

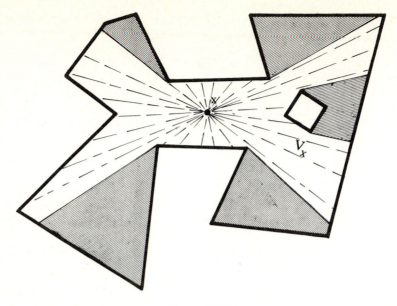

Figure 11.7.7.2

The sets V_x are compact, and by 11.1.8.7 so are their convex hulls $\mathcal{E}(V_x)$. The hypothesis of the corollary is exactly that any $(d + 1)$-element family of sets V_x, whence also of sets $\mathcal{E}(V_x)$, has non-empty intersection. Thus by Helly's theorem (11.7.1) there is a point $y \in \bigcap_{x \in A} \mathcal{E}(V_x)$. We will show that A is star-shaped at y, the difficulty being, of course, that we do not know *a priori* that y is in the intersection of the V_x (observe that we cannot apply Helly's theorem to the V_x, which are not necessarily convex).

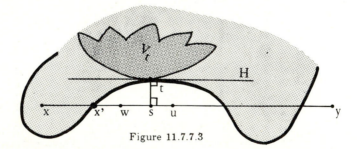

Figure 11.7.7.3

The proof is by contradiction. Let $x \in A$ be such that $[x, y] \not\subset A$. Introduce a Euclidean structure on X. Let x' be the farthest point from x on $[x, y]$ such that $[x, x'] \subset A$, and let u be an arbitrary point in $[x, y] \setminus A$. By continuity, there exists $w \in]x', u]$ such that $d(w, x') < d(u, A)$. Now take $s \in [u, w]$ and $t \in A$ such that $d(x, t) = d([u, w], A)$. We have $s \neq u$ by construction. Applying 11.1.7.2 twice, once to s and A and once to t and $[u, w]$, we see that $(\overrightarrow{st} \mid \overrightarrow{sy}) \leq 0$ and that V_t (whence also $\mathcal{E}(V_t)$) lies in the half-space separated from x by the hyperplane H containing t and orthogonal to $\overrightarrow{st}$. But this implies $y \notin \mathcal{E}(V_t)$. □

11.7.8. For other applications of Helly's theorem, see [VE, part 6], [EN, chapter 2] and [D–G–K].

11.8. Convex functions

11.8.1. We recall that a function $f : I \to \mathbf{R}$, where I is an interval of $\mathbf{R}$, is called *convex* if, for any $\lambda \in [0, 1]$ and any $x, y \in I$, we have

$$f\big(\lambda x + (1 - \lambda)y\big) \leq \lambda f(x) + (1 - \lambda)f(y).$$

We can generalize this definition as follows:

11.8.2. DEFINITION. *Let $A \subset X$ be a convex set in an affine space X. A map $f : A \to \mathbf{R}$ is called* convex *if, for every $\lambda \in [0, 1]$ and any $x, y \in A$, we have*

$$f\big(\lambda x + (1 - \lambda)y\big) \leq \lambda f(x) + (1 - \lambda)f(y).$$

Such a condition only makes sense if A is convex. An equivalent definition requires the concept of the epigraph of a function, that is, everything above and including its graph:

11.8.3. DEFINITION AND PROPOSITION. *Let the epigraph of a map $f : Y \to \mathbf{R}$, where Y is arbitrary, be the set*

$$\operatorname{Epigr}(f) = \big\{ (x, t) \in Y \times \mathbf{R} \mid x \in A \text{ and } t \geq f(x) \big\}.$$

A necessary and sufficient condition for a map $f : A \to \mathbf{R}$ on a convex set A to be convex is that its epigraph be convex.

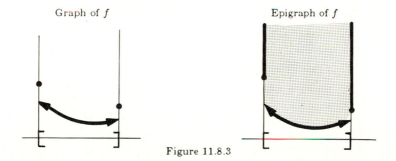

Graph of f Epigraph of f

Figure 11.8.3

An easy induction argument shows that, for a convex function f,

11.8.4
$$f\left(\sum_i \lambda_i x_i\right) \leq \sum_i \lambda_i f(x_i)$$

for any non-negative scalars λ_i such that $\sum_i \lambda_i = 1$ and any points x_i in A.

11.8.5. DEFINITION. *A function* $f : A \to \mathbf{R}$, *where A is convex, is called strictly convex if, for any $\lambda \in \,]0,1[$ and any $x, y \in A$ with $x \neq y$, we have*

$$f\big(\lambda x + (1 - \lambda)y\big) < \lambda f(x) + (1 - \lambda)f(y).$$

11.8.6. DEFINITION. *A function* $f : A \to \mathbf{R}$ *is called concave if the function $-f$ is convex, that is, if*

$$f\big(\lambda x + (1 - \lambda)y\big) \geq \lambda f(x) + (1 - \lambda)f(y)$$

for any $\lambda \in [0, 1]$ and any $x, y \in A$.

11.8.7. EXAMPLES.

11.8.7.1. Figure 11.8.3 shows that a convex function is not necessarily continuous. In fact, a convex function on a compact set does not even have to be bounded: just take $A = B(0,1) \subset \mathbf{R}^2$ and make $f = 0$ in the interior of A and f positive, but arbitrary, on the frontier. On the continuity of convex functions, see 11.8.10.4.

11.8.7.2. Affine forms $f : X \to \mathbf{R}$ (cf. 2.4.8.3) are convex, but certainly not strictly convex, since equality always holds in 11.8.4, by 3.5.1.

11.8.7.3. The function $x \mapsto x^2$, defined on $\mathbf{R}$, is strictly convex, but $x \mapsto |x|$ is only convex.

11.8.7.4. If $f_i : A \to \mathbf{R}$ is convex and $c_i \geq 0$ for all i, the linear combination $\sum_i c_i f_i$ is still convex.

11.8.7.5. If $f_i : A \to \mathbf{R}$ is convex for all i, the supremum of the f_i is convex. This is seen by using 11.8.3, since the epigraph of the supremum is the intersection of the epigraphs of the f_i.

11.8.7.6. If x, y, z, t are four points in a Euclidean affine space, the function

$$[0,1] \ni \lambda \mapsto d\big(\lambda x + (1 - \lambda)y, \lambda z + (1 - \lambda)t\big) \in \mathbf{R}$$

is convex. In fact,

$$d\big(\lambda x + (1 - \lambda)y, \lambda z + (1 - \lambda)t\big) = \|\lambda \overrightarrow{xz} + (1 - \lambda)\overrightarrow{yt}\|,$$

and the desired inequality follows from the proof of 9.2.2.

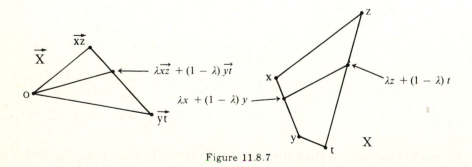

Figure 11.8.7

We now present two somewhat less trivial constructions for convex functions. These examples will be essential later: see 12.11.3 and 15.5.9.

11.8.8. The theorem of Brunn–Minkowski

11.8.8.1. Theorem (Brunn–Minkowski). *Let A and B be compact sets in a d-dimensional affine space X, and let $\mathcal{L}$ be a Lebesgue measure on X. The function*

$$[0,1] \ni \lambda \mapsto \mathcal{L}\big(\lambda A + (1-\lambda)B\big)^{1/d} \in \mathbf{R}$$

is concave (cf. 11.1.3 for the definition of $\lambda A + (1-\lambda)B$).

Proof. Consider on X a Euclidean structure with Lebesgue measure $\mathcal{L}$, and an orthonormal frame for this Euclidean structure. Let $\mathcal{F}$ be the family of open parallelepipeds with edges parallel to the axes in our frame.

11.8.8.2. First case. The sets A and B are in $\mathcal{F}$. Denote by a_i, b_i ($i = 1,\ldots,d$) the sides of A and B. The set $\lambda A + \mu B$ is in $\mathcal{F}$ and has sides $\lambda a_i + \mu b_i$, for any $\lambda, \mu \geq 0$. We thus have

$$\mathcal{L}(A) = \prod_i a_i, \quad \mathcal{L}(B) = \prod_i b_i, \quad \mathcal{L}(\lambda A + \mu B) = \prod_i (\lambda a_i + \mu b_i).$$

Set $u_i = \dfrac{a_i}{\lambda a_i + \mu b_i}$ and $v_i = \dfrac{b_i}{\lambda a_i + \mu b_i}$, so that $\lambda u_i + \mu v_i = 1$; by 11.8.11.6 we have

$$\frac{\lambda \mathcal{L}(A)^{1/d} + \mu \mathcal{L}(B)^{1/d}}{\mathcal{L}(\lambda A + \mu B)^{1/d}} = \lambda \prod_i u_i^{1/d} + \mu \prod_i v_i^{1/d}$$

$$\leq \frac{\lambda}{d} \sum_i u_i + \frac{\mu}{d} \sum_i v_i = \frac{1}{d} \sum_i (\lambda u_i + \mu v_i) = 1.$$

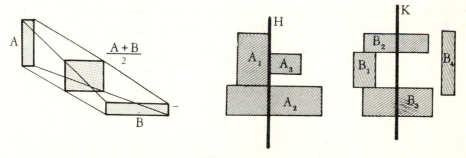

Figure 11.8.8

11.8.8.3. Second case. Assume that $A = \bigcup_{i=1}^{m} A_i$ and $B = \bigcup_{j=1}^{n} B_j$, where $A_i, B_j \in \mathcal{F}$ for all i, j and $A_i \cap A_{i'} = \emptyset$, $B_j \cap B_{j'} = \emptyset$ for all $i \neq i'$, $j \neq j'$. The proof is by induction on $m+n$. Assume, for example, that $m > 1$; then there exists a hyperplane H, parallel to a coordinate hyperplane, and strictly separating at least two sets A_i. We can write $A = A^+ \cup A^-$, where A^+ and A^- are of the same type as A, but composed of only m^+ and m^-

little parallelepipeds, where $m^+ < m$ and $m^- < m$. By continuity, there exists a hyperplane K, parallel to H and splitting B into pieces B^+, B^- such that $\dfrac{\mathcal{L}(A^+)}{\mathcal{L}(A^-)} = \dfrac{\mathcal{L}(B^+)}{\mathcal{L}(B^-)}$. Further, B^+ and B^- are of the same type as B, with $n^+ \leq n$ and $n^- \leq n$. We can now apply the induction hypothesis to $\lambda A^+ + \mu B^+$ and $\lambda A^- + \mu B^-$, obtaining

$$
\begin{aligned}
\mathcal{L}(\lambda A + \mu B) &= \mathcal{L}(\lambda A^+ + \mu B^+) + \mathcal{L}(\lambda A^- + \mu B^-) \\
&\geq (\lambda \mathcal{L}(A^+)^{1/d} + \mu \mathcal{L}(B^+)^{1/d})^d + (\lambda \mathcal{L}(A^-)^{1/d} + \mu \mathcal{L}(B^-)^{1/d})^d \\
&= (\lambda \mathcal{L}(A)^{1/d} + \mu \mathcal{L}(B)^{1/d})^d.
\end{aligned}
$$

11.8.8.4. Third case. Now let A and B be arbitrary compact sets. It is a standard result in Lebesgue measure theory that A and B can be approximated by sets A_n and B_n as in 11.8.8.3 and satisfying $|\mathcal{L}(A) - \mathcal{L}(A_n)| < \epsilon$, $|\mathcal{L}(B) - \mathcal{L}(B_n)| < \epsilon$. This leads to the desired conclusion. □

11.8.8.5. Corollary. *For arbitrary compact sets A, B we have*

$$
\mathcal{L}(A + B)^{1/d} \geq \mathcal{L}(A)^{1/d} + \mathcal{L}(B)^{1/d}. \qquad □
$$

11.8.8.6. Remark. One can show that the function in 11.8.8.1 is strictly concave if and only if A and B are not homothetic. The proof is more delicate, cf. [EN, 97] or [HR, 187].

11.8.9. THE FUNCTION OF LOEWNER–BEHREND

11.8.9.1. Recall the notation introduced in 8.2.5.2: $Q(E)$ is the space of Euclidean structures on a d-dimensional real vector space E, and forms a convex open set in the vector space $P_2^\bullet(E)$. Fix on E a Lebesgue measure $\mathcal{L}$. Given $q \in Q(E)$, consider the ellipsoid $q^{-1}(1)$ (cf. 15.3.3.3), whose convex hull is the *solid ellipsoid* $\mathcal{E}(q) = q^{-1}([0,1])$. We wish to calculate the volume $\mathcal{L}(\mathcal{E}(q))$ of this ellipsoid (cf. 9.12.4).

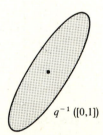

$q^{-1}([0,1])$

Figure 11.8.9

11.8.9.2. Let A (resp. A') be the matrix of q with respect to a basis $\mathcal{B}$ (resp. $\mathcal{B}'$) of E (cf. 13.1.3.6); then $A' = {}^t S A S$, where S is the matrix with respect to $\mathcal{B}$ of the map $f \in \mathrm{Isom}(E)$ such that $f(\mathcal{B}) = \mathcal{B}'$. In particular, if $\mathcal{B}$ and $\mathcal{B}'$ are *unitary bases* of E with respect to $\mathcal{L}$, that is, if their associated isomorphisms

$\mathbf{R}^d \to E$ are as in 2.7.4.2, we have $\det S = 1$ and $\det A = \det A'$. Thus it makes sense to define the *determinant of q with respect to $\mathcal{L}$*:

11.8.9.3 $\det_\mathcal{L} q = \det A,$

where A is the matrix of q in any unitary basis $\mathcal{B}$. We will show that

11.8.9.4 $\mathcal{L}\big(\mathcal{E}(q)\big) = \beta(d)(\det_\mathcal{L} q)^{-1/2},$

where $\beta(d)$ is defined in 9.12.4.7.

Proof. Consider a Euclidean structure on E with canonical measure $\mathcal{L}$ (cf. 9.12), and let $f \in \mathrm{Isom}(E)$ be such that $f\big(B(0,1)\big) = \mathcal{E}(q)$, that is, such that $f^*q = \| \cdot \|^2$, the norm of the Euclidean structure (cf. 13.1.3.9). In particular (13.1.3.10), since the matrix of $\| \cdot \|^2$ is I, the matrix of A satisfies $I = {}^t U A U$, where U is the matrix of f in an arbitrary orthonormal basis. We thus have $\det A (\det f)^2 = 1$. By 2.7.4.3 we can write

$$\mathcal{L}\big(\mathcal{E}(q)\big) = |\det f| \, \mathcal{L}\big(B(0,1)\big) = |\det f| \, \beta(d) = \beta(d)(\det_\mathcal{L} q)^{-1/2}. \qquad \square$$

The theorem of Loewner–Behrend will be demonstrated in 11.8.10.7 using the following proposition:

11.8.9.5. Proposition. *The function* $Q(E) \ni q \mapsto \mathcal{L}\big(\mathcal{E}(q)\big) \in \mathbf{R}$ *is strictly convex.*

Proof. Take $q, q' \in Q(E)$; by 13.5.5, there exists a basis, which we can assume to be $\mathcal{L}$-unitary, diagonalizing q and q' simultaneously. Thus we can write

$$q = \sum_i a_i x_i^2, \qquad q' = \sum_i a'_i x_i^2.$$

For determinants calculated in this basis, we have $\det q = \prod_i a_i$, $\det q' = \prod_i a'_i$, and, for $\lambda, \lambda' \geq 0$ with $\lambda + \lambda' = 1$,

$$\det(\lambda q + \lambda' q') = \prod_i (\lambda a_i + \lambda' a'_i).$$

Applying 11.8.11.4, we can write

$$\big(\det(\lambda q + \lambda' q')\big)^{-1/2} = \prod_i (\lambda a_i + \lambda' a'_i)^{-1/2} \leq \prod_i (a_i^\lambda a_i'^{\lambda'})^{-1/2}$$

$$= \left(\left(\prod_i a_i\right)^{-1/2}\right)^\lambda \left(\left(\prod_i a'_i\right)^{-1/2}\right)^{\lambda'}$$

$$\leq \lambda (\det q)^{-1/2} + \lambda' (\det q')^{-1/2}.$$

Equality can only hold if it holds for each pair (a_i, a'_i), that is, if $a_i = a'_i$ for every i. But this implies $q = q'$.

11.8.10. PROPERTIES OF CONVEX FUNCTIONS. We have seen in 11.8.7.1 that a convex function, even on a compact set, is not necessarily continuous or bounded; however, we will see that complications can only arise at the boundary and for upper bounds, as hinted in 11.8.7.1.

11.8.10.1. Proposition. *A convex function $f : A \to \mathbf{R}$ is bounded above on every compact subset of the relative interior of A (cf. 11.2.8), and bounded below on every bounded subset of A.*

Proof. As usual, we reduce to the case $\overset{\circ}{A} \neq \emptyset$. Let $x \in \overset{\circ}{A}$ be arbitrary, let H be a supporting hyperplane to $\mathrm{Epigr}(f)$ at $(x, f(x))$ (cf. 11.8.3 and 11.5.2), and take an affine form $g : X \to \mathbf{R}$ such that $H = g^{-1}(0)$. We have $f \geq g$ by construction; but an affine form is bounded over bounded sets.

Let $K \subset \overset{\circ}{A}$ be compact; for every $x \in K$, there exists a simplex S_x of X contained in $\overset{\circ}{A}$ and such that $x \in \mathrm{Int}(\mathcal{E}(S_x))$. By compactness, we can cover K with a finite number of such simplices; by 11.8.4, the value of $f(x)$ at any point $x \in K$ is bounded above by the values of f at the vertices of those simplices, and there are only finitely many such vertices.

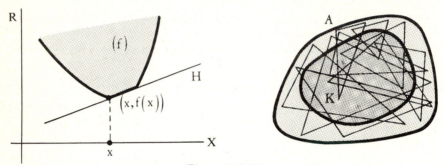

Figure 11.8.10.1

11.8.10.2. Lemma. *Let $f : [a, b] \to \mathbf{R}$ be convex, and take $c \in \,]a, b[$.*

i) *The function $[a, b] \setminus c \ni t \mapsto \dfrac{f(t) - f(c)}{t - c}$ is increasing. The function f has derivatives to the right and to the left, denoted by $f'_l(c)$ and $f'_r(c)$, and $f'_l(c) \leq f'_r(c)$; in particular f is continuous at c.*

iii) *If f is differentiable on $]a, b[$, its derivative is increasing; if f is twice differentiable at c, we have $f''(c) \geq 0$.*

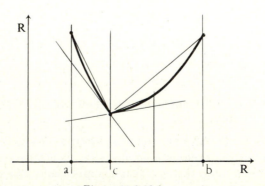

Figure 11.8.10.2

Proof. Part (i) follows from the definition (see figure 11.8.10.2). For every $t > c$, the function $t \mapsto \dfrac{f(t) - f(c)}{t - c}$ is increasing and bounded below by $\dfrac{f(a) - f(c)}{a - c}$, so the limit

$$f'_r(c) = \lim_{t \to c} \frac{f(t) - f(c)}{t - c}$$

exists; the same is true of $f'_l(c)$, and $f'_l(c) \leq f'_r(c)$. As to f', if it exists, it must be increasing because $f'(c) \leq \dfrac{f(d) - f(c)}{d - c} \leq f'(d)$ for every $c, d \in \left]a, b\right[$. The last assertion in (iii) is well-known.   □

11.8.10.3. Notes. The last proof can be easily extended to show that

$$\{ c \in \left]a, b\right[\mid f'_l(c) \neq f'_r(c) \}$$

is countable. For finer properties of convex functions, and the infinite-dimensional case, see [R–V, chapter IV].

Lemma 11.8.10.2 implies that a convex function $f : A \to \mathbf{R}$, where $A \subset X$ has arbitrary dimension, f has a directional derivative on any half-line starting at a point $a \in \overset{\circ}{A}$. We apply this idea as follows:

11.8.10.4. Proposition. *Let $f : A \to \mathbf{R}$ be convex; then f is continuous at every point in the interior of A.*

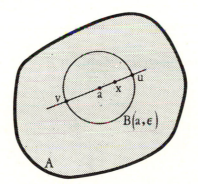

Figure 11.8.10.4

Proof. As always, we reduce to the case $\overset{\circ}{A} \neq \emptyset$. Take $a \in \overset{\circ}{A}$ and $\epsilon > 0$ such that $B(a, \epsilon) \subset \overset{\circ}{A}$ (the ball is measured with respect to an arbitrary Euclidean structure). Using 11.8.10.1, define

$$M = \sup \{ f(x) \mid x \in B(a, \epsilon) \}.$$

For arbitrary $x \in B(a, \epsilon)$, $x \neq a$, let u, v be the points where the line $\langle a, x \rangle$ intersects $S(a, r)$. We apply 11.8.10.2 to the restriction of f to the affine line $\langle v, u \rangle$:

$$\frac{f(x) - f(a)}{\|x - a\|} \leq \frac{f(u) - f(a)}{\|u - a\|} \leq \frac{M - f(a)}{\epsilon}$$

and
$$\frac{f(x) - f(a)}{-\|x - a\|} \le \frac{f(v) - f(a)}{\|v - a\|} \le \frac{M - f(a)}{\epsilon},$$
whence
$$|f(x) - f(a)| \le \frac{M - f(a)}{\epsilon}\|x - a\|. \qquad \square$$

11.8.10.5. Theorem. *Let $f : A \to \mathbf{R}$, be a convex function on an open convex set A. Then f is differentiable almost everywhere, and its derivative is continuous where defined.* $\qquad \square$

We will not have the time to demonstrate this important and natural theorem; the reader can consult [R–V, 116–117], or see 11.9.14. The matter of second and higher derivatives of convex functions is still a practically open research question ([R–V, 120]). Theorem 11.8.10.5 play an essential role in 12.10.11.1.

The remainder of 11.8.10 is devoted to some extremum properties of convex functions. We start with the following

11.8.10.6. Proposition. *Let $f : A \to \mathbf{R}$ be strictly convex. If f achieves its minimum, it does so at a unique point.*

Proof. If $m = \inf_A f$ and $f(a) = m = f(b)$ we have
$$m \le f\left(\frac{a + b}{2}\right) \le \frac{f(a) + f(b)}{2} = m,$$
whence $a = b$ by 11.8.5. $\qquad \square$

We now apply this elementary proposition to the question we started studying in 11.8.9.

11.8.10.7. Theorem (Loewner–Behrend). *Let E be a finite-dimensional real vector space, endowed with a Lebesgue measure. If K is a compact subset of E with non-empty interior, there exists a unique least-volume solid ellipsoid containing K.*

Proof. By 11.8.10.5, 11.8.9.5 and the continuity of $q \mapsto \mathcal{L}\big(\mathcal{E}(q)\big)$, it is enough to show that we can find a minimum over an appropriately chosen compact convex subset of $Q(E)$. Since K is bounded, there exists at least one solid ellipsoid $\mathcal{E}(q_0)$ containing K. The set we want is
$$A = \big\{\, q \in Q(E) \mid \mathcal{E}(q) \supset K \text{ and } \det_{\mathcal{L}} q \ge \det_{\mathcal{L}} q_0 \,\big\},$$
and we have to show that this is a compact convex subset of $Q(E)$. Convexity for the condition $\det_{\mathcal{L}} q \ge \det_{\mathcal{L}} q_0$ follows from 11.8.9.5; as for the condition $\mathcal{E} \supset K$, we have
$$\mathcal{E}(q) = \big\{\, x \in X \mid q(x) \le 1 \,\big\}, \qquad \mathcal{E}(q') = \big\{\, x \in X \mid q'(x) \le 1 \,\big\},$$
so if $q, q' \in A$ and $\lambda \in [0, 1]$ we indeed have
$$\big(\lambda q + (1 - \lambda q')\big)(x) = \lambda q(x) + (1 - \lambda)q'(x) \le 1.$$

The set A is closed in $P_2^\bullet(E)$, being the intersection of inverse images of intervals under continuous maps, but we must watch out for the fact that

$Q(E)$ is an open subset of $P_2^\bullet(E)$. But the boundary of $Q(E)$ is made up of degenerate quadratic forms, whose $\det_{\mathcal{L}}$ is zero, so A is in fact closed. To see that it is bounded, observe first that $\mathcal{E}(q) \supset K$ implies $\mathcal{E}(q) \supset \mathcal{E}(K \cup (-K))$, where $-K$ is the reflection of K through the origin, and that $\mathcal{E}(K \cup (-K))$ has non-empty interior because K does. Consider on E an arbitrary Euclidean structure and let $\epsilon > 0$ be such that $B(0, \epsilon) \subset \mathcal{E}(K \cup (-K))$; the eigenvalues λ_i of $q \in Q(E)$, with respect to the Euclidean structure under consideration, must satisfy $\lambda_i < 1/\epsilon$ for all i if we are to have $\mathcal{E}(q) \supset B(0, \epsilon)$; and this shows that A is bounded. $\qquad\square$

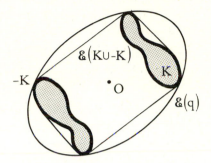

Figure 11.8.10.7

We shall use 11.8.10.7 in 15.5.9, but we remark from now that it provides the promised third proof for 8.2.5 (a proof not involving integration theory, since volumes of ellipsoids are mere determinants!).

11.8.10.8. Corollary. *If G is a compact subgroup of* $\mathrm{GL}(E)$, *there exists* $q \in Q(E)$ *invariant under* G.

Proof. Let H be an arbitrary compact subset of E with non-empty interior, $K = G(H)$ be the orbit of H under the action of G, and $\mathcal{E}(q)$ the least-volume ellipsoid containing K. The set $\mathcal{E}(q)$ is invariant under any $g \in G$, since, by construction, $g(K) = K$ and $g(\mathcal{E}(q)) \supset K$. Finally,

$$\mathcal{L}\big(g(\mathcal{E}(q))\big) = \mathcal{L}(\mathcal{E}(q))$$

since $\det g = 1$ for any g (otherwise G is not compact, as $\det g^n = (\det g)^n$). Since the least-volume ellipsoid is unique, we have $g(\mathcal{E}(q)) = \mathcal{E}(q)$, and we have shown that the form q satisfies $g^*q = q$ for all $g \in G$. $\qquad\square$

Now back to extremum values of convex functions.

11.8.10.9. Proposition. *If $f : A \to \mathbf{R}$ is convex and continuous, f achieves its minimum at at least one extremal point of A, that is (cf. 11.6.8)*

$$\sup_A f = \sup_{\mathrm{Extr}(A)} f.$$

Proof. By 11.1.8.6 and 11.6.8, every $a \in A$ is a finite barycenter of extremal points; thus, if $M = \sup_{\mathrm{Extr}(A)} f$, we have $f(a) \leq M$ for all $a \in A$, by 11.8.4. $\qquad\square$

11.8.10.10. Note. Proposition 11.8.10.9 is of paramount importance in the practical search for maxima, as in game theory and linear programming, for example ([KE, 86 ff.]): if the function whose maximum we seek is convex, it suffices to know its values at points of $\mathrm{Extr}(A)$. Such points are finite in number if A is a polyhedron, for instance (cf. 12.1.9). For details and examples, see [KE] and [R–V, chapter V].

11.8.11. CONVEXITY CRITERIA. EXAMPLES

11.8.11.1. Proposition. *Let $I \subset \mathbf{R}$ be an interval and $f : I \to \mathbf{R}$ a twice differentiable function. In order for f to be convex it is necessary and sufficient that $f''(x) \geq 0$ for all $x \in I$. For f to be strictly convex it is necessary and sufficient that $f''(x) > 0$ for all $x \in I$.*

Proof. Set, for $a, b \in I$,

$$g(x) = f(x) - f(b) - f(b)\frac{x - a}{f(b) - f(a)};$$

we have to show that $g \leq 0$ over $[a, b]$. But $g'' = f''$, so g' is increasing; since g' vanishes at some $\alpha \in [a, b]$ (at the point where g is minimal, for example), g' is non-positive to the left of α and non-negative to the right of α. Thus g decreases starting at $g(a) = 0$, then it increases again till $g(b) = 0$. This shows that $g \leq 0$. □

	a	α	b
g	0	↘	↗ 0
g'	−	0	+
g"		+	

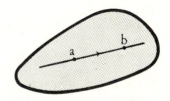

Figure 11.8.11

11.8.11.2. Corollary. *Let $f : A \to \mathbf{R}$ be of class C^2, where A is an open subset of X. In order that f be convex it is necessary and sufficient that its second derivative*

$$f''(x) : \vec{X} \times \vec{X} \to \mathbf{R}$$

be a positive semidefinite quadratic form for every x, that is, $f''(x)(y, y) \geq 0$ for every $y \in \vec{X}$. For f to be convex it is necessary and sufficient that $f''(x)$ be positive definite, that is $f''(x)(y, y) > 0$ for every $y \in \vec{X} \setminus 0$.

Proof. Given $a, b \in A$, it is enough to apply 11.8.11.1 to the map $t \to f(a + t\overrightarrow{ab})$, whose second derivative is equal to $f''(a + t\overrightarrow{ab})(\overrightarrow{ab}, \overrightarrow{ab})$. □

11.8.11.3. Example. The function $-\log x$ is strictly convex on $\mathbf{R}_+^*$, since $(-\log x)'' = 1/x^2$, whence

$$-\log(\lambda a + \lambda' a') \leq -\lambda \log a - \lambda' \log a'$$

for $\lambda, \lambda' \geq 0$, $\lambda + \lambda' = 1$. Since the log is increasing, we get

11.8.11.4
$$\lambda a + \lambda' a' \geq a^\lambda a'^{\lambda'}$$

for any $\lambda, \lambda' \geq 0$ with $\lambda + \lambda' = 1$. More generally,

11.8.11.5
$$\sum_i \lambda_i a_i \geq \prod_i a_i^{\lambda_i}$$

for any set of $\lambda_i \geq 0$ with $\sum_i \lambda_i = 1$. In particular, we obtain the well-known inequality between geometric and arithmetic means:

11.8.11.6
$$a_1 \cdots a_n \leq \left(\frac{a_1 + \cdots + a_n}{n} \right)^n,$$

where equality holds if and only if all the a_i are equal, since $-\log x$ is strictly convex. For $n = 2$, we recover the elementary inequality $a + b \geq 2\sqrt{ab}$.

11.8.11.7. Example. For every fixed real number $p > 1$, the function

$$\mathbf{R}_+^* \ni x \mapsto x^p \in \mathbf{R}$$

is strictly convex, since $f''(x) = p(p-1)x^{p-2}$. We thus have

11.8.11.8
$$\left(\sum_i \lambda_i a_i \right)^p \leq \sum_i \lambda_i a_i^p$$

for any $a_i > 0$ and $\lambda_i \geq 0$ with $\sum \lambda_i = 1$. This leads to the *Hölder inequality*

11.8.11.9
$$\sum_i x_i y_i \leq \left(\sum_i x_i^p \right)^{1/p} \left(\sum_i y_i^q \right)^{1/q}$$

$(p > 1$, $1/p + 1/q = 1$, $x_i \geq 0$, $y_i \geq 0)$, of which 8.1.3 is a particular case $(p = 2)$. Hölder's inequality, however, is much less elementary than 8.1.3; to deduce it from 11.8.11.8, one must find numbers λ_i, a_i and k such that

$$\lambda_i a_i = x_i y_i, \qquad \lambda_i a_i^p = k x_i^p, \qquad \sum_i \lambda_i = 1.$$

The right numbers are

$$\lambda_i = \left(\sum_i y_i^q \right)^{-1} y_i^q, \qquad a_i = \left(\sum_i y_i^q \right) x_i y_i^{1-q}$$

11.8.11.10. Lastly, one deduces from 11.8.11.9 that, for every $p > 1$,

$$\mathbf{R}^d \ni (x_1, \ldots, x_n) \mapsto \left(\sum_i |x_i|^p \right)^{1/p}$$

defines a norm on $\mathbf{R}^d$. This requires proving the *Minkowski inequality:*

11.8.11.11
$$\left(\sum_i |x_i + y_i|^p \right)^{1/p} \leq \left(\sum_i |x_i|^p \right)^{1/p} + \left(\sum_i |y_i|^p \right)^{1/p};$$

this is done by using 11.8.11.9 twice, as follows:

$$\sum_i |x_i|\,|x_i + y_i|^{p-1} \le \left(\sum_i |x_i|^p\right)^{1/p}\left(\sum_i |x_i + y_i|p\right)^{(p-1)/p},$$

$$\sum_i |y_i|\,|x_i + y_i|^{p-1} \le \left(\sum_i |y_i|^p\right)^{1/p}\left(\sum_i |x_i + y_i|p\right)^{(p-1)/p},$$

$$\sum_i |x_i + y_i|^p \le \sum_i (|x_i| + |y_i|)(|x_i + y_i|^{p-1})$$

$$\le \left(\sum_i |x_i + y_i|^p\right)^{(p-1)/p}\left(\left(\sum_i |x_i|^p\right)^{1/p} + \left(\sum_i |y_i|^p\right)^{1/p}\right).$$

11.8.11.12. Notes. Inequalities 11.8.11.9 and 11.8.11.11 are absolutely essential in analysis, as they allow the definition of L_p-spaces.

For more convexity inequalities, see, for example, [DE3, 40 ff.], [R–V, chapter VI] and the references therein, especially the classical book [H–L–P] and the modern text [B–B].

11.8.12. CONVEX FUNCTIONS VERSUS CONVEX SETS. We have already encountered a connection between convex functions and convex sets in 11.8.3. Two more general relationships are discussed below, but not in detail: the reader can consult [EN, 54].

11.8.12.1. Throughout 11.8.12 X denotes a Euclidean vector space. A *gauge* on X is any convex function $f : X \to \mathbf{R}$ satisfying

$$f(\lambda x) = \lambda f(x)$$

for any $x \in X$ and any $\lambda \ge 0$. Norms, for example, are gauges.

11.8.12.2. If f is a gauge, the ball $C(f) = \{\, x \in X \mid f(x) \le 1 \,\}$ is a convex set. Conversely, if C is a convex compact set containing 0 in its interior, the function given by

$$f_C(x) = \inf\{\, \lambda \mid \lambda > 0,\ x \in \lambda C \,\}$$

is a gauge, but will only be a norm if $C = -C$, that is, if C is symmetric. The boundary of C is $f_C^{-1}(1)$; we say that f_C is the *distance function* on C.

11.8.12.3. If C is bounded in X, the function $h_C(x) = \sup\{\, (y\,|\,x) \mid y \in C \,\}$ is also a gauge, called the *supporting function* of X.

11.8.12.4. The relationship between h_C and f_C is a consequence of 11.1.5 and 11.4.8; if C is a convex compact set with $0 \in \overset{\circ}{C}$, we have

$$f_{C^*} = h_C, \qquad h_{C^*} = f_C.$$

11.8.12.5. A gauge is determined by its restriction to the unit sphere $S = S(0, 1)$. In general, it is not possible to explicitly recover C from $h = h_C$ (given on S). If h is differentiable, one can say that C is the *envelope* of the hyperplanes $H_t = \{\, x \in X \mid (x\,|\,t) = 1 \,\}$ for t ranging over S, or again that C

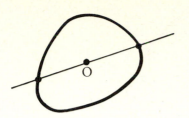
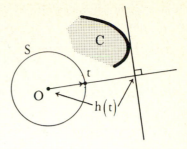

Figure 11.8.12

is the transform of $h^{-1}(1)$ under reciprocal polarities with respect to S. In the case of the plane, h corresponds to the *Euler equations* of the curve $\operatorname{Fr} C$ (see 12.12.14 and 12.12.15).

11.8.12.6. Supporting functions are a powerful tool in the geometric study of convex sets: see, for example, [EN, 54 and the whole of chapter V] and [BU1, chapter II, § 6]. See also 11.9.14.

11.9. Exercises

11.9.1. Show that Minkowski addition satisfies the relations
$$(A + B) + C = A + (B + C),$$
$$(A \cup B) + C = (A + C) \cup (B + C),$$
$$(A \cap B) + C = (A + C) \cap (B + C),$$
$$H_{a.\lambda}(A + B) = H_{a.\lambda}(A) + H_{a.\lambda}(B),$$
$$H_{a.\lambda+\mu}(A) \subset H_{a.\lambda}(A) + H_{a.\mu}(A).$$

When does equality hold in the last relation?

11.9.2. True or false: For every bounded set A, $\operatorname{diam}(\mathcal{E}(A)) = \operatorname{diam} A$.

11.9.3. Let X be a d-dimensional affine space and A a subset of X with at most d connected components. Show that every point of $\mathcal{E}(A)$ is the barycenter of d points in A.

* **11.9.4.** HILBERT GEOMETRY. Let A be a convex compact subset of X whose interior is non-empty. Given two distinct points x, y in $\overset{\circ}{A}$, set $d(x, y) = \left| \log[x, y, u, v] \right|$, where u, v are the two points where the line $\langle x, y \rangle$ meets the frontier of A (figure 11.9.4). Show that $d : \overset{\circ}{A} \times \overset{\circ}{A} \to \mathbf{R}$, defined by the equation above and $d(x, x) = 0$ for all $x \in \overset{\circ}{A}$, is a metric. Show that this metric is excellent (9.9.4.4). Study the relation between the strict triangle inequality and the nature of the boundary points of A. For a notion of area in this geometry, see [B–K, 167]. See also 6.8.14.

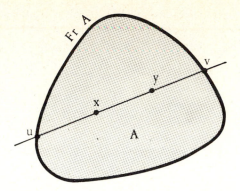

Figure 11.9.4

11.9.5. Let X be an affine space of dimension greater than 1, and $f : X \to X$ a bijection. Show that if, for any convex set A, the image $f(A)$ is convex, then f is affine.

11.9.6. Consider in $\mathbf{R}^2$ the sets

$$B = \big\{ (x,0) \mid x \geq 1 \text{ or } x \leq -1 \big\}, \qquad A = \mathbf{R}^2 \setminus B.$$

The set A is star-shaped (figure 11.3.6.0.3). For the homeomorphism $F :$ $A \to \mathbf{R}^2$ constructed in the proof of 11.3.6.1, draw the inverse images of circles around the origin.

11.9.7. Let E be the vector space of $n \times n$ real matrices, and K the subset of E consisting of *bistochastic* matrices, that is, those satisfying

$$a_{ij} \geq 0 \text{ for all } i, j, \qquad \sum_i a_{ij} = 1 \text{ for all } j, \qquad \sum_j a_{ij} = 1 \text{ for all } i.$$

Show that K is a compact polyhedron in E and that the vertices are the *permutation matrices*, that is, those whose every row or column have exactly one non-zero element, equal to 1.

* **11.9.8.** Prove that the extremal points of a convex set in the plane form a closed set.

11.9.9. Consider in $\mathbf{R}^3$ the closed convex cone C defined by the inequalities $x \geq 0$, $y \geq 0$, $z \geq 0$ and $z^2 \leq xy$. Show that the line D with equations $x = 0$, $z = 1$ does not intersect C, but that there exists no plane containing D and not intersecting C.

11.9.10. KIRCHBERGER'S THEOREM. Let A and B be finite subsets of a d-dimensional space X. If, for every $(d+2)$-element subset of X, it is possible to find a hyperplane that strictly separates $A \cap Y$ and $B \cap Y$, there exists a hyperplane that strictly separates A and B.

11.9.11. Let A_i $(i = 1, \ldots, r, r > d+1)$ be convex subsets of $\mathbf{R}^d$ such that any $r - 1$ of them have non-empty intersection. Show that the intersection of the A_i is non-empty, as follows: take $x_i \in \bigcap_{j \neq i} A_i$ $(i = 1, \ldots, r)$, and find scalars λ_i, not all zero, such that $\sum_{i=1}^{r} \lambda_i = 0$ and $\sum_{i=1}^{r} \lambda_i x_i = 0$. Finish off by splitting the λ_i into two groups, one non-negative and the other negative.

11.9.12. MAXIMAL AND MINIMAL WIDTHS. Let C be a compact convex set. The *maximal* (resp. *minimal*) width of C, denoted by $D(C)$ (resp. $d(C)$), is the supremum (resp. infimum) of the width of C in the direction ξ, as ξ ranges over the set of directions of lines (cf. 11.5.6.3). Show that $D(C) = \operatorname{diam}(C)$, the diameter of C.

The *interior radius* of C, denoted by $r(C)$, is the supremum of the radii of all spheres contained in C. Show that, for d the dimension of the ambient space, we have

$$r(C) \geq \frac{d(C)}{2\sqrt{d}} \quad \text{for odd } d,$$

$$r(C) \geq \frac{\sqrt{d+2}}{2(d+1)} d(C) \quad \text{for even } d.$$

See [EN, 112–114] if necessary. Are these inequalities sharp?

11.9.13. Let A be an open convex set and $f : A \to \mathbf{R}$ a function satisfying

$$f\left(\frac{x+y}{2}\right) \leq \frac{f(x) + f(y)}{2}$$

for every $x, y \in A$. Show that A is convex. See [VE, 130] if necessary.

11.9.14. SUPPORTING FUNCTIONS. Find an explicit formula for the supporting function of a point, of a ball $B(a, r)$, and of a cube centered at the origin. Show that if f (resp. g) is the supporting function of A (resp. B), the supporting function of the Minkowski sum $\lambda A + \mu B$ is simply $\lambda f + \mu g$.

11.9.15. DIFFERENTIABILITY OF CONVEX FUNCTIONS. Let A be an open convex set and $f : A \to \mathbf{R}$ a convex function. Show that if the partial derivatives $\partial f / \partial x_i$ (with respect to some frame $\{e_i\}$ for the ambient space) exist at a point $a \in A$, the function f is differentiable at a. Deduce, using 11.8.10.2, that f is differentiable almost everywhere (use Fubini's theorem).

11.9.16. CONVEX SETS AND THE GEOMETRY OF NUMBERS. Let $A = \mathbf{Z} \times \mathbf{Z}$ be the unit lattice in $\mathbf{R}^2$, and C a compact convex subset of $\mathbf{R}$ satisfying $-C = C$. Show that if the area of C (cf. 9.12 or 12.12) satisfies $\mathcal{L}(C) \geq 4$, then C intersects $A \setminus (0, 0)$, that is, A contains points with integer coordinates not both zero (Minkowski). Generalize for dimension d. Deduce that if $f(x, y) = ax^2 + 2bxy + cy^2$ is a positive definite quadratic form with discriminant $D = ac - b^2$, there exist integers x, y, not both zero, such that $f(x, y) < 4/\pi\sqrt{D}$. For elaborations on this result, the first in the so-called "geometry of numbers", see the excellent work [CS]. See also [LE] and [DE3, 202–203].

11.9.17. Show that for every compact convex subset A of a Euclidean affine space X we have $\operatorname{Is}_A(X) = \operatorname{Is}_{\operatorname{Extr}(A)}(X)$.

11.9.18. Is there a connection between Jung's theorem (11.5.8) and Helly's theorem (11.7.1)?

11.9.19. KAKUTANI'S LEMMA. Let C_1 and C_2 be disjoint convex sets, and $x \notin C_1 \cup C_2$ a point. Denote by Γ_i the convex hull of $\{x\} \cup C_i$ $(i = 1, 2)$. Show that at least one of the sets $\Gamma_1 \cap C_2$ and $\Gamma_2 \cap C_1$ is empty.

* **11.9.20.** Given a subset A of an affine space X, consider the set $N(A)$ of points a such that A is star-shaped with respect to a (cf. 11.1.2.4). Show that $N(A)$ is convex. Find $N(A)$ for a number of shapes of A.

* **11.9.21.** THE THEOREM OF LUCAS. Let P be a polynomial with complex coefficients, and let P' be its derivative. Show that in the affine space $\mathbf{C}$, all the roots of P' belong to the convex hull of the roots of P. When P has degree three and distinct roots a, b, c, show that there is an ellipse inscribed in the triangle $\{a, b, c\}$ whose foci are the roots u, v of P'.

* **11.9.22.** Find all partitions of the plane into two convex sets.

11.9.23. THE THEOREM OF KRITIKOS. Let C be the frontier of a compact convex subset K of a Euclidean space E, and assume K has non-empty interior. Given $m \in \overset{\circ}{K}$, denote by $R(m)$ the radius of the smallest sphere with center m containing K, and by $r(m)$ the radius of the largest sphere with center m contained in K.

Show that there exists a unique point $m \in \overset{\circ}{K}$ such that the difference $R(m) - r(m)$ is minimal. Show that for this point the associated containing and contained spheres have at least two points in common which C.

Deduce that for every compact convex differentiable hypersurface C of E (that is, the differentiable frontier of a compact convex set) there exists at least one point from which one can draw four normals to C.

[CH2] Henri Cartan. *Elementary theory of analytic functions of one or more complex variables*. Hermann, Paris, 1963.

[CH–GR] S. S. Chern and P. Griffiths. Abel's theorem and webs. *Jahresberichte der Deutschen Math. Vereinigung*, 80:13–110, 1978.

[CL1] Robert Connelly. A counterexample to the rigidity conjecture for polyhedra. *Publ. Math. I.H.E.S*, 47, 1978.

[CL2] Robert Connelly. A flexible sphere. *Mathematical Intelligencer*, 1:130–131, 1978.

[CL3] Robert Connelly. An attack on rigidity (I and II). Preprint, Cornell University.

[C–M] H. S. M. Coxeter and W. O. J. Moser. *Generators and Relations for Discrete Groups. Ergebnisse der Mathematik, 14*, Springer-Verlag, Berlin, fourth edition, 1980.

[CN] R. Cuénin. *Cartographie Générale*. Eyrolles, Paris.

[CN–GR] G. D. Chakerian and H. Groemer. Convex bodies of constant width. In P. M. Gruber and J. M. Wills, editors, *Convexity and Its Applications*, Birkhäuser, Basel, 1983.

[COO] Julian L. Coolidge. *A Treatise on the Circle and the Sphere*. Clarendon Press, Oxford, 1916.

[CP] Christophe. *L'idée fixe du savant Cosinus*. Armand Colin, Paris.

[CR1] H. S. M. Coxeter. *Introduction to Geometry*. Wiley, New York, second edition, 1969.

[CR2] H. S. M. Coxeter. *Regular Polytopes*. Dover, New York, third edition, 1973.

[CR3] H. S. M. Coxeter. *Non-Euclidean Geometry*. University of Toronto Press, Toronto, third edition, 1957.

[CR4] H. S. M. Coxeter. *Regular Complex Polytopes*. Cambridge University Press, London, 1974.

[CR5] H. S. M. Coxeter. The problem of Apollonius. *American Math. Monthly*, 75:5–15, 1968.

[CS] J. W. S. Cassels. *An Introduction to the Geometry of Numbers*. Springer-Verlag, Berlin, 1959.

[CT] Gustave Choquet. *Cours d'Analyse, II: Topologie*. Masson, Paris, 1964.

[CY] Claude Chevalley. *Theory of Lie Groups*. Princeton University Press, Princeton, 1946.

[CZ] Jean-Pierre Conze. *Le théorème d'isomorphisme d'Ornstein et la classification des systèmes dynamiques en théorie ergodique. Lecture Notes in Mathematics, 383*, Springer-Verlag, Berlin, 1974 (Séminaire Bourbaki, November 1972).

[D–C1] Robert Deltheil and Daniel Caire. *Géométrie*. J. B. Baillière, Paris.

[D–C2] Robert Deltheil and Daniel Caire. *Compléments de Géométrie*. J. B. Baillière, Paris.

[DE1] Jean Dieudonné. *La Géométrie des Groupes Classiques. Ergebnisse der Mathematik, 5*, Springer-Verlag, Berlin, third edition, 1971.

[DE2] Jean Dieudonné. *Linear Algebra and Geometry*. Hermann and Houghton, Mifflin, Paris and Boston, 1969.

[DE3] Jean Dieudonné. *Infinitesimal Calculus*. Hermann, Paris, 1971.

[DE4] Jean Dieudonné. *Treatise on Analysis. Pure and Applied Mathematics, 10*, Academic Press, New York, 1969–.

[DE5] Jean Dieudonné. *History of Algebraic Geometry*. Wadsworth, Belmont, Ca., 1985.

[D–G–K] L. Danzer, B. Grünbaum, and V. Klee. Helly's theorem and its relatives. In *Convexity*, pages 101–180, AMS, Providence, R.I., 1963.

[DI] P. Dembowski. *Finite Geometries. Ergebnisse der Mathematik, 44*, Springer-Verlag, Berlin, 1968.

[DI–CA] Jean Dieudonné and James B. Carrell. *Invariant Theory, Old and New*. Academic Press, New York, 1971.

[DO] Heinrich Dorrie. *100 Great Problems of Elementary Mathematics*. Dover, New York, 1965.

[DP] Baron Charles Dupin. *Géométrie et Mécanique des Arts et Métiers et des Beaux-Arts*. Bachelier, Paris, 1825.

[DQ] Ernest Duporcq. *Premiers Principes de Géométrie Moderne*. Gauthier-Villars, Paris, third edition.

[DR] Jacques Dixmier. *Cours de Mathématiques du premier cycle, deuxième année*. Gauthier-Villars, Paris, 1968.

[DV] Aryeh Dvoretzky. Some results on convex bodies an Banach spaces. In *International Symposium on Linear Spaces*, pages 123–160, Jerusalem Academy Press, Jerusalem, 1961.

[DX] Gaston Darboux. *Principes de Géométrie Analytique*. Gauthier-Villars, Paris, 1917.

[G-W] J. C. Gibbons and C. Webb. *Circle-preserving maps of spheres.* Preprint, Illinois Institute of Technology, Chicago.

[GZ] Heinz Götze. *Castel del Monte: Gestalt und Symbol der Architektur Friedrichs II.* Prestel-Verlag, München, 1984.

[HA] M. Hall. *The Theory of Groups.* Macmillan, New York, 1959.

[HA-WR] G. H. Hardy and E. M. Wright. *An Introduction to the Theory of Numbers.* Clarendon Press, Oxford, 1945.

[H-C] D. Hilbert and S. Cohn-Vossen. *Geometry and the Imagination.* Chelsea, New York, 1952.

[HD] Jacques Hadamard. *Leçons de Géométrie Elémentaire.* Armand Colin, Paris, fifth edition, 1911-15.

[HG] Hans Haug. *L'art en Alsace.* Arthaud, Grenoble, 1962.

[H-K] O. Haupt and H. Künneth. *Geometrische Ordnungen. Grundlehren der Mathematischen Wissenschaften, 133,* Springer-Verlag, Berlin, 1967.

[HL] Alan Holden. *Shapes, space and symmetry.* Columbia Univ. Press, New York, 1971.

[H-L-P] G. H. Hardy, J. E. Littlewood, and G. Pólya. *Inequalities.* Cambridge University Press, Cambridge [Eng.], 1934.

[HM] P. Hartman. On isometries and a theorem of Liouville. *Mathematische Zeitschrift,* 69:202–210, 1958.

[HN] Sigurdur Helgason. *Differential Geometry, Lie Groups and Symmetric Spaces.* Academic Press, New York, 1978.

[HOL] Raymond d'Hollander. *Topologie Générale.* Volume 1, Eyrolles, Paris.

[HO-PE] W. V. D. Hodge and D. Pedoe. *Methods of Algebraic Geometry.* Cambridge University Press, Cambridge [Eng.], 1947–1954.

[H-P] Daniel R. Hughes and Fred C. Piper. *Projective Planes. Graduate Texts in Mathematics, 6,* Springer-Verlag, New York, 1973.

[HR] Ludwig Hadwiger. *Vorlesungen über Inhalt, Oberfläche und Isoperimetrie. Grundlehren der Mathematischen Wissenschaften, 93,* Springer-Verlag, Berlin, 1957.

[HS] Joseph Hersch. Quatre propriétés des membranes sphériques homogènes. *Comptes Rendus Acad. Sci. Paris,* 270:1714–1716, 1970.

[HU] Dale Husemoller. *Fibre Bundles. Graduate Texts in Mathematics, 20,* Springer-Verlag, New York, second edition, 1975.

[H-W] P. J. Hilton and S. Wylie. *Homology Theory.* Cambridge University Press, Cambridge [Eng.], 1960.

[H-Y] John G. Hocking and Gail S. Young. *Topology.* Addison-Wesley, Reading, Mass., 1961.

[HZ] M. A. Hurwitz. Sur quelques applications géométriques des séries de Fourier. *Annales Ec. Norm.*, 19:357–408, 1902.

[I-B] I. M. Îaglom and V. G. Boltyanskii. *Convex Figures.* Holt, Rinehart and Winston, New York, 1961.

[I-R] G. Illiovici and P. Robert. *Géométrie.* Eyrolles, Paris.

[JE] Jürgen Joedicke. *Shell Architecture.* Reinhold, New York, 1963.

[KE] Paul Krée. *Introduction aux Mathématiques et à leurs applications fondamentales, M.P. 2.* Dunod, Paris, 1969.

[KF] Nicholas D. Kazarinoff. *Geometric Inequalities.* Random House, New York, 1961.

[KG1] Wilhelm Klingenberg. *A Course in Differential Geometry. Graduate Texts in Mathematics, 51,* Springer-Verlag, New York, 1978.

[KG2] Wilhelm Klingenberg. Paarsymmetrische alternierende Formen zweiten Grades. *Abhandl. Math. Sem. Hamburg,* 19:78–93, 1955.

[KH] A. G. Kurosh. *Lectures on General Algebra.* Pergamon, Oxford, 1965.

[KJ1] Marie-Thérèse Kohler-Jobin. Démonstration de l'inégalité isopérimétrique. *Comptes Rendus Acad. Sci. Paris,* 281:119–120, 1976.

[KJ2] Marie-Thérèse Kohler-Jobin. Une propriété de monotonie isopérimétrique qui contient plusieurs théorèmes classiques. *Comptes Rendus Acad. Sci. Paris,* 284:917–920, 1978.

[KM] Tilla Klotz-Milnor. Efimov's theorem about complete immersed surfaces of negative curvature. *Advances in Math.,* 8:474–543, 1972.

[KN1] Felix Klein. *Lectures on the Icosahedron.* Paul, London, 1913.

[KN2] Felix Klein. *Vorlesungen über Höhere Geometrie.* Chelsea, New York, third edition, 1949.

[KO-NO] Shoshichi Kobayashi and Katsumi Nomizu. *Foundations of Differential Geometry. Tracts in Mathematics, 15,* Interscience, New York, 1963–1969.

[KS] Gabriel Koenigs. *Leçons de Cinématique.* Hermann, Paris, 1897.

[KT] Herbert Knothe. Contributions to the theory of convex bodies. *Michigan Math. Journal,* 4:39–52, 1957.

[KY] George P. Kellaway. *Map Projections.* Methuen, London, 1970.

[LB1] Henri Lebesgue. *Leçons sur les Constructions Géométriques.*
Gauthier-Villars, Paris, 1949.

[LB2] Henri Lebesgue. *Les coniques.* Gauthier-Villars, Paris, 1955.

[LB3] Henri Lebesgue. Octaèdres articulés de Bricard. *L'Enseignement
Mathématique (ser. 2)*, 13:175–185, 1967.

[LE] Cornelis G. Lekkerkerker. *Geometry of Numbers.* Wolters-Noordhoff,
Groningen, 1969.

[LEV] Paul Lévy. Le problème des isopérimètres et les polygones articulés.
Bull. Soc. Math. France, 90:103–112, 1966.

[LF1] Jacqueline Lelong-Ferrand. *Géométrie Différentielle.* Masson, Paris,
1963.

[LF2] Jacqueline Lelong-Ferrand. Transformations conformes et
quasi-conformes des variétés riemanniennes compactes. *Mémoires Acad.
Royale Belg., Cl. Sci. Mém. Coll.*, 5, 1971.

[LF3] Jacqueline Lelong-Ferrand. Invariants conformes globaux sur les
variétés riemanniennes. *J. of Diff. Geometry*, 8:487–510, 1973.

[LF–AR] J. Lelong-Ferrand and J.-M. Arnaudiès. *Géometrie et Cinématique.
Cours de mathématiques, vol. 3*, Dunod, Paris, 1972.

[LG1] Serge Lang. *Elliptic Functions.* Addison-Wesley, Reading, Mass.,
1973.

[LG2] Serge Lang. *Analysis II.* Addison-Wesley, Reading, Mass., 1969.

[LM1] J. Lemaire. *Hypocycloïdes et Epicycloïdes.* Blanchard, Paris, 1967.

[LM2] J. Lemaire. *L'hyperbole Equilatère.* Vuibert, Paris, 1927.

[LS] Jean-Jacques Levallois. *Géodésie Générale.* Volume 2, Eyrolles, Paris,
1969.

[LU] Lazar A. Lûsternik. *Convex Figures and Polyhedra.* Dover, New York,
1963.

[LW] K. Leichtweiß. *Konvexe Mengen.* Springer-Verlag, Berlin, 1980.

[LY] Harry Levy. *Projective and Related Geometries.* Macmillan, New York,
1964.

[LZ] V. F. Lazutkin. The existence of caustics for a billiard problem in a
convex domain. *Math. USSR: Izvestia*, 7:185–214, 1973.

[MA] Paul Malliavin. *Géométrie Différentielle Intrinsèque.* Hermann, Paris,
1972.

[MB] Benoît B. Mandelbrot. *Fractals: Form, Chance and Dimension.* W. H. Freeman, San Francisco, 1977.

[MD] A. Marchaud. Les surfaces du second ordre en géométrie finie. *J. Math. pures et appl.*, 9–15:293–300, 1936.

[ME] Ricardo Mañé. *Ergodic Theory. Ergebnisse der Mathematik,* Springer-Verlag, Berlin, 1986.

[MG] Caroline H. MacGillavry. *Fantasy and Symmetry: The Periodic Drawings of M. C. Escher.* Harry N. Abrahams, New York, 1976.

[MI] John Milnor. A problem in cartography. *Amer. Math. Monthly,* 76:1101–1102, 1969.

[ML] Charles Michel. *Compléments de Géométrie Moderne.* Vuibert, Paris, 1926.

[M–P] P. S. Modenov and A. S. Parkhomenko. *Géométrie des Transformations.* Volume 1, Academic Press, New York, 1965.

[MR] John Mather. The nice dimensions. In *Liverpool Singularities Symposium,* pages 207–253. *Lecture Notes in Mathematics,* 192, Springer-Verlag, New York, 1971.

[M–T] André Martineau and François Trèves. *Eléments de la théorie des espaces vectoriels topologiques et des distributions.* Centre de Documentation Universitaire, Paris, 1962–1964.

[MW1] G. D. Mostow. *Strong Rigidity of Locally Symmetric Spaces. Annals of Mathematics Studies, 78,* Princeton University Press, Princeton, 1973.

[MW2] G. D. Mostow. Discrete subgroups of Lie groups. *Advances in Mathematics,* 15:112–123, 1975.

[MY] William S. Massey. *Algebraic Topology: an Introduction. Graduate Texts in Mathematics, 56,* Springer-Verlag, New York, 1977.

[NA] Rolf Nevanlinna. On differentiable mappings. In *Analytic Functions,* pages 3–9, Princeton University Press, Princeton, 1960.

[NU] Jason John Nassau. *Practical Astronomy.* McGraw-Hill, New York, second edition, 1948.

[NW] P. E. Newstead. Real classification of complex conics. *Matematika,* 28:36–53, 1981.

[OA] M. Obata. The conjecture on conformal transformations of riemannian manifolds. *J. of Diff. Geometry,* 6:247–258, 1972.

[OM] O. T. O'Meara. *Introduction to Quadratic Forms. Grundlehren der Mathematischen Wissenschaften, 117,* Springer-Verlag, Berlin, 1963.

[SE3] Jean-Pierre Serre. *Algèbres de Lie Semisimples Complexes.* W. A. Benjamin, New York, 1966.

[SE-TH] H. Seifert and W. Threlfall. *A Textbook of Topology.* Academic Press, New York, 1980.

[SF1] Hans Schwerdtfeger. Invariants of a class of transformation groups. *Aequationes Math.*, 14:105–110, 1976.

[SF2] Hans Schwerdtfeger. Invariants à cinq points dans the plan projectif. *Comptes Rendus Acad. Sci. Paris*, 285:127–128, 1977.

[SG] A. Seidenberg. *Lectures in Projective Geometry.* Van Nostrand, Princeton, 1962.

[SI] Ya. G. Sinai. *Introduction to Ergodic Theory.* Princeton University Press, Princeton, 1976.

[SK] Michael Spivak. *A Comprehensive Introduction to Differential Geometry.* Publish or Perish, Berkeley, Ca., second edition, 1979.

[SL1] G. T. Sallee. Maximal area of Reuleaux polygons. *Canadian Math. Bull.*, 13:175–179, 1970.

[SL2] G. T. Sallee. Reuleaux polytopes. *Mathematika*, 17:315–323, 1970.

[SO1] Luis Antonio Santaló Sors. *Introduction to Integral Geometry.* Hermann, Paris, 1953.

[SO2] Luis Antonio Santaló Sors. *Integral Geometry and Geometric Probability.* Addison-Wesley, New York, 1976.

[SR] Edwin H. Spanier. *Algebraic Topology.* McGraw-Hill, New York, 1966.

[S-T] Ernst Snapper and Robert J. Troyer. *Metric Affine Geometry.* Academic Press, New York, 1971.

[ST-RA] E. Steinitz and H. Rademacher. *Vorlesungen über die Theorie der Polyeder. Grundlehren der Mathematischen Wissenschaften, 41,* Springer-Verlag, Berlin, 1934.

[SU] Pierre Samuel. Unique factorization. *American Math. Monthly*, 75:945–952, 1968.

[SY] J. L. Synge and B. A. Griffith. *Principles of Mechanics.* McGraw-Hill, New York, 1942.

[SW] Ian Stewart. *Galois Theory.* Chapman and Hall, London, 1973.

[TM] René Thom. Sur la théorie des enveloppes. *J. de Math. Pures et Appl.*, 16:177–192, 1962.

[TS] Jacques Tits. *Buildings of spherical type and finite BN-pairs. Lecture Notes in Mathematics, 386,* Springer-Verlag, Berlin, 1974.

[VE] Frederik A. Valentine. *Convex Sets*. McGraw-Hill, New York, 1964.

[VG1] H. Voderberg. Zur Zerlegung eines ebenen Bereiches in Kongruente. *Jahresberichte d. Deutschen Math. Ver.*, 46:229–231, 1936.

[VG2] H. Voderberg. Zur Zerlegung der Ebene in kongruente Bereiche in Form einer Spirale. *Jahresberichte d. Deutschen Math. Ver.*, 47:159–160, 1937.

[VL] Patrick du Val. *Homographies, Quaternions and Rotations*. Clarendon Press, Oxford, 1964.

[VN] Georges Valiron. *The Geometric Theory of Ordinary Differential Equations and Algebraic Functions. Lie Groups: History, Frontiers and Applications, XIV*, Math Sci Press, Brookline, Mass., 1984.

[V–Y] O. Veblen and J. W. Young. *Projective Geometry*. Ginn and Co., Boston, 1910–1918.

[WF] Joseph A. Wolf. *Spaces of Constant Curvatures*. Boston, 1974.

[WK] R. J. Walker. *Algebraic Curves*. Princeton University Press, Princeton, 1950.

[WL] Hermann Weyl. *Symmetry*. Princeton University Press, Princeton, 1952.

[WN] Magnus J. Wenninger. *Polyhedron Models*. Cambridge University Press, Cambridge [Eng.], 1971.

[WO] Yung-Chow Wong. *Isoclinic n-planes in Euclidean 2n-space, Clifford parallels in elliptic $(2n-1)$-space and the Hurwitz matrix equations. AMS Memoirs, 41*, American Math. Society, Providence, R.I., 1961.

[WR] Frank Warner. *Foundations of Differentiable Manifolds and Lie Groups*. Scott, Foresman, Greenville, Ill., 1971.

[ZN] Michel Zisman. *Topologie Algébrique Elémentaire*. Armand Colin, Paris, 1972.

[ZR] C. Zwikker. *The Advanced Geometry of Plane Curves and their Applications*. Dover, New York, 1963.

Index of notations

Index

Entries in italics refer to definitions. Entries consisting of a roman numeral and a number refer to volume and page number.

Acknowledgements

We are pleased to acknowledge the permission of various publishers and institutions to reproduce some of the figures appearing in this book:

Figs.   9.12.7.2, 12.10.9.2.1, 18.1.1.3, 18.1.6.1: ©Armand Colin Editeur, Paris

Fig.   10.12.1.2: ©Librairie Arthaud, Paris

Figs.   1.7.4.8, 1.7.6.13, page 108 (vol. I), 19.6.12.5:
        ©BEELDRECHT, Amsterdam/BILD-KUNST, Bonn 1982

Figs.   1.7.4.1 - 1.7.4.5, 1.7.6.1 - 1.7.6.12: ©Cedic, Paris

Figs.   12.1.3.2.1, 12.1.3.2.2, 12.6.10.5.3, 12.6.10.5.4:
        ©Cambridge University Press, Cambridge

Figs.   4.3.9.11, 4.3.9.12, 15.3.3.3.2, 18.1.8.6, 20.7.3:
        ©Chelsea Publishing Company, New York

Figs.   12.1.1.5, 12.1.1.6: ©1971 Columbia University Press, New York

Fig.   18.9.4.2: ©Conference Board of the Mathematical Sciences, Washington

Figs.   12.5.6.1, 12.5.6.2.1 - 12.5.6.2.4: ©Dover Publications, Inc., New York

Fig.   18.1.8.7: ©Eyrolles, Paris

Figs.   4.7.4.1, 4.7.4.2: ©Institut Géographique National, Paris
        (Autorisation no. 99-0175)

Figs.   1.8.5, 19.6.12.4: ©John Wiley & Sons Inc., New York

Figs.   14.4.4.6.3, 14.4.6.4, 15.3.3.3.3, 15.3.3.3.4:
        ©Karl Krämer Verlag, Stuttgart

Fig.   page 22 (vol. II): ©Prestel-Verlag, München

Fig.   12.5.5.7: ©Princeton University Press, Princeton

Fig.   12.10.5.2: ©Vandenhoeck & Ruprecht, Göttingen